Dahlem Workshop Reports
Life Sciences Research Report 36

The goal of this Dahlem Workshop is:
towards a new understanding of large-scale evolutionary change by integrating paleobiological and biological approaches

Life Sciences Research Reports
Editor: Silke Bernhard

Held and published on behalf of the
Stifterverband für die Deutsche Wissenschaft

Sponsored by:
Senat der Stadt Berlin
Stifterverband für die Deutsche Wissenschaft

Patterns and Processes in the History of Life

D. M. Raup, D. Jablonski,
Editors

Report of the Dahlem Workshop on
Patterns and Processes in the History of Life
Berlin 1985, June 16–21

Rapporteurs: K. W. Flessa · D. Järvinen ·
J. S. Levinton · D. B. Wake

Program Advisory Committee:
D. M. Raup and D. Jablonski, Chairpersons
B. Charlesworth · H. K. Erben · K. J. Hsü · W. Nagl
A. J. de Ricqlès · R. M. Rieger · D. Simberloff
S. C. Stearns

Springer-Verlag
Berlin Heidelberg New York
London Paris Tokyo 1986

Copy Editors: K. Geue, J. Lupp

Photographs: E. P. Thonke

With 4 photographs, 36 figures, and 8 tables

ISBN-13:978-3-642-70833-6 e-ISBN-13: 978-3-642-70831-2
DOI: 10.1007/978-3-642-70831-2

Library of Congress Cataloging-in-Publication Data
Dahlem Workshop on Patterns and Processes in the History of Life
(1985: Berlin, Germany)
Patterns and processes in the history of life.
(Dahlem workshop reports) (Life sciences research report; 36)
Includes indexes.
1. Evolution–Congresses.
2. Paleontology–Congresses.
I. Raup, David M.
II. Jablonski, David.
III. Title.
IV. Series.
V. Series: Life sciences research report; 36
QH359.D34 1985 575 86-26005
ISBN 0-387-15965-7 (U.S.)

Softcover reprint of the hardcover 1st edition 1986

Typesetting, printing and bookbinding: Brühlsche Universitätsdruckerei, Giessen
2131/3020-543210

Table of Contents

The Dahlem Konferenzen

Founders

Recognizing the need for more effective communication between scientists, especially in the natural sciences, the Stifterverband für die Deutsche Wissenschaft*, in cooperation with the Deutsche Forschungsgemeinschaft**, founded Dahlem Konferenzen in 1974. The project is financed by the founders and the Senate of the City of Berlin.

Name

Dahlem Konferenzen was named after the district of Berlin called "Dahlem", which has a long-standing tradition and reputation in the arts and sciences.

Aim

The task of Dahlem Konferenzen is to promote international, interdisciplinary exchange of scientific information and ideas, to stimulate international cooperation in research, and to develop and test new models conducive to more effective communication between scientists.

Dahlem Workshop Model

Dahlem Konferenzen organizes four workshops per year, each with a limited number of participants. Since no type of scientific meeting proved effective enough, Dahlem Konferenzen had to create its own concept. This concept has been tested and varied over the years, and has evolved into its present form which is known as the *Dahlem Workshop Model*. This model

---

* The Donors Association for the Promotion of Sciences and Humanities
** German Science Foundation

provides the framework for the utmost possible interdisciplinary communication and cooperation between scientists in a given time period.

The main work of the Dahlem Workshops is done in four interdisciplinary discussion groups. Lectures are not given. Instead, selected participants write background papers providing a review of the field rather than a report on individual work. These are circulated to all participants before the meeting to provide a basis for discussion. During the workshop, the members of the four groups prepare reports reflecting their discussions and providing suggestions for future research needs.

Topics

The topics are chosen from the fields of the Life Sciences and the Physical, Chemical, and Earth Sciences. They are of contemporary international interest, interdisciplinary in nature, and problem-oriented. Once a year, topic suggestions are submitted to a scientific board for approval.

Participants

For each workshop participants are selected exclusively by special Program Advisory Committees. Selection is based on international scientific reputation alone, although a balance between European and American scientists is attempted. Exception is made for younger German scientists.

Publication

The results of the workshops are the Dahlem Workshop Reports, reviewed by selected participants and carefully edited by the editor of each volume. The reports are multidisciplinary surveys by the most internationally distinguished scientists and are based on discussions of new data, experiments, advanced new concepts, techniques, and models. Each report also reviews areas of priority interest and indicates directions for future research on a given topic.

The Dahlem Workshop Reports are published in two series:
1) Life Sciences Research Reports (LS), and
2) Physical, Chemical, and Earth Sciences Research Reports (PC).

Director

Silke Bernhard, M.D.

Address

Dahlem Konferenzen
Wallotstrasse 19
1000 Berlin (West) 33

Patterns and Processes in the History of Life,
eds. D. M. Raup and D. Jablonski, pp. 1–5. Dahlem Konferenzen 1986
Springer-Verlag Berlin, Heidelberg

Introduction

D. M. Raup and D. Jablonski
Dept. of Geophysical Sciences
University of Chicago
Chicago, IL 60637, USA

Much of the current excitement in evolutionary thought lies at the conceptual interface between biology and paleobiology. Research questions now attracting special attention include the interplay between cladogenetic and phyletic evolution, the origin of major taxonomic groups, mass extinction, the role of competition and other biotic interactions in the evolution of lineages and communities, and the relationship between rates of evolution at the genomic and organismic levels. These are fundamental questions about the evolution of life that demand an interdisciplinary approach. Unfortunately, there is still surprisingly little communication between biologists and paleobiologists. The lag time between a research development in one of the fields and its absorption by the other may stretch into years or even decades. Serious misunderstandings can arise owing to differing vocabularies and conceptual differences dictated by the respective kinds of available data and techniques. The main purpose of this Dahlem Workshop in June of 1985 was to reduce the gap between biology and paleobiology and to help establish more effective communication and interaction between the two disciplines.

Ever since Darwin, evolutionists have recognized the unique value of the fossil record as a historical document of the evolution of life, a document providing access to phenomena over time spans far greater than ever available to the field biologist. Unfortunately, the fossil record has been used primarily as a source of evolutionary anecdote almost to the exclusion of its potential as a laboratory for framing and testing hypotheses. At the same time, biologists have tended to concentrate on problems at the scale of individual organisms and local populations or have been obliged to assume that questions regarding greater time scales and higher taxa were amenable to extrapolation of theory and results based on shorter-term and smaller-scale phenomena. Despite general recognition that the two fields share a common concern with the patterns and processes of evolution, many basic questions about this common ground remain.

Current controversies on the microevolution-macroevolution question are a case in point. On the one hand, it has been argued that the patterns of change at and above the species level and over millions or tens of millions of years (known, descriptively at least, as macroevolution) are simply the result of the accumulation of small-scale changes among individuals within populations (microevolution). On the other hand, it has been argued that the larger patterns involve additional processes such as selection or other sorting processes at the species level. Advances in this area would benefit from true collaboration between the theory and data of paleontology and of biology, as distinguished from the excessively contentious interaction that has dominated much of the discussion thus far. One welcome step at this meeting was agreement on the need to explore causes and consequences of differential production and persistence of higher-level taxa, labelled species selection when dictated by species-level processes, but deserving other terms when causation or effect resides at other levels in the biological hierarchy.

On the other hand, far less rapprochement was achieved regarding the empirical assessment of the macroevolutionary role of speciation, a phenomenon generally too slow for direct biological study but too rapid to be recorded in most paleontological sequences. Given the lack of one-to-one correspondence between the achievement of reproductive isolation and the phenotypic manifestation of that event, analysis of individual speciation events in the fossil record requires very special conditions and a great deal of knowledge of the biology of the organisms involved. Paleontologists remained considerably more optimistic than some of the biologists that indirect assessment of relative differences in speciation rates might still be possible, thus allowing testing of carefully constructed hypotheses on the role of morphological stasis and speciation (as opposed to other mechanisms of phenoptypic change) in large-scale evolutionary patterns. At the same time, biologists were considerably more optimistic than some of the paleontologists on the degree to which ecological and population genetic theory could explain species-level patterns. Reliance on tight species packing and competitive interactions as the cause for stasis across many marine and terrestrial phyla, for example, appeared to be pushing the role of compensation in structuring communities beyond that conceded by many ecologists and paleoecologists. Disagreements thus persist, but we hope they will be fruitful ones as collaboration between the two disciplines increases.

Hypothesis testing is not a straightforward matter in the fossil record and here, too interactions with biology can be extremely profitable. Quite simply, predictions regarding long-term consequences of processes observed in living organisms can be tested directly using paleontological data if those living organisms have an adequate fossil record, thus avoiding the pitfalls of extrapolative approaches. We hope to see a burgeoning of this interactive effort in the coming years. Framing and testing of hypotheses in paleontological subjects inevitably raises the problem of inferring process from pattern, and the consideration and elimination of a broad range of rival hypotheses is an essential procedure here. In a historical science such as paleontology, the problem often arises that the events that are of most interest are unique in the history of life. For example, replication of the metazoan radiation at the beginning of the Cambrian is not feasible. However, decomposition of such problems into component hypotheses may at least in part alleviate this difficulty. For example, hypotheses built upon the role of species packing might be tested by comparing evolutionary dynamics (both morphological and taxonomic) during another global diversification, such as the biotic rebound from the end-Permian extinction, which removed perhaps 95% of the marine species (see Valentine, this volume).

The subject of extinction, and mass extinction in particular, has become important in both paleobiology and biology. On the paleobiologic side, our knowledge of the basic facts of past extinctions has now become sufficiently robust to permit rigorous statistical analysis of the extinction phenomenon over a wide range of spatial and temporal scales. Success in this area will not only tell us something about the geological history of physical environmental stress (by shedding light on the proximate causes of extinction events) but will have bearing on the larger evolutionary significance of mass extinctions (by tracking large-scale evolutionary patterns that originate, transcend, or terminate near extinction events). Paleontological extinction studies in turn have relevance to the pressing problem of present-day extinctions, as in tropical forest and reef habitats. The feedback loop comes full circle as biologists begin to explore new data and theory regarding patterns and mechanisms of species contraction and extinction in living organisms, where detailed information is available on population structures, genetical processes, and other factors not accessible in the fossil record.

We are, therefore, much impressed by the possibilities for the interweaving of paleontological and biological approaches. There is good precedent for such an undertaking. In the late nineteenth century, exchange between paleontology and evolutionary was at a high pitch (see [2, 3, 4, 6)]. The term "paleobiology" was invented soon thereafter by the great Viennese paleontologist Othenio Abel in 1912 [1], who founded a journal with that title in 1928. *Palaeobiologica* rapidly developed an international audience centered around an extremely productive central European research program (see [5]). Communication between biologists and paleontologists and the biological investigation of the fossil record, fell into some quiescence in the decades immediately following the Second World War, but the past few years have seen a rebirth of, and a building upon, this earlier vigor. Such renewed exchange can only aid in the pursuit of the ultimate object, a better synthetic theory of the evolution of life.

This volume contains 17 background papers written by individual workshop participants. The first chapter, "The Nature of the Fossil Record," is intended to be a brief "Paleobiology for the Evolutionist" introduction – an attempt to outline some of the problems that can, and cannot, be addressed in the fossil record, and to provide an entry into the paleobiological literature for the uninitiated. The second chapter, "Natural Selection and Fitness, Adaptation and Constraint," performs the reciprocal task for the paleontologist. As is customary for Dahlem Workshops, all the background papers were distributed before the meeting, to be discussed during our week in Berlin but not formally presented. Following the meeting each of the background papers was revised by its author in response to comments received, but of course the papers remain personal statements and reviews of segments of the larger subject.

By contrast, the four group reports represent statements prepared during and shortly after the meeting by the four subgroups that operated throughout the conference. These papers also are not designed as definitive statements but rather as summaries of the views and viewpoints of small, but we hope representative, subsets of researchers in the several subfields. We urge that readers view the group reports simply as state-of-the-art bulletins at an instant in time in an array of disciplines and subdisciplines that are in period of exceptional activity and, we are convinced, a new level of cross-fertilization.

We are most grateful to Silke Bernhard and the Dahlem Konferenzen organization for providing a forum so conducive to the exploration of the interdisciplinary potential of our science. Our successful discussions owe much to the skillful and gracious work of Silke Bernhard and her staff. K. Geue and J. Lupp provided sympathetic and exceptionally able editorial assistance as the book was completed. We thank S. Suter and S. M. Kidwell, both of the University of Chicago, for reviews and much-needed second opinions.

References

1. Abel O (1912) Grundzüge der Paläobiologie der Wirbeltiere. Stuttgart: Schweizerbart
2. Bowler PJ (1976) Fossils and progress: Paleontology and the idea of progressive evolution in the nineteenth century. New York: Science History Publications
3. Bowler PJ (1983) The eclipse of Darwinism: Anti-Darwinian evolution theories in the decades around 1900. Baltimore; Johns Hopkins University
4. Desmond A (1982) Archetypes and Ancestors: Palaeontology in victorian London 1850–1875. London: Frederick Miller Ltd. (1984 Chicago: University of Chicago Press)
5. Reif W-E (1980) Paleobiology today and fifty years ago. N Jb Geol Paläont Mh 1980:361–382
6. Rudwick MJS (1972) The meaning of fossils: Episodes in the history of palaeontology. New York: American Elsevier (2nd Ed., 1976, New York: Science History Publications, reprinted 1985 by University of Chicago Press)

Patterns and Processes in the History of Life,
eds. D. M. Raup and D. Jablonski, pp. 7–22. Dahlem Konferenzen 1986
Springer-Verlag Berlin, Heidelberg

The Nature of the Fossil Record: A Biological Perspective

D. Jablonski*, S. J. Gould**, and D. M. Raup*

* Dept. of Geophysical Sciences
University of Chicago
Chicago, IL 60637, USA

** Museum of Comparative Zoology
Harvard University
Cambridge, MA 02138, USA

Abstract. Paleontology can be a rich source of theory and data on evolutionary and ecological processes at more inclusive hierarchical levels and greater time scales than those available to the neontologist. Although the generation-by-generation record of ancient populations and communities is obscured by inconstancy of sedimentation and bioturbation, the record of lager-scale patterns of intra- and interspecific morphological change (and stasis) can be analyzed with confidence when sampling schemes – and the questions being asked – allow for the discontinuous nature of the rock record and the difficulty of precise time correlation among localities. These considerations permit the analysis of such diverse problems as the evolutionary consequences of different genetic population structures (e.g., fragmented vs. panmictic) or genetical systems (e.g., sexual vs. asexual); the intrinsic and extrinsic factors giving rise to a bias towards stasis or gradualism within the species of a given group; and the origin, persistence, and cohesiveness of different ecological communities and community types. The rich morphological data of paleontology (generally only hardparts, although some extraordinary localities provide a far more complete record) yield insights into the pattern of occupation of morphospace through geological time and into the ways in which differential speciation and extinction generate evolutionary trends in groups dominated by morphological stasis. The smooth extrapolation of microevolutionary processes evidently also fails to explain the evolutionary consequences of mass extinction events, which are more frequent, more disruptive, and more important as agents of faunal replacement than previously thought. In its direct study of inclusive levels and long time intervals and of the large and rare events that only repeat sufficiently often in the fulness of geological time, paleontology yields a wealth of phenomena not accessible to neontology and promises a fruitful union with microevolution and ecology for a more comprehensive theory of organic change and stability.

Introduction

Paleontology paid a heavy price for admission into the modern evolutionary synthesis. In abjured its long and sorry history of such "special" macroevolutionary nations as orthogenesis and racial life cycles but accepted the stricture that its phenomena, at all amounts of change and lengths of time, might be encompassed as extrapolated results of natural selection working within populations. Paleontology became a kind of gigantic tapestry or playing field for the extension of microevolutionary processes.

The contemporary rebirth in the biological application of paleontological data – celebrated by Maynard Smith [38] as a return of paleontology to the high table of evolutionary theory – includes a challenge to the purely extrapolative paradigm. The fossil record offers the means to assess directly the extent to with selection within populations can smoothly build to trends within monophyletic groups, and the competitive struggles of individuals in normal times can explain relays of faunas across the entire Phanerozoic. In this chapter we review briefly what can and cannot be expected from the fossil record: we underscore the great breadth of information that can be retrieved and draw cautionary attention to some question that are as yet intractable to paleontological approaches. Thus, we must consider both evolutionary issues and taphonomy in its broadest sense, i.e., "the study of how preservation controls information in the fossil record" (see [3]; also [45, 46, 54] for summary statements on the general quality of the fossil record). We hope to encourage a broader appllication of paleontological data to evolutionary problems, and to draw attention to the new *sorts* of questions that can be addressed when the data of the fossil record – biases and all – are taken on their own terms.

Genetics

Nucleic acids and isozymes have minimal fossilization potential, but genetical approaches to the fossil record hold rich possibilities. Even though only a fraction of an organism's phenotype is fossilized in most instances, indirect tests for longterm consequences of particular genetic mechanisms or processes are often possible.

a) The paleogenetics of certain aspects of morphology can be analyzed, on rare occasions using simple Mendelian traits, and with greater generality and power using the techniques of quantitative genetics (see [48]). Such analyses can yield estimates of the roles of drift and selection in shaping morphological time series extending over millions of years in a variety of settings and higher taxa [37]. The few analyses available thus far have dealt with isolated individual species or lineages, but a study of several co-occurring lineages might permit a better assessment of the roles of environmental factors and intrinsic biological traits such as effective population sizes in the observed phenotypic changes.

All population biological approaches to the fossil record are seriously limited by problems of resolution. Special conditions of mass mortality are required to produce a sample comprising a single contemporaneous population. The great majority of paleontological samples – including those on individual bedding planes – are time-averaged, representing an accumulation of the remains of individuals that lived years, decades, or centuries apart, admixed by an insufficiency of diluting sedimentation and/or by the churning activities of later organisms (see [18]). This may not be a problem when comparing samples themselves separated by hundreds of thousands of years, but it must be kept in mind when attempting to apply genetic models of any kind.

b) The significance of certain population genetical models can be tested, if such models generate unique predictions of long-term, morphological evolution. For example, gastropod lineages composed of species with subdivided genetic population structures exhibit significantly higher speciation rates than lineages of relatively panmictic species (see [28, 30] for review). Jablonski et al [31] found environmental trends in the production of major evolutionary novelties that appeared consistent with speciation models derived from population genetics [10, 57]; results were preliminary, but they point to new lines of cooperative research.

c) Perhaps the most exciting possibilities lie in the comparative study of macroevolutionary patterns in taxa having radically different genetic systems. Major evolutionary differences have been postulated for groups in which the maternal genome is heavily involved in early development and those in which the zygote's genome directs early development from the outset [12, 13]; groups in which the germ line is sequestered early in ontogeny and those in which somatic mutations may be incorporated into the germ line (a category that includes most of life on Earth, i.e., plants and many invertebrate phyla [9]; groups in which genetic recombination occurs and groups in which reproduction is clonal [56]; groups in which the horizontal transfer of hereditary information among species (or higher taxa) is common and those in which it is rare [20]. If assumptions are made regarding the genetic systems of fossil organisms on the basis of their extant relatives, the fossil record offers a testing ground for an intriguing array of genetical hypotheses.

Morphology

The fossil record is incredibly rich in morphological data, although certain kinds of organisms and certain habitats are better represented than others because of the vagaries of preservation. Organisms with hard skeletons, particularly shelled invertebrates, are much more likely to be preserved as fossils than soft-bodied organisms. As a general rule, terrestrial habitats are poorly represented, as are deep-sea environments. But important exceptions occur in the so-called Lagerstätten: those geologic flukes where a much greater proportion of the original biota is preserved due to exceptional conditions. Classic Lagerstätten include the La Brea tarpits of the Pleistocene, Solnhofen Limestone (with its *Archaeopteryx*) of the Jurassic, Mazon Creek of the Carboniferous, and the Burgess Shale of the Cambrian. These and others like them provide invaluable windows into the evolutionary past.

Morphological data from the fossil record make two main contributions to evolutionary biology. First, they greatly increase the available sample by recording a large number of now-terminated evolutionary experiments. Extinct organisms such as flying reptiles may be used to develop a broad picture of what has and has not happened in several billion years of evolution, and of alternative solutions to problems of design that enrich and constrain hypotheses on the origin of major adaptations (e.g., vertebrate flight [42]; see also the remarkable patterns of iterative evolution in Tertiary planktonic foraminifera, Jurassic ammonites, and Cambrian trilobites). Second, the array of morphologies in the fossil record provides a chronology of events in evolution: who begat whom and when. It is thus possible to describe and interpret the history of the occupation of morphospace, that hyperspace including all organisms that could theoretically exist. Without this history, evolutionary biologists would have no way of really knowing, for example, whether the present-day biota represents a steady-state condition or not. The extraordinary range of bizarre morphologies in early Paleozoic Lagerstätten suggests that a steady state has not prevailed.

For many years paleontologists were content to describe morphologies and to construct loose scenarios for the functional anatomy and evolution of extinct organisms. But the past twenty years have seen important advances in methodology (see [21]). The approach initiated by Rudwick in the 1960s and applied by him to the interpretation of enigmatic morphologies in extinct brachiopods [49] set a standard for functional analysis that has been extended by other to living organisms. More recently, constructional morphology [50, 52] has developed as a highly rigorous approach which, again, started from analysis of extinct forms.

The power of constructional morphology lies in its broadening of hypotheses for the causes of form [24]. We need no longer assume adaptation a priori, either as direct selective response to immediate environment or as a phyletic legacy of past adaptations passively inherited (Darwin's argument for the ultimate primacy of adaptation over "laws of form"). Constructional morphology provides a third, nonadaptive axis to Seilacher's "triangle" of causes: nonselected covarying consequences of other changes and inherent properties of building materials developed for other reasons. Examples include Seilacher's [51] argument for a constructional basis of divaricate ornament in mollusks and brachiopods (often co-opted secondarily for an immediate use), and Gould's [23] claim that eight separate lineages of dwarf *Cerion* evolved a shape foreclosed to the vastly more common lineages of normal size and developed as a nonselected consequence of the dwarfing itself.

Origination and Extinction

Speciation

With increasing interest among evolutionary biologists in the speciation process, it was inevitable that attempts would be made to document speciation events in the fossil record. At first glance, this would seem appropriate in the light of paleontological time: knowledge of the distribution of populations in space and time should make it possible, for example, to follow the course of geographic speciation from initial isolation through to the development of reproductive barriers. Yet most detailed studies with this objective have failed to add significantly to our knowledge of the speciation process – principally because time resolution is such a problem. Preservation is largely dependent in the pattern of accumulation of the entombing sediments. Not only does time-averaging blur together many generations, but sedimentation over slightly longer time scales is almost invariably episodic and patchy in the shallow-marine environments that constitute the bulk of the accessible fossil record [1, 19], resulting in a history that is commonly more gap than record at the requisite scale of resolution. Thus, although several hundred thousand speciation events are recorded in the fossil record, they are known only in the sense that we recover the species that are produced: the events leading to the formation of these species cannot generally be seen.

Probabilistic models have been developed to put confidence limits on the resolving power of the stratigraphic record (reviewed in [3, 4, 17]). These should only be applied in conjunction with other taphonomic and sedimentologic evidence, however; Velbel [59] showed that a 1 cm shale bed could represent from a few hours to 1200 years of sedimentation – a span of six orders of magnitude that fortunately can be narrowed using independent physical evidence. In light of these problems, it is particularly important that sampled populations (which, again, will almost inevitably encompass several to hundreds of generations) be collected from taphonomically comparable deposits [36].

Two promising settings for tracking the speciation process are the deep-sea record of shelled microplankton [7] and the freshwater deposits of large and ancient lakes [8, 40, 62]. Even here, however, some of the major difficulties of paleontological documentation of speciation arise. Fine-scale time resolution is rarely sufficient to identify a geographic array of populations known to have existed at precisely the same time see [11] for an exception. This problem blocks efforts to document the fates of individual populations and makes it difficult to distinguish ecophenotypic and evolutionary change (but see [5, 34]). Furthermore, we can deal only with morphospecies, not all reproductively isolated units; the imprecise correlation between speciation and morphologic change prevents a truly exhaustive paleontological tally of speciation even within readily fossilized taxa [37a]. However, other questions related to speciation and evolutionary change are available to paleontologists; these include tracing patterns of phyletic transformation within and between species and determining how these patterns in turn relate to large-scale evolutionary trends.

Phyletic Transformation

It has been assumed in evolutionary biology that species lineages undergo gradual transformation over time through directional selection. Given enough time in an ancestor-descendant sequence of populations, changing allelic frequencies and occasional introduction of genetic novelties should lead to sufficient change to make the descendant populations different species from the ancestral form. For evolutionary biologists this scheme has been more a matter of faith than observation because of the lack of long time spans over which direct observation can be made. A relatively small number of cases, e.g., the evolution of industrial melanism during the past century, have been extrapolated from time spans of tens of years to millions of years. However, pure extrapolation – which may indeed be appropriate in some instances – inevitably limits the breadth of phenomena being sought and tested.

The fossil record, on the other hand, allows us to analyze directly the stability or lack of stability of species over time (using morphological rather than genomic data, of course). Gradualistic change can be detected even among relatively widely spaced samples (although the difficulties mentioned above in interpreting fossil morphologies still obtain), and the absence of morphological change – stasis – and the temporal overlap of putative ancestor and descendant is readily documented even in relatively widely spaced samples. Some trends in monophyletic groups show a punctuational pattern of stasis and rapid change ([61] on African pigs; [16] on Pleistocene ostracodes), others a more gradualistic scheme [7, 37a] or a mixture of the two modes [47]. But one thing is clear: stasis is far more prevalent than would have been predicted from population-level studies of living organisms. Evidently, in many groups the potential for change inherent in the genetic system is not generally realized owing to canalizing effects of stabilizing selection and other forces. Combined paleontological-neontological research programs could shed much light on the phenomenon of stasis and its prevalence in some groups but not in others. For example, interesting preliminary suggestions identify biases toward one mode or the other correlated with ecological status or genetic systems ([9, 35]; see [15] for a new collection of paleontological case histories).

Trends

In groups exhibiting a high degree of within-species stasis, morphologic change will tend to be concentrated at times of lineage branching, i.e., speciation. Given the difficulties in examining the process and consequences of speciation in extant organisms, the fossil record may be the only basis for determining the origin of evolutionary trends (character gradients of substantial magnitudes occuring within monophyletic groups over substantial periods of time). Trends in punctuational groups, in which the course of evolution follows a stepping stair or rectangular pattern [56] at usual scales of geological resolution, will be a matter of differential speciation annd extinction events among species, not (usually) a simple long-term extension of natural selection within populations. This differential may itself still be reducible to selection at the organismal level, for example, with species longevity as a consequence of organismal adaptations built in competition (see [33] on trends in evolution of brain size). Conversely, trends, particularly those driven by differential speciation, may be caused by true species selection acting on irreducible properties of populations (individuals, after all, do not speciate). As Vrba [60] emphasizes, these alternatives are testable with an appropriate combination of paleontological and neontological data. Vrba has supported extrapolated individual success for trends in African antelopes (but producing unexpected results in among-species patterns via her "effect hypothesis"). Hansen's [26] trend in early ontogenetic lifestyles of Tertiary volutid gastropods seems to imply true species selection [22, 28].

Trends driven by differential origination and extinction are higher-level analogs (within monophyletic groups) to conventional birth and death within populations. But if trends occur because species tend to arise with one kind of morphology rather than others, then we encounter (again with monophyletic groups) an analog to mutation pressure within populations – an unconventional process in microevolution, but a promissing and as yet unevaluated macroevolutionary force. Possible candidates for paleontological testing of the generality of such "directional speciation" [56] include the tendency to originate at larger or smaller size than ancestors [55] and, especially, ontogenetic channeling where new species either back down (progenesis or neoteny) or climb up (hypermorphosis) their ontogenetic trajectories (see [39] for a possible example in Tertiary scallops). Many fossil invertebrates contain a wealth of ontogenetic data in their accretionary or exuviated shells, but these data have not yet been fully exploited because of the need for more precise markers of ontogenetic state than simple size measurements. More detailed work on skeletal ontogenesis in living invertebrates and vertebrates should improve calibration and will open the door to a fuller exploration of the phylogenies of ontogenies.

Testing of hypotheses regarding the origin of large-scale patterns through differential speciation and extinction is a daunting task. Large volumes of data must be collected in a rigorous phylogenetic framework, and the biases discussed earlier must be overcome or circumvented. We do argue paleontological access only to morphospecies need not be a crippling problem so long as a comparative aproach is adopted. For closely related groups, relative differences in morphospeciation rates will reflect real differentials in overall speciation rates unless sibling or cryptic species are overwhelmingly more common in the lineage that otherwise exhibits the lower morphospeciation rate. More research is needed here, but the few data available give little reason to suspect that such a bias prevails for most pairs of fossil taxa that have been compared in this fashion. For example, the proportion of sibling species in marine invertebrates having high dispersal abilities and low morphospeciation rates is evidently no higher than in groups having low dispersal and high morphospeciation rates [28]. The limited phenotypic data of the fossil record and its much-lamented incompleteness make the delineation of ancestral-descendent relationships difficult, indeed futile according to some [43]. But for well skeletonized marine organisms in particular, the record of higher taxa and their approximate times of origination and extinction is better than generally appreciated [37a, 44]. Major problems remain, however: the quality of the fossil record improves with decreasing age, and an additional bias, the "pull of the Recent," is imposed by the completeness of our knowledge of the present biota – which engenders the extension of geologic ranges of young but rarely preserved forms from their occasional fossil discoveries to the Recent, but cannot provide the same extension to older, now extinct taxa (see [45, 46]). Increasing appreciation of these biases, and attempts to analyze them statistically, has engendered a more realistic assessment of patterns in origination for higher taxa in the fossil record and new approaches to treatment of extinction as an important phenomenon in its own right.

Extinction

Virtually all species that ever lived are extinct; thus, the number of extinction events is approximately equal to the number of speciation events over the total span of geologic time. For this reason, extinction must be an important element in the total evolutionary process. For an evolutionary biologist to ignore extinction would be about as reasonable as for a demographer to consider birth not death.

The fossil record provides the only data base for analyzing the distribution of extinction events in space and time. Comparisons among large samples of fossil taxa can yield generalizations on the interplay between intrinsic biological traits (from body size to dispersal ability) and external forcing mechanisms in determining vulnerability to extinction of genealogical groups and biotas of particular time intervals. Van Valen [58] compiled taxonomic durations within a large number of higher taxa and found that extinction appeared to be stochastically constant – that is a taxon's probability of survival was not dependent upon how long it had been in existence. From this observation he extrapolated ecological processes into evolutionary time to propose his famous Red Queen Hypothesis, that the effective environment of a species is continually deteriorating owing to evolutionary advances by the other species around it. Testing the Red Queen model requires paleontological time scales but has proven difficult to accomplish, in large part because physical environments rarely remain quiescent for sufficient periods of time to allow purely biological interactions to work to their conclusions [27].

The Red Queen and her rivals are concerned primarily with times of normal, background levels of extinction. Most intense discussion is now focussing on the role of mass extinctions in setting new patterns (or simply disrupting those arising at lower levels), rather than just providing a more intense refiner's fire of conventional competition. If mass extinctions are more frequent, extensive, rapid, and different in effect than we had previously imagined, then they may either impose important randomness on patterns of faunal replacement through time, or sort taxa by rules different from those operating in normal times (see Jablonski, this volume). This is primarily the territory of the paleontologist, of course, although sadly today's human-generated acceleration in extinction rates may soon come to match in magnitude and biological consequences some of the major events in the geologic past.

The fine structure of mass extinctions are difficult to resolve because, ironically, these very interesting events tend to coincide (with the question of causation still hotly debated) with marine regressions, which tend to obscure and destroy the record. Sampling schemes are difficult to design, and the temptations are great to overextend data from a single depositional site with a relatively complete record, but models are becoming available to place confidence limits on patterns of extinction and survival around extinction events (Jablonski, this volume). These approaches, along with more rigorous analyses of synoptic data sets and a new generation of mesoscale studies encompassing tens of millions of years surrounding an extinction event, may be leading towards new evolutionary paradigms, with extinction as a major force perturbing the evolutionary system away from a steady state.

A search for major trends in the overall history of life must be an integration of both mass and background extinction processes. These may therefore transcend not only the adaptive struggles of individuals in conventional Darwinian ("wedging") competition, but also the simple extension of trends within monophyletic groups to the origin and persistence of major new designs. Sepkoski's [53] three-fauna theory suggests an intrinsic patterning of directions in the history of marine invertebrates, not simply produced by fortuitously coordinated responses to mass extinctions – in other words, recording some kind of ecological "coherence" among taxonomically disparate members of the faunas (see also the remarkably similar patterns revealed by analysis of the plant record [41]).

Community Evolution

The discrimination of biological communities in the fossil record is fraught with problems: the soft-bodied organisms that are often overwhelmingly important in the structure and function of living communities are rarely preserved (although their traces sometimes are), and time-averaging can merge not only temporally disjunct populations but discrete communities. Different environments are subject to different rates of sedimentation and nondeposition, and hardparts of different mineralogies and microstructures are subject to different rates and patterns of preservation. These difficulties are compounded in terrestrial plant and animal communities, where rich accumulations tend to consist of transported rather than in situ remains. Most ecological processes, such as succession, are vanishingly brief when viewed on geological time scales, and attempts to recognize them in the fossil record have generally been misleading. (One exception that holds considerable promise lies in reefs and other hard-bottom marine communities, in which the temporal superposition involved in successional processes is maintained at the appropriate scale; see [32].) All these obstacles notwithstanding, paleontologists have been able to recognize recurrent associations that appear to have at least some ecological significance in a host of marine and nonmarine environments.

The time scales of the fossil record are more suited to studying the persistence, cohesiveness, and replacement of these biotic associations, rather than their inner workings. For example, we now have a large body of fossil evidence (reviewed in [25]) that the end of the last glaciation saw a profound fragmentation of terrestrial communities, and many species that had been sympatric are today distantly allopatric and seemingly ecologically incompatible. This millennium-long perspective places limits on the concept of stable, tightly bound ecological units, and it is interesting that any necessary readjustments to new ecological associates (predators, prey, competitors) has occurred with little detectable morphologic change – stasis has been pervasive, even in groups with short generation times, such as insects [14].

On a larger scale, the fossil record can yield insights into patterns of decline and replacement among ecologically equivalent taxa, although again, contemporaneity or dynamics among individual populations is generally beyond resolution. For example, Benton [6] has argued that dinosaurs did not competitively displace therapsid reptiles but, harkening back to one of our earlier themes, radiated after the therapsids had gone into decline, presumably due to climatic and vegetational changes. Among marine invertebrates, as mentioned above, Sepkoski's [53] three major faunas appear to exhibit ecological as well as temporal discreteness: these and other major innovations originate in near-shore habitats and expand offshore over tens of millions of years [31]. Differences in species richness within these successive major faunas are not simply accommodated by addition of species to communities with roughly the same pattern of organization, but have evidently resulted from changes in community organization and number of adaptive types [2]. Such paleontological studies can again test the limits of extrapolation for ecological phenomena and detect additional processes not manifested at lower levels and at smaller time scales. We look forward to a fruitful extension with differences, not a confutation, of Darwinian principles.

Conclusion: Strengths and Weaknesses of the Fossil Record

The fossil record contains a wealth of biological information, but we face strong limitations and biases to the extraction of that information. We can only briefly summarize them here.

Quality of Preservation

Usually hard skeletons are the parts preserved, with little or no information on soft tissues, biochemistry, or genome. However, occasional Lagerstätten can preserve remarkable detail on soft tissues, including internal anatomy, and even certain kinds of behavioral information.

Quantity of Preservation

Only a very small fraction of the original biomass or taxonomic diversity is preserved (perhaps 250 000 preserved species vs. the several billion that surely lived in the past). Because fossilization and subsequent preservation require extraordinary conditions, the record is biased and certain environments and modes of life are vastly over- or under-represented. But our available sample is nonetheless concentrated on the well skeletonized marine invertebrates, permitting a comparative approach among clades in many instances. We face temporal biases as well, with certain geologic eras being disproportionately represented in the stratigraphic record; and superimposed on this is the additional time-dependent bias of improving preservation in increasingly younger rocks, with the present-day biota constituting an extraordinarily good, qualitatively different sample. Controlling for these difficulties presents a major challenge to the paleobiologist. New approaches are becoming available for appropriate choice of taxa, time intervals, sampling procedures, and the setting of realistic confidence limits.

Time Resolution

Under certain circumstances, a local stratigraphic section can preserve even a year-by-year chronology. But in the general case, we deal with separated packages of time, which add up to but a fraction of total elapsed time. Within those packages, bioturbation and other agents of time-averaging often limits time resolution. Sedimentation is spatially patchy, so that there is no guarantee, and rarely any way to demonstrate, that the same instant in time is being sampled in widely separated localities. Models are becoming available for environment-specific assessment of resolution within and among stratigraphic sections, and these will greatly improve our ability to choose the scale of the question to be asked. The fine structure of the speciation process will not be retrievable from most paleontological sequences, but the fossil record richly documents the origin and demise of a host of higher taxa and major adaptations.

Geographic Resolution

Normally it is not possible to map distributions at an instant in time due to problems in correlation; at best, tens or hundreds of thousands of years' uncertainty must be allowed. Analysis is further limited by the fact that the rock record for a given time interval may be restricted to relatively few scattered sites around the world. Therefore, biogeographic studies must be at a coarser scale than the biologist would use. This disadvantage is offset by the possibility of examining changes in diversity and distribution through successive slices of time (which admittedly become coarser with increasing age).

Drifting continents make biogeography more difficult except in the younger parts of the record. But drifting continents also suggest a new diversity of interesting questions, from the role of changing provinciality in governing global diversity to the biotic effects exhibited "when provinces collide" (e.g., [29]). As with so many other biological questions, the fossil record allows us to address these phenomena directly, rather than as extrapolations from small-scale, short-term observations of living systems.

Acknowledgements. We thank S.M. Kidwell and many of the participants of this Dahlem Workshop for helpful comments and criticism.

References

1. Ager DV (1980) The nature of the stratigraphical record, 2nd ed. Wiley, New York
2. Bambach RK (1983) Ecospace utilization and guilds in marine communities through the Phanerozoic. In: Tevesz MJS, McCall PL (eds) Biotic Interactions in Recent and Fossil Benthic Communities. Plenum Press, New York, pp 719–746
3. Behrensmeyer AK, Kidwell SM (1985) Taphonomy and paleobiology. Paleobiology 11:105–119
4. Behrensmeyer AK, Schindel D (1983) Resolving time in paleobiology. Paleobiology 9:1–8
5. Belyea PR, Thunnell RC (1984) Fourier shape and planktonic foraminiferal evolution: The *Neogloboquadrina-Pulleniatina* lineage. J Paleontol 58:1026–1040
6. Benton MJ (1983) Dinosaur success in the Triassic: A noncompetitive ecological model. Q Rev Biol 58:29–55
7. Berggren WA, Casey RE (eds) (1983) Symposium on tempo and mode of evolution from micropaleontological data. Peleobiology 9:326–428
8. Büttner D (1982) Biometrie und Evolution der *Viviparus*-Arten (Mollusca, Gasteropoda) aus der Plio-Pleistozän-Abfolge von Ost-Kos (Dodekanes, Griechenland). Berliner Geowiss Abh A 42:1–79
9. Buss LW (1983) Somatic variation and evolution. Paleobiology 9:12–16
10. Carson HL, Templeton AR (1984) Genetic revolutions in relation to speciation phenomena: The founding of new populations. Ann Rev Ecol Syst 15:97–131
11. Cisne JL, Chandlee GO, Rabe BD, Cohen JA (1982) Clinal variation, episodic evolution, and possible parapatric speciation: The trilobite *Flexicalymene senaria* along an Ordovician depth gradient. Lethaia 15:325–341
12. Cohen J, Massey BD (1983) Larvae and the origins of major phyla. Biol J Linn Soc 19:321–328
13. Cooke J, Webber JA (1983) Vertebrate embryos: Diversity in developmental strategies. Nature 306:423–424
14. Coope GR (1979) Late Cenozoic fossil Coleoptera: Evolution, biogeography, and ecology. Ann Rev Ecol Syst 10:247–268

15. Cope JCW, Skelton PW (eds) (1985) Evolutionary case histories from the fossil record. Spec Pap Palaeontol 33:1–203
16. Cronin TM (1985) Speciation and stasis in marine Ostracoda: Climatic modulation of evolution. Science 227:60–63
17. Dingus L, Sadler PM (1982) The effects of stratigraphic completeness on estimates of evolutionary rates. Syst Zool 31:400–412
18. Dodd JR, Stanton RJ Jr (1981) Paleoecology, concepts and applications. Wiley, New York
19. Einsele G, Seilacher A (eds) (1982) Cyclic and event stratification. Springer, Berlin
20. Erwin DH, Valentine JW (1984) "Hopeful monsters," transposons, and Metazoan radiation. Proc Natl Acad Sci USA 81:5482–5483
21. Fisher DC (1985) Evolutionary morphology: Beyond the analogous, the anecdotal, and the ad hoc. Paleobiology 11:120–138
22. Gould SJ (1982) The meaning of punctuated equilibrium and its role in validating a hierarchical approach to macroevolution. In: Milkman R, (ed) Perspectives on Evolution. Sinauer, Sunderland, MA, pp 83–104
23. Gould SJ (1984) Morphological channeling by structural constraint: Convergence in styles of dwarfing and gigantism in *Cerion,* with a description of two new species and a report on the discovery of the largest *Cerion*. Paleobiology 10:172–194
24. Gould SJ, Lewontin RC (1979) The spandrels of San Marco and the Panglossian paradigm: A critique of the adaptationist programme. Proc R Soc Lond B 205:581–598
25. Graham RW, Lundelius EL Jr (1984) Coevolutionary disequilibrum and Pleistocene extinctions. In: Martin PS, Klein RG (eds) Quaternary Extinctions. University of Arizona Press, Tucson, pp 223–249
26. Hansen TA (1980) Influence of larval dispersal and geographic distribution on species longevity in neogastropods. Paleobiology 6:193–207
27. Hoffman A, Kitchell JA (1984) Evolution in a pelagic planktic ecosystem: A paleobiologic test of models of multispecies evolution. Paleobiology 10:9–33
28. Jablonski D (1986) Larval ecology and macroevolution in marine invertebrates. Bull Mar Sci
29. Jablonski D, Flessa KW, Valentine JW (1985) Biogeography and paleobiology. Paleobiology 11:75–90
30. Jablonski D, Lutz RA (1983) Larval ecology of marine benthic invertebrates: paleobiological implications. Biol Rev 5:21–89
31. Jablonski D, Sepkoski JJ Jr, Bottjer DJ, Sheehan PM (1983) Onshore-offshore patterns in the evolution of Phanerozoic shelf communities. Science 222:1123–1125
32. Jackson JBC (1983) Biological determinants of present and past sessile animal distribution. In: Tevesz MJS, McCall PL (eds) Biotic Interactions in Recent and Fossil Benthic Communities. Plenum Press, New York, pp 39–120
33. Jerison HJ (1973) Evolution of the brain and intelligence. Academic Press, New York
34. Johnson ALA (1981) Detection of ecophenotypic variation in fossils and its application to a Jurassic scallop. Lethaia 14:277–285
35. Johnson JG (1982) Occurence of phyletic gradualism and punctuated equilibria through geological time. J Paleontol 56:1329–1331
36. Kidwell SM, Aigner TA (1985) Sedimentary dynamics of complex shell beds: Implications for ecologic and evolutionary patterns. In: Bayer U, Seilacher A (eds) Cycles in Sedimentation and Evolution. Springer, Berlin, pp 382–395

37. Lande R (1976) Natural selection and random genetic drift in phenotypic evolution. Evolution 30:314–334
37a. Levinton JS (1983) Stasis in progress: The empirical basis of macroevolution. Ann Rev Ecol Syst 14:103–137
38. Maynard Smith J (1984) Paleontology at the high table. Nature 309:401–402
39. Miyazaki JM, Mickevich MF (1982) Evolution of *Chesapecten* (Mollusca: Bivalvia, Miocene-Pliocene) and the biogenetic Law. Evol Biol 15:369–409
40. Mensink H (1984) Die Entwicklung der Gastropoden im miozänen See des Steinheimer Beckens (Süddeutschland). Palaeontographica Abt. A 183:1–63
41. Niklas KJ, Tiffney BH, Knoll AH (1985) Patterns in vascular land plant diversification: A factor analysis at the species level. In: Valentine JW (ed) Phanerozoic Diversity Patterns: Profiles in Macroevolution. Princeton University Press, Princeton, NJ
42. Padian K (1983) A functional analysis of flying and walking in pterosaurs. Paleobiology 9:218–239
43. Patterson C (1981) Significance of fossils in determining evolutionary relationship. Ann Rev Ecol Syst 12:195–223
44. Paul CRC (1982) The adequacy of the fossil record. In: Joysey KA, Friday AE (eds) Problems of Phylogenetic Reconstruction. Syst Assoc Spec Vol 21. Academic Press, London, pp 75–117
45. Raup DM (1972) Taxonomic diversity during the Phanerozoic. Science 177:1065–1071
46. Raup DM (1979) Biases in the fossil record of species and genera. Bull Carnegie Mus Nat Hist 13:85–91
47. Raup DM, Crick RE (1981) Evolution of single characters in the Jurassic ammonite *Kosmoceras*. Paleobiology 7:200–215
48. Reyment RA (1983) Phenotypic evolution in microfossils. Evol Biol 16:209–254
49. Rudwick MJS (1964) The inference of function from structure in fossils. Brit J Phil Sci 15:27–40
50. Seilacher A (1970) Arbeitskonzept zur Konstruktionsmorphologie. Lethaia 3:393–396
51. Seilacher A (1972) Divaricate patterns in pelecypod shells. Lethaia 5:325–343
52. Seilacher A, Reif WE, Westphal F (eds) (1982) Studies in paleoecology. N Jb Geol Paläont Abh 164:1–305
53. Sepkoski JJ Jr (1984) A kinetic model of Phanerozoic taxonomic diversity. III. Post-Paleozoic families and mass extinctions. Paleobiology 10:246–267
54. Simpson GG (1960) The history of life. In: Tax S (ed) Evolution After Darwin, vol. 1. University of Chicago Press, Chicago, pp 117–180
55. Stanley SM (1973) An explanation for Cope's Rule. Evolution 27:1–26
56. Stanley SM (1979) Macroevolution: Pattern and Process. W. H. Freeman, San Francisco
57. Templeton AR (1980) Modes of speciation and inferences based on genetic distances. Evolution 34:719–729
58. Van Valen L (1973) A new evolutionary law. Evol Theory 1:1–30
59. Velbel DB (1984) Sedimentology and taphonomy in a clastic Ordovician sea. Geol Soc Am Abstr Progr 16:204 (Abstract)
60. Vrba ES (1984) Patterns in the fossil record and evolutionary processes. In: Ho M-W, Saunders PT (eds) Beyond Neo-Darwinism. Academic Press, London, pp 115–142
61. White TD, Harris JM (1977) Suid evolution and correlation of African hominid localities. Science 198:13–21
62. Williamson PG (1981) Paleontological documentation of speciation in Cenozoic molluscs from Turkana Basin. Nature 293:437–443

Patterns and Processes in the History of Life,
eds. D. M. Raup and D. Jablonski, pp. 23–44. Dahlem Konferenzen 1986
Springer-Verlag Berlin, Heidelberg

Natural Selection and Fitness, Adaptation and Constraint

S. C. Stearns
Zoologisches Institut
4051 Basel, Switzerland

Abstract. Biologists often use abstractions developed in one context to describe material mechanisms observed in another, where the connection is not as certain. This article discusses four abstractions: "natural selection," "fitness," "adaptation," and "constraint." *Natural selection* operates simultaneously on all levels of the biological hierarchy, but its strength, determined by cycle time and proportion of variation accounted for, varies by orders of magnitude across levels. For traits varying at all levels, selection within populations must dominate. For traits fixed within but varying among clades, only selection at higher levels can have any impact. Here selection will be slow but can produce major change. *Fitness* is not a trait but a technical embodiment of assumptions about selection. Fitness definitions are either short-term of long-term and measure either abundance or risk minimization, and are either absolute (intrinsic) or relative (extrinsic). Within specialties one can often use different fitness definitions to reach the same prediction. Because fitness definitions are simply mental tools, we are free to change definitions if by so doing we gain predictive and explanatory power. No universal definition of fitness can yet be made, nor is one ever likely to be made, because different problems *define* different fitness definitions. *Adaptations* are polished products of natural selection with particularly clear relationships to particular problems faced by the organisms that posses them. They have not often been produced by selection among clades, which has produced the appearance of "history" – rough, clade-specific patterns – because selective events have been few. Constraint usually means "an explanation imported from outside the local context to explain the limits on the patterns observed." Constraints generated by a particular set of processes, here called "intermediate structure," are clade-specific and intervene between genes and phenotype to produce comparative biology. These structures are currently either formal (e.g., genetic covariance matrices) or material (e.g., hormonal integration). If we could make the formal aspects of intermediate structure material, precise, concrete, and subject to observation and experiment, we could explain much of the stasis and clade-specificity seen in the fossil record.

Introduction

This article aims to provide an overview of evolution broad enough to encompass historical and contemporary biology, macroevolution and microevolution, and comparative and experimental analysis. Its goal is to clarify discussions among specialists who might otherwise use the same terms to describe different processes and different terms to describe the same processes. Each speciality interprets key evolutionary mechanisms in ways often appropriate only for local problems. For example, evolution has been viewed as optimizing the phenotype, as the mechanical consequences of populations dynamics and meiosis, as the product of historical accident and developmental constraint, or as the interplay of speciation, extinction, and the diversity of extant Baupläne. Each view brings with it a certain enlightenment and a certain misrepresentation.

The device used to produce the overview is analysis of the basic elements of the theory – natural selection, fitness, adaption, and constraint. The argument is motivated by the contrast between the analytical modes, types of data available, and time scales considered in each specialty.

The distinction between macroevolutionary pattern and macroevolutionary process will prove useful. By macroevolutionary patterns I mean large-scale patterns detectable in the fossil record and patterns detectable by comparison of existing higher taxa. Much of this paper is devoted to the proposition that most such macroevolutionary patterns are consistent with microevolutionary explanations. One of my main points, however, is that natural selection acts simultaneously on all levels of the biological hierarchy. This leaves open the possibility that differential extinction and speciation rates of different clades, or selection for certain types of survivors during mass extinction events, by which I mean macroevolutionary processes, could have affected the patterns under study.

Natural Selection

For natural selection to have an evolutionary effect on a population consisting of some arbitrary entities, there are three necessary conditions. The trait under selection must be heritable in a broad sense, the entities must vary with respect to the trait, and this heritable variability must be associated with differential reproductive success (reproduction and/or survival [11, 23]). If these conditions are satisfied, then the population will evolve and the trait *may* change over time in a manner that can be interpreted as "increasing in adaptation." The mechanism is purely mechanical, by which I mean: in organic evolution, all that happens is that organisms are born, reproduce, and die. We see the ones that survive and try to make sense of the patterns they display. To do so, we invent concepts such as "natural selection," "fitness," "adaptation," and "constraint." These mental tools, incorporated in many abstract formalisms, are abbreviated representations of the dynamics implied by perfectly concrete, finely detailed mechanism, played out across the levels of the biological hierarchy: gene, organism, population, species...

Any collection of things with the requisite properties of heritability, variability, and differential reproductive success will, in principle, experience natural selection. Such entities include but are not limited to genes, clones, phenotypes, demes, species, and clades (23). Statically described, a clade is some level of supraspecific organization, from genera through kingdoms. Dynamically described, a clade is the trace through time of a character state or a collection of character states shared by common descent, defined by the organisms that possess it and in contrast to other clades containing organisms that do not posses it. Clades are most easily recognized by invariant traits, while also being characterized by patterns in the variance of the other traits.

We do not yet know why some traits are invariant while others vary in given clades. This is another face of the problem of stasis and constraint, discussed below.

For purposes of convenience, I refer to selection operating on "higher" or "lower" taxonomic levels or on "genera," "families," or other supraspecific levels of organization. No static description such as this can capture a dynamic pattern; the phyla that we see might have been genera to a Precambrian taxonomist. Where possible, I will use the neutral term "clade" to avoid the static prejudices with which the standard system of biological nomenclature molds our view of evolution.

Many would claim, with considerable justification, that the species is the highest level of biological organization that can be treated as a unit of selection, and that all supraspecific taxa are artifacts that we create strictly for our convenience. From this point of view, there can be no selection among clades, which are not themselves units. While granting the force of this remark, I wish to leave open the possibility of natural selection acting on collections of organisms that we might call orders, classes, or phyla. The mechanism would be mass extinction. My reason for making this move is that such collections of organisms possess the first two of the three required conditions of heritability [35], variability and differential reproductive success, and I would like to stimulate someone to evaluate the third.

Ecologists and geneticists usually think of natural selection as operating within populations. It in fact operates simultaneously on all levels in the biological hiearchy that possess the three necessary conditions. (Slatkin [35] has defined population-level heritability; his extension of the concept can be generalized, with complications, to any level in the taxonomic hierarchy.) Natural selection has different impact on different levels. The force of natural selection on a given level depends on three things [23, 31, 40]: the amount of variability for the trait in question among units within that level relative to total variability over all levels, the cycle time of those units, and the degree to which the reproductive success of those units is related to variation in the trait among the units.

For genes and individuals the cycle time is the generation time. For units at all higher levels cycle times are longer, often by orders of magnitude. (By the cycle time of species, for example, I mean the average duration of species in time: roughly 10000 to 10000000 individual generations.) The individual organism plays a key role in this hierarchy of levels of selection. Genes themselves are not units of selection but are selected through their phenotypic effects on the organisms that carry them [12]. The partitioning of variance among levels of organization should always be assessed by comparison with the amount of variance expressed as differences among individuals.

Therefore, if a trait varies heritably among individuals within populations, among populations within species, among species within genera, and so forth over all levels, then selection acting on genes and individuals within populations will usually be strong enough to resist any *counteracting* selection at higher levels simply because generation times are much shorter than the cycle times at higher levels of organization.

For the special case contrasting selection on individuals and on demes (within species), selection on demes can only by effective given an unlikely population structure (small demes with restricted migration [21, 25]. This condition is relaxed for species selection, where migration is not important because it has no consequences for the mixing of gene pools between species.

If, on the other hand, selection acts in the *same* direction at all levels, then an analysis of variance of how much variation in the trait could be accounted for by processes occuring at each of the levels would normally indicate that selection operating within populations was the dominant contributor.

A second factor amplifies the rate of evolution within sexual species: recombination. Above the level of the species, selection acts on units that behave like clones with no potential for recombination. Species selection cannot generate new character combinations but can only eliminate ones that have been generated by lower-level processes ([39] and Charlesworth, personal communication). Within species, recombination speeds the fixation and elimination of alleles under selection. This process is modified by numerous factors – ploidy, dominance, and linkage, among others – but even so proceeds more rapidly within sexual species than within clones or among species. Thus two powerful factors suggest in principle that microevolution should dominate over macroevolution: generation time and recombination.

Now consider what happens when a trait becomes invariant within a species, for whatever reason. (That some traits do become invariant is a fact. Why they become invariant is not well understood. See below, under stasis and constraint). If a trait is fixed within a species but varies among species, then selection *cannot* act on differences among individuals within species, which do not exist. It can only act, slowly, on differences among species. If a trait is fixed within a family but varies among families within an order, then selection *cannot* act on differences among genera, which do not exist. It can only act, *very* slowly, on differences among families. Thus the overall pattern of variation across levels of organization is critical to the determination of the relative importance of selection acting on each level.

It is easy enough to claim that natural selection could act in principle at all levels of the biological hierarchy, but it is quite difficult in practice to demonstrate that it has effects on any units of selection larger than the gene or the individual. Since the general importance of group selection and kin selection have not been unambigously demonstrated after two decades of work, it would be unrealistic to expect an early verdict on species selection. As Vrba ([39], p 322) clearly states, "If one can explain sorting among species solely by comparision of characters and dynamics at the levels of organisms and genomes (the effect hypothesis), then there is no need to invoke species selection. The argument amounts to a plea for recognition that not all nonrandom sorting among species need be caused by species selection... the earlier claims of species selection were not about species selection at all."

The calculation of actual forces of selection is rendered complex by the fixation of traits at different levels in different clades. For example, within one bird order, the Procellariformes, clutch size is fixed at one egg; within all other bird orders it varies. Almost all gekkos lay two eggs. The trait is nearly fixed within the family Gekkonidae, while all lizards in the genus *Anolis* lay one egg. In all other lizard families clutch size varies. In most bird and lizard species clutch size varies among individuals. Thus the calculation of selection pressures on traits fixed within larger clades must take into account the fact that often the trait is not fixed within other clades at the same level.

Are there processes that eliminate variation within levels and transfer the strength of selection up the biological hierarchy? Wade [40] found that selection on individuals within populations tended to exhaust heritable variability within populations, fixing the values of quantitative traits at different levels in different populations. This created the precondition that makes group selection effective: no variation within groups, but considerable variation among groups. The result in this case depended on a high degree of isolation among groups, and was therefore special, but it suggests what the general prerequisites would be.

The prerequisite can be extended to higher levels of organization. Speciation cuts off the gene flow between populations that helps to maintain intraspecific and intrapopulational variability. Within species selection often fixes certrain traits, Once a trait is fixed at different values for each species, the only possible evolutionary change will occur through differential rates of speciation and extinction of species correlated with the fixed values of the trait.

However, intense selection, or drift in small isolated populations, is a poor general explanation for the elimination of variation within a species, because as soon as selection is relaxed, the population has grown, or gene flow begins, genetic variation will increase again. What might be a good general explanation? We know that morphological stasis can accompany dramatic genetic change (in plethodontid salamanders [42]; in Tetrahymena, where "proteins responsible for cell architecture have diverged greatly within the genus while cellular architecture itself has remained unchanged" [45]; in all cases of canalization). This suggests that the key process in the stabilization of a complex trait *must* be sought in modifications to the phenotypic expression of genetic information: canalization, thresholds, plasticity, and so forth (G. Wagner, personal communication).

Fitness

Fitness is a technical tool used to represent some model of natural selection [38]. Fitness is not a natural property of genes, organisms, or species, and it is not a property of a phenotype. It characterizes a specific phenotype-environment interaction [12]. When the environment changes, relative and absolute fitnesses also change.

Different definitions of fitness are used for different problems. The criteria used to choose fitness definitions reflect the goals of specialties as much as they do the process of natural selection. Because the biological hierarchy and the action of natural selection on it are complicated, no single definition of fitness captures the essence of the entire process. (Some population geneticists disagree with me on this point, arguing that relative reproductive success in the next generation is adequate for all cases. I agree that this is the most broadly successful definition but beg to differ on its universality. Important research programs could be closed off before they have a chance to get started if alternative fitness definitions are ruled out.)

Even within a given level of the hierarchy, which often means within a given specialty, different definitions of fitness are used when different assumptions are made about the problem under analysis. It seems that any single fitness definition can only represent a special aspect of natural selection. For example, the definitions of fitness used in life-history evolution are particularly effective at drawing attention to the *causes* of differences in fitness among the units under selection, whereas the definitions of fitness used in populations genetics are particularly effective in analyzing the *consequences* of those fitness differences [38]. Differences in age-specific schedules of fecundity and mortality are the causes of changes in gene frequencies.

The taxonomy of fitness definitions is in its infancy. Most definitions can be usefully ordered in a three-way table defined by a first distinction between single-generation and long-term definitions, a second distinction between measures of abundance and measures of risk minimization, and a third distinction between absolute (intrinsic) measures defined in an isolated system and relative (extrinsic) measures defined in terms of competition against other types (genes, clones, species). The usual fitness measure is a single-generation measure of abundance. Examples include lifetime reproductive success in any of its forms (the W of population genetics, the Malthusian parameter r for age-structured populations) and the carrying capacity (K) of population dynamics. (Energy uptake per unit time, used in foraging theory, is a token for fitness defined over much less than a generation.) A single-generation measure of risk minimization expresses the idea of minimizing the probability of leaving no offspring that survive to maturity, for example, while a long-term measure of risk minimization expresses the idea of minimizing the probability that a gene, clone, or species would go extinct.

We know much about when to use which fitness definition for single loci where there are two sexes with age structure (5). In most cases, either r, K, or a combination of each term defined for separate sexes will suffice. For whole organisms, r has led to some success in life-history theory [37], and "lifetime reproductive success" has had considerable influence, in ordering research on animal behavior [9] and sex allocation [7]. Inclusive fitness [17] is "lifetime reproductive success" extended to take account of genetic relationships within groups of interacting individuals and has had some success in sociobiology [18, 20]. Paleontologists, implicitly, or explicitly, normally employ long-term measures of risk minimization applied to species, but long-term measures of abundance have also been suggested [33].

One striking feature of fitness definitions is the different uses to which they are put. When a population geneticist says he understands what fitness is, he means that he knows which measure to pick to predict changes in gene frequencies under certain assumptions. When a life-history theorist says he understands what fitness is, he means that he knows which measure to pick to predict the equilibrium properties of the phenotype. When the sociobiologist says he knows what fitness is, he means that he knows which measure to pick to explain apparently self-sacrificial behavior. When the paleontologist says he understands what fitness is, he means that he knows which measure to pick to explain a pattern of duration or diversity in the fossil record.

In all these research programs it is possible for the same prediction to be made or the same pattern explained, by models using different definitions of fitness. In most cases, the information at hand is not sufficient to discriminate among the alternative definitions available. For example, in life-history evolution one gets the same predictions for the optimal equilibrium properties of the phenotype over a rather broad range of conditions of one defines fitness as increasing *either* with increasing r *or* with decreasing probability of extinction. In sociobiology, behavioral patterns that for a decade had been interpreted as caused by kin selection are now being reinterpreted as the straightforward expression of the lifetime reproductive success of unrelated individuals calculated as a problem in game theory [2, 27].

In general most success has been achieved when fitness has been defined as a single-generation measure of abundance: relative contribution to the next generation. There are two reasons for this. First, models of abundance are normally deterministic and are easier to analyze than the stochastic models of risk minimization. Second, predictions made for single generations are easier to test than predictions made for the long term. In only three cases known to this writer are models based on long term measures of risk minimization used to interpret important biological patterns; the evolution of sex (recombination [3, 26, 44], the evolution of repeated reproduction over a long lifetime (iteroparity), and the evolution of dispersal [36]. In all three cases, alternative explanations based on single-generation measures of abundance have not been refuted and continue to multiply. There is no reason at present to conclude that the patterns we perceive have had a single cause. They probably were generated by a combination of processes some of which can be usefully associated with abundance measures of fitness, others with measures of risk minimization [19].

To summarize, different definitions of fitness are used in different contexts; in some contexts different definitions lead to the same predictions; in all contexts the use of a particular definition of fitness involves assumptions about how natural selection works and what units it works on. Most discussions of fitness definitions that turn into arguments do so because the implicit assumptions are not clearly articulated. If one recognizes that fitness definitions are simply tools to be applied to different problems, then there should be little controversy over using different definitions, so long as care is taken to demonstrate that there are good grounds for selecting one rather than another.

Adaptations

The word "adaptation" is used in biology to refer to three different things: a process, a product, and a state [15]. Here I ignore the historical precedence of adaptation to the theory of evolution and adopt a special, narrow definition: I take an adaptation to be a *product* of natural selection.

Adaptations can be recognized by a combination of complexity, economy, precision, and efficiency with broad taxonomic distribution [43], convergent evolution, or point-for-point correlation with environmental factors [10]. Adaptations are *not* the cause of natural selection; that notion leads directly to an empty circle. Adaptation is a difficult concept, easily misused [15, 16, 24, 43], in part because its several meanings are sometimes conflated. It has not proved easy to develop a general procedure that will allow us to discriminate clearly between those patterns that can be ascribed with certainty to some form of constraint. The difficulty arises because both selection and constraint have been involved in the production of all organic phenomena. Therefore the problem is not to decide whether this phenomenon was produced by selection or by constraints, but how much of each was involved, and precisely how they interacted. In concrete cases within a local problem area this separation can be achieved.

Is adaptation a term that can be properly employed in discussions of macroevolution? Two extreme positions are possible. One answers, "No. To do so would be confusing, because practically all examples of adaptation in which there is some hope of establishing evolutionary causation have been produced by microevolution. They are at least potentially analyzable in terms of genetics, life histories, physiology, experimental embryology, and so forth. To attribute adaptation to processes not within the reach of experiment is to risk circular reasoning." The other answers, "Yes. By the terms of your argument, natural selection proceeds simultaneously across all levels of the biological hierarchy, and adaptation is the product of natural selection according to some definition of fitness. Therefore, all we have to do in attributing some adaptation to a macroevolutionary process is to establish the level at which selection occurred and the definition of fitness appropriate to the problem at hand. Once those terms are clear, then the sense in which we use the concept of adaptation will also be clear."

I prefer the first position, am not willing to rule out the second, but find it unlikely. It takes many selective events to produce adaptations. As one goes up the taxonomic scale, one encounters fewer and fewer possible selective events since the origin of life. Only a few kingdoms have ever originated or gone extinct, a few tens of phyla, a few hundreds of classes, and a few thousands of orders, whereas billions upon billions of organisms have reproduced and died. The higher we go up the biological hierarchy, the longer we go back in time to the point where the trait was fixed, the less the chance that selection will have produced the polished appearance of adaptations. The law of large numbers diminishes in importance, and the luck of the finite draw takes over, producing patterns that are more rough-hewn and harder to interpret, precisely because the sample size is small.

Basic differences in body plan were originally produced and have been subsequently refined by microevolution. Diversity in the Precambrian was, however, low, and a small number of extinctions had unusually large effects on the sample of surviving body plans. No essential differences in natural selection at higher levels produced this difference in pattern. The number of events sampled was smaller, random events were more important, and the opportunity for selection on a large sample to produce an appearance of polish was not available. A relatively small number of extinctions in a Precambrian fauna of unusually low diversity generated clade-specific patterns of long duration and gave us the impression of "history" – the effects of a small sample. The production of the condition we call adaptation by macroevolutionary events is not impossible in principle but does seem highly unlikely.

Constraints

"Constraint" is a covering term hiding a diversity of meanings, just as "cancer" conceals a diversity of diseases. Here I list some ways the term is used and some types of constraints on evolutionary processes. These suggest meanings of constraint sharing a common feature that resembles the 19th century concept of Unity of Type (11) and that I call *intermediate structure*. It may help to connect micro- and macroevolution.

The concept of constraint originated independently in different biological traditions. In classical morphology, one defined major clades by their different basic body plans, Baupläne. These clades evidently have different potentials for producing certain roles – terrestrial herbivores, intertidal filterfeeders, nocturnal aerial raptors. One is tempted to explain different evolutionary potentials by differences in the constraints implicit in different Baupläne. For example, the locomotion of echinoderms is based on a water-vascular system. There are no terrestrial or freshwater echinoderms. This suggests that the commitment to this particular type of hydraulic locomotion rules out evolving into certain habitats. Similarly, the chitinous exoskeleton of arthropods places limits on body size, especially on land, lower than those imposed by calcareous endoskeletons, while the use of gills simultaneously for feeding and breathing may have kept bivalves from occupying the land. Here "constraint" evidently means "a plausible explanation for something that did *not* happen, based on the functions of certain structures and the evident difficulty of retaining function while changing structure."

In reductionist biology, "constraint" describes both assumptions and predictions. Explanations are couched in terms of constraints flowing directly from physical and chemical principles. Organisms are seen as reflecting both the properties of the materials out of which they are built and the principles used to organize these properties into functioning systems. For example, in biomechanics one predicts that the relationship of limb cross section to body weight is *constrained* by physical laws. In population genetics one predicts that the Mendelian mechanism *constrains* the possible rates of gene frequency change in different ways for dominant and recessive genes. In molecular biology one explains the selection of a genetic code based on triplets of base pairs in terms of *constraints* on the numbers of amino acids necessary to build functioning enzymes – more than 16, less than 64. In developmental biology one notes that the embryos of advanced organisms tend to pass through stages in which they resemble the embryos of ancestors. This suggests that the evolution of new adult morphologies is *constrained* by the possible modifications of ancestral embryos. Here "constraint" means "the properties invoked at a lower level of analysis, such as physics or chemistry, in order to explain limits on the patterns seen at some higher level, such as development or evolution" – it is practically synonymous with "reductionist explanation."

Each of these examples is presented on its own level of the biological hierarchy, but in many cases constraints cut across levels. Water has the following unusual or extreme physical and chemical properties: unusual chemical stability, very high heat capacity, very high melting and boiling points, high heat conductivity, capacity to dissolve an extremely broad range of compounds and to ionize many of them, unusually high surface tension, and lower density in the solid than in the liquid state. The charge distribution and geometry of the water molecule lead to higher-order interactions among molecules and arrays of molecules; these produce the properties listed. Because of these properties, water is used universally in living organisms. Biochemistry is the chemistry of aqueous solutions and the hydrophobic surfaces bordering them. Much of physiology represents designs based on the properties of water: water as a heat buffer, water as a medium for circulating oxygen and nutrients and removing waste products, water's high heat conductivity used in countercurrent exchange. The Reynold's number of water changes at a scale of about 1 mm, with drastic consequences for the locomotion and morphology of creatures below that size, living in a viscous medium, and above that size, living in a more fluid one.

Much of ecology is determined by the availability of water and its effect on temperatures in the environment and in the organism. The temperature of the planet itself is regulated within relatively narrow bounds by heat storage in the world's oceans and the modulation of incoming and outgoing radiation by water in the atmosphere. If we consider evolution as an engineer designing organisms, he clearly must know what water will do for you and what it prevents you from doing.

In optimality theory constraint is at times loosely used in the sense of "something I have to invoke to explain why my theory didn't work," but it has a rigorous and constructive use as well. When one formulates an optimality problem, one must state mathematical boundary conditions. These are usually derived, explicitly of implicitly, from clade-specific biological constraints. For example, of one seeks to predict optimal age-at-maturity, one must make assumptions about how fecundity increases with size and how the juvenile mortality of offspring is related to the parent's age-at-maturity. These relationships differ from clade to clade in a manner that reflects, at least in part, the different constraints implicit in different body plans. The optimality predictions are only successful if these biological constraints are properly translated into mathematical boundary conditions. The problem will only be well posed if the constraints are clearly stated and sufficient to force a solution. Thus optimality theory, far from being in conflict with the notion of constraint, depends directly on it for its success. Here constraint means "the manner in which biology sets limits on the kinds of functions that can be used and the values their parameters can take."

To say that something is constrained is simply to say that one can analyze it with reduction and that when one does so, one discovers good causal reasons why certain things are not found. There is nothing mysterious about constraints; biologists have been using them as explanations for a long time. The problem has not been with the concept of constraint, or with knowing in principle how to use it, but with the fact that the particular constraints relevant to evolutionary patterns have not been easy to identify because they arise in a poorly explored region of biology.

One could describe all of biology as the consequence of constraints – not just reductionist biology, but natural selection itself, viewed as mechanism. The meaning of the word would vanish. We can preserve it in a relative sense if we recognize that it only has meaning in a local context where one concentrates on the possibilities latent in certain processes and views the limitations on those possibilities as arising from outside that context. One could always stand the problem on its head, switch contexts, and what had been constraints would become a set of possibilities, and what had been possibilities would become constraints. Thus life histories can constrain the evolution of behavior, and behavior can constrain the evolution of life histories. Development can constrain the evolution of biochemistry, and biochemistry can constrain the evolution of development.

It is important to distinguish clearly between selective constraints and systemic constraints. Selective constraints come into play when a change in the phenotype producing a gain in one component of fitness also produces a loss in others. Because the constraint here is not on which mutations are possible but on how the possible mutations affect fitness, selective constraints show up in genetic covariances. Systemic constraints come into play when the present state of the organism constraints the possible range of phenotypes producible by mutation. Such constraints do not appear in genetic covariances, because there is no variation available on which to base a covariance estimate, and are therefore beyond the reach of standard genetic analysis.

Lack of variation in single characters, such as clutch size or number of digits, is readily explained by stabilizing selection leading to canalization [6]. Lack of variation in characters that are made up of integrated sub-characters, especially characters that perform two functions at once (such as the gills of bivalve molluscs), plausibly results from the difficulty of preserving function while changing form. It is not completely clear, however, how we are to identify a single character or conclude that it only has one function. I suggest, moreover, that canalization is normally achieved by a progressive integration of the phenotype that inevitably involves other characters and their functions. Thus lack of variation in "single" characters may have the same underlying cause as lack of variation in complex characters: organismal integration.

In evolutionary biology some constraints are seen as arising from genetics, development, physiology, and demography. Listing a few obvious constraints in each of these areas will suggest how the concept can be used locally to define new problems.

Genetic constraints flow from the use of DNA as the genetic material, from its organization into chromosomes, from diploidy and meiosis, and from their consequences for populations. Diploidy makes possible dominance, and dominance makes possible deleterious recessives. Diploidy also leads to heterozygotes, thus making possible heterozygote advantage. Inbreeding depression is caused by deleterious recessives and the superiority of heterozygotes, lost when alleles are fixed. Thus inbreeding depression can be viewed as a constraint on the evolution of mating systems that stems indirectly from diploidy.

Linkage is also an important constraint. In few-locus systems the location of genes on the same chromosome can force a genetic equilibrium in which fitness is not maximized [28]. This constraint may have consequences for the evolution of the genetic system – single-locus, few-locus, polygenic.

The evolutionary potential of a population is not guaranteed by the existence of additive genetic variance, as a naive reading of Fisher [14] would suggest. It also depends on the relationship between the genetic and phenotypic covariance matrices and the geometry of the fitness landscape. Genetic and phenotypic covariance matrices, which describe organismal integration, must evolve [6, 34] but not under all circumstances and not in all directions [4].

Developmental constraints limit possible adult morphologies and life histories [1, 4, 32]. Development proceeds as a cascading sequence of events containing sensitive points where small perturbations – either genetic or environmental – can trigger large changes in the outcome. Each step in the sequence is often embodied in a few local rules capable of generating a globally coherent pattern. The genes code for the local rules, not for the global pattern [30]. This means that the global pattern – the adult morphology, the life history – is not under point-for-point genetic control, at least not in all its important aspects. (Most traits can be fine-tuned by modifier genes.) Developmental processes intervene between genes and phenotypes in such a way that only a restricted set of phenotypes is presented to the environment [1]: The intermediate phenotypes cannot be produced and are never "seen" by selection.

Another sort of developmental constraint arises when a change in the timing of events is favored. For changes in timing to evolve, processes that were previously integrated must become uncoupled in such a way that function is preserved [32]. This is probably not possible for all processes, but we do not yet know in general what separates developmental processes that can be uncoupled from those that cannot.

Not all developmental processes function as constraints on the phenotype. We know, for example, that major rearrangements of *early* development can take place (in larval stages, in embryos of viviparous species) without any essential change in the adults. Many developmental mechanisms are better viewed as making possible opportunities for adaptation rather than constraining the phenotype. Any balanced treatment of developmental constraint should take this into account. Similar comments could be made about all other classes of constraints.

Prominent physiological constraints include the different implications of heterothermy and endothermy and the long list of consequences of different surface area: volume ratios. Equally significant is the selection of a regulatory mode: hormones or nerves. Once a regulatory mechanism is chosen, further development and refinement of that mechanism is constrained in ways strongly tied to the initial choice.

The basic demographic constraint is generation time, which is closely related to developmental rate and age-at-maturity. It limits the rate at which many ecological and all evolutionary processes can occur. Other demographic constraints include the different consequences of discrete vs. overlapping generations, of stable vs. unstable age distributions, of simple vs. complex life cycles, and of different general patterns (long life vs. short life, many reproductive events vs. one reproductive event). These constraints affect population dynamics and thus influence both rates of gene frequency change and patterns at the community level.

Lists of constraints do not help much unless they suggest new problems or new ways of thinking about old problems, for such lists could be extended almost without limit. To focus attention on what is essential, the following questions may be useful: What is the minimal abstraction of the organism nesessary and sufficient to explain the patterns of interest? What mechanisms *must* we acknowledge?

Just as with the definition of fitness, one gets different answers when these questions are asked in different contexts. If one is interested in explaining differences in the design of whole organisms, then one answer leads to the concept of *intermediate structure*. This is a new term, introduced with reluctance, but necessary because it describes a collection of processes that share important effects and because these processes are not accurately described by terms such as "developmental and physiological constraints" – they span practically all levels of the biological hierarchy, and their proper analysis will require a redefinition of organismal structure that ignores the traditional boundaries of specialties.

Intermediate Structure

The term "intermediate structure" describes the processes intervening between genotype and phenotype that produce local, clade-specific constraints. Darwin ([11], Ch. 6) recognized such processes as Unity of Type but was not in a position to analyze them. The constraints inherent in genetics, population genetics, in most of biochemistry and cell biology, and in demography are usually general. They do not often depend on what *kind* of organism one is talking about. Intermediate structure, on the other hand, depends very much on clade. It consists mostly of those principles of developmental biology, physiology, and functional morphology that constrain the set of phenotypes that can be poduced within a clade. Specialists in these disciplines have not usually seen their results as principles applicable to problems in a larger context.

For example, consider the following ways of describing a facet of "organismal integration." This population geneticist sees pleiotropy, the impact of one gene on two or more traits. The quantitative geneticist summarizes many pleiotropic effects in one genetic covariance matrix, which is thought to play a major role in constraining the rate and direction of evolution (cf. papers in [13]). The genetic covariance matrix, however, is a formal representation of developmental and physiological constraints [8]. The life-history theorist sees the cost of reproduction – increased reproduction associated with decreased growth and survival. The physiologist sees endocrine control shifting rates of energy transfer to reproduction, growth, and maintenance.

Some of these descriptions are formal, as is the case with the genetic covariance matrix, with pleiotropy, and with the cost of reproduction. Others are material, as is the case with the effect of a particular hormone on the reproductive organs. The formal descriptions have arisen as a provisional attempt to grasp the key features of a reductionist programm that is not yet complete. To connect the material to the formal and thereby make the formal precise, concrete, and subject to experiment is a natural goal.

To attain a rigorous comparative biology, one must have in hand and a clear summary of how intermediate structure differs concretely among clades. To appreciate the full impact of historical processes on what we now see, we must first have a rigorous comparative biology of living organisms. The results can then be used to infer the impact of the contemporary biology of times long past on the generation of historical patterns. Clearly both research programs can only advance together, since they depend on each other's results.

How does intermediate structure, and thus clade-specific constraint, arise? If there is a short answer, it is, throught the different methods of *organismal integration* that evolve in each clade. Some of these differences are potentially predictable from inherent differences in body plan, mode of development, and life cycle, but others represent strictly opportunistic divergences. Evolution works with the materials at hand and takes the route of least resistance. The results, in terms of organismal design, can be wildly nonlinear (Brenner, quoted by Lewin [22]).

Once this process has begun within a clade, one cannot return to the beginning and recover the range of opportunities present at an earlier stage. Both the organism and the environment will have changed irreversibly. The progressive building up of layers of integration, all based on materials and interactions actually existing in the organisms in each intermediate stage, produces a progressive commitment to a certain design and its limitations. These can be expressed as limitations on the directions in which the genetic covariance matrix can change, as constraints implicit in the commitment to a certain feeding mode or type of skeleton, or as some phrase in the language of practically any biological specialty. They almost all reduce to some aspect of organismal integration.

Misunderstandings

This perspective makes clear one way of solving two misunderstandings about evolution.

First, it is often argued, particularly by mathematicians making initial contact with evolutionary theory, that natural selection could not possibly work because it claims that mutations are random, and there is no way that a random process could produce so many complex, efficient, polished structures in the time available. Such arguments miss the point that mutations are only random with respect to fitness; they are constrained by, indeed defined by, clade-dependent intermediate structure and can only modify already-existing phenotypes [29, 34]. Selection necessarily acts on modifications of what is already there. These phenotypes occupy a tiny fraction of the space of potential phenotypes and have been moving for a long time through that space in the direction of some sort of increasing adaptation. Thus the claim of the mathematicians is wrong because they have assumed an inappropriate definition of "random".

Second, the concept of "regulatory genes" often comes up in discussions of developmental constraints. It is used, for example, to explain why large genetic differences can be found with small morphological differences, and why large morphological differences can at times be associated with small genetic differences [46]. There is, however, no such thing as a regulatory gene, only regulatory systems with genes embedded in them. Some genes make products that have direct control over the expression of other genes, and some genes make products that have indirect control over the expression of other genes. At times, gene products may be primarily structural in their impact. At other times, the same products may be primarily regulatory in their impact. Since the same gene product can play various context-dependent roles, it is a mistake to assign to the genes a property that is properly applied to the different roles their products can play in different contexts (G. Wagner, personal communication). The technical distinction between regulatory and structural in worth preserving, so long as one remembers that they refer to context-dependent roles, not to a property of a gene as fixed as its DNA sequence.

Discussion

From my perspective, there is no inconsistency between microevolution and macroevolution, and no break in causation between the two. The process of natural selection proceeds in the same fashion within populations and among species. The essential difference is that selection above the level of the population is slow and only effective on traits that are fixed within species or clades, where it is the only possible kind of selection. The recombination and short generation time within populations normally mean that microevolution will dominate. In contrast, selection at and above the species level is effectively clonal.

The interactions between microevolution and macroevolution are most interesting. The microevolutionary emphasis is on processes that maintain genetic heterogeneity within populations, because heritable variability is a prerequisite for effective selection. The macroevolutionary emphasis is on processes that eliminate genetic heterogeneity from populations, species, genera, etc., fixing traits within clades and transferring the relative strength of selection up the hierarchy. The process that fixes a trait permanently within a species is probably selection acting on intermediate structure itself, leading to canalization, thresholds, or plasticity. The fixation of a trait will be followed by further organismal reintegration based on the constancy of the trait: all subsequent changes will "assume" it is fixed. Thus changes in intermediate structure followed by further integration based on those changes can maintain traits in a fixed state in spite of subsequent mutations, thus creating the preconditions for (much slower) species selection. Processes tending to fix traits within genera, families, and so forth are less well understood and may not exist. (Species selection is the logical candidate.) If they do not, then all of stasis could be traced back to microevolutionary modifications of intermediate structure.

Microevolutionary processes may produce adaptations that have the effect of reducing the normal – but not the catastrophic – extinction probability. Candidate traits include iteroparity (a long life with many reproductive events), sex (recombination and diverse offspring), broad dispersal, many offspring, and variation within a single brood of the requirements for breaking diapause and leaving a resistant stage. Such adaptations can be grouped under the heading of "risk-minimizing traits."

These traits appear to do two things at once: first, minimize the probability that an individual organism will *not* leave offspring that survive to maturity; second, decrease the probability that a species will go extinct. There are sound microevolutionary explanations for the evolution of each of these traits, but they have macroevolutionary impact in the sense that species possessing them tend to go extinct less frequently than species that to not. This is, of course, a form of clade selection that generates a pattern in the taxonomic distribution of the traits. Should we attribute that pattern to macroevolution or to microevolution? Vrba (39) would choose microevolutionary on ground of parsimony and explanatory power. I think it is caused by the interaction of both. If risk-minimizing traits in fact have such effects, it is interesting to note that many lineages do not evolve these properties. They constitute cases in which microevolution is in apparent conflict with avoidance of extinction. To explore this idea more thoroughly, we need information on the relative rates of origination and extinction of lineages that possess these traits versus their opposites, using well controlled comparisons that explicitly acknowledge and deal with the effect of confounding traits not involved in the issue.

A basic question focuses attention on how microevolution and macroevolution interact. How much of the pattern under discussion is explained by the organism' relationship with the external world, and how much by the necessity to coordinate processes within the organisms? What is the relative importance of the outside and the inside of organisms?

Acknowledgements. G. Wagner, J. Tuomi, E. Haukioja, M. Wade, and D. Raup stimulated many of these ideas. W. Hamilton explained Price's paper to me. S. Stevenson, D Wake, G. Wagner, A. van Noordwijk, H. Reichert, B. Stearns, J. Koella, T. Bürgin, P. Schmidt, B. Charlesworth, A. Panchen, D. Futuyma, M. Ghiselin, J. Dzik, M. Andersson, K. Linsenmair, E. Vrba, and V. Mosbrugger read a draft and improved it with their comments. My thanks to all.

References

1. Alberch P (1980) Ontogenesis and morphological diversification. Am Zool 20:653–667
2. Axelrod R, Hamilton WD (1981) The evolution of cooperation. Science 211:1390–1396
3. Bell G (1982) The Masterpiece of nature: The evolution and genetics of sexuality. Berkeley, University of California Press
4. Bonner JT (ed) (1982) Evolution and Development. Dahlem Konferenzen. Springer, Berlin Heidelberg New York
5. Charlesworth B (1980) Evolution in Age-structured Populations. Cambridge University Press, Cambridge
6. Charlesworth B, Lande R, Slatkin M (1982) A neo-darwinian commentary on macroevolution. Evolution 36:474–498
7. Charnov EL (1982) The theory of sex allocation. Princeton University Press, Princeton
8. Cheverud JM (1984) Quantitative genetics and developmental constraints on evolution by selection. J Theoret Biol 110:155–1712
9. Clutton-Brock TH, Guinness F, Albon SD (1982) Red Deer: Behavior and ecology of two sexes. University of Chicago Press, Chicago
10. Curio E (1973) Towards a methodology of teleonomy. Experientia 29:1045–1058
11. Darwin C (1859) On the origin of species by means of natural selection. John Murray, London
12. Dawkins R (1982) The extended phenotype. W.H. Freeman, San Francisco
13. Dingle H, Hegmann JR (eds) (1982) Evolution and genetics of life histories. Springer, Berlin Heidelberg New York
14. Fisher RA (1930) The genetical theory of natural selection. Oxford University Press, Oxford
15. Ghiselin M (1966) On semantic pitfalls of biological adaptation. Phil Sci 33:147–153
16. Gould SJ, Lewontin RC (1979) The spandrels of San Marco and the Panglossian paradigm: a critique of the adaptationist program. Proc Roy Soc Lond B 205:581–598
17. Hamilton WD (1964) The genetical evolution of social behavior, I, II. Theoret Biol 7:1–52
18. Hamilton WD (1972) Altruism and related phenomena, mainly in social insects. Ann Rev Ecol Syst 3:193–232
19. Hilborn R, Stearns SC (1982) On inference in ecology and evolutionary biology: the problems of multiple causes. Acta Biotheor 31:145–164
20. King's College Sociobiology Study Group (1982) Current problems in sociobiology. Cambridge University Press, Cambridge
21. Leigh EG Jr (1983) When does the good of the group override the advantage of the individual? Proc Natl Sci USA 80:2985–2989
22. Lewin R (1984) Why is development so illogical? Science 224:1327–1329
23. Lewontin RC (1970) The units of selection. Ann Rev Ecol Syst 1:1–18
24. Lewontin RC (1978) Adaptation. Sci Am 239:156–169
25. Maynard Smith J (1976) Group selection. Q Rev Biol 51:277–283
26. Maynard Smith J (1978) The evolution of sex. Cambridge University Press, Cambridge

27. Maynard Smith J (1982) Evolution and the theory of games. Cambridge University Press, Cambridge
28. Moran PAP (1964) On the nonexistence of adaptive landscapes. Ann Hum Genet 27:383–393
29. Muller HJ (1949) Reintegration of the symposium on genetics, paleontology, and evolution. In: Jepsen GL, Simpson GG, Mayr E (eds) Genetics, Paleontology, and Evolution. Princeton University Press, Princeton, pp 421–445
30. Oster G, Alberch P (1982) Evolution and bifurcation of developmental programs. Evolution 36:444–459
31. Price GR (1972) Extension of covariance selection mathematics. Ann Hum Genet 35:485–490
32. Raff RA, Kaufman TC (1983) Embryos, genes, and evolution. MacMillan, New York
33. Raup D (1984) Evolutionary radiations and extinctions. In: Holland HD, Trendall AF (eds) Patterns of Change in Earth Evolution, Dahlem Konferenzen. Springer, Berlin Heidelberg New York Tokyo pp 5–14
34. Simpson GG (1953) The major features of evolution. Columbia University Press, New York
35. Slatkin M (1981) Populational heritability. Evolution 35:859–871
36. Stearns SC, Crandall RE (1981) Bet-hedging and persistence as adaptations of colonizers. In: Scudder GGE, Reveal JL (eds) Evolution Today. Hunt Institute, Pittsburg, pp 371–383
37. Stearns SC, Crandall RE (1984) Plasticity for age and size at sexual maturity: A life-history response to unavoidable stress. In: Potts G, Wootton R (eds) Fish Reproduction. Academic Press, New York, pp 13–33
38. Tuomi J, Haukioja E (1979) An analysis of natural selection in models of life-history. Savonia 3:9–16
39. Vrba ES (1984) What is species selection? Syst Zool 33:328–335
40. Wade M (1977) An experimental study of group selection. Evolution 31:134–153
41. Wagner G (1984) Coevolution of functionally constrained characters: prerequisites for adaptive versatility. Biosystems 17:51–55
42. Wake DB, Roth G, Wake MH (1983) On the problem of stasis in organismal evolution. J Theoret Biol 101:211–224
43. Williams GC (1966) Adaptation and natural selection. Princeton University Press, Princeton
44. Williams GC (1975) Sex and evolution. Princeton University Press, Princeton
45. Williams NE (1984) An apparent disjunction between the evolution of form and substance in the genus Tetrahymena. Evolution 38:25–33
46. Wilson AC, Carlson SS, White TJ (1977) Biochemical evolution. Ann Rev Biochem 46:573–636

Standing, left to right:
Reinhard Rieger, Dan Fisher, Stephen Gould, Jerzy Dzik, Volker Mosbrugger

Seated (center)), left to right:
Wolf-Ernst Reif, Duane Meeter, Armand de Ricqlès, Dolf Seilacher

Seated (front), left to right:
Mike LaBarbera, Ed Connor, David Wake
(Not shown) Günter Wagner

Patterns and Processes in the History of Life,
eds. D. M. Raup and D. Jablonski, pp. 47–67. Dahlem Konferenzen 1986
Springer-Verlag Berlin, Heidelberg

Directions in the History of Life

Group Report

D. B. Wake, Rapporteur
E. F. Connor
A. J. de Ricqlès
J. Dzik
D. C. Fisher
S. J. Gould
M. LaBarbera
D. A. Meeter
V. Mosbrugger
W.-E. Reif
R. M. Rieger
A. Seilacher
G. P. Wagner

Introduction

The discussions of this group centered on the oldest topic in evolutionary biology – patterns in the history of life and their causes. We live in an apparently ordered world, and because evolutionary biology developed as a discipline in the context of western intellectual history, it was perhaps inevitable that perceived patterns in the history of life would be interpreted in terms of some of the major hopes and aspirations of Europeans. Thus we see the wide prevalence of ideas relating such concepts as directionality, progress, determinism, and adaptationism to diverse patterns and trends, ranging from the taxonomy of life (Linnaeus) to the Scala Naturae [6, 19]. We now reject simple notions of order such as special creation, foreordination, or that life is just the manifestation of the postulated law-like structure of the universe. But we are still left with the facts that there are perceived patterns and directional trends in the history of life. We must first recognize and define patterns in order to help us detect and identify underlying structure, cause, or both. Patterns demand explanation and careful analysis of them may lead to the identification, recognition, and understanding of underlying processes. In turn, we may generate hypotheses concerning both pattern and process, and attempt tests.

Our job, then, has been to develop an analytical approach to patterns in evolution. Patterns are first detected, then characterized and defined. When we are satisfied that a pattern is real, we seek cause(s). Directionality is especially intriguing, for we have no a priori expectation from current theory that evolution should be directional or progressive.

Patterns in the History of Life

Nineteenth-century biologists perceived patterns in the history of life (e.g., Cope's Rule, Williston's Principle, Dollo's Law) [18]. While these evolutionary laws were based on empiricisms from the fossil record, all were framed using intuitive ideas and non-quantitative approaches. Our group in discussion arranged patterns and directional trends according to levels of organization in tabular form for purposes of discussion (Table 1, a selective list). In analyzing potential trends, problems of hierarchy and genealogy and matters related to scaling arise. We largely restricted our discussion to general patterns in order to avoid narrow discussion of trends that appear to be group- or lineage-specific. We also limited our discussion primarily to patterns and trends that could be detected in the fossil record.

Identification of Patterns

The reality of the trends selected remains controversial, and thus a good deal of our group's discussion centered on recognition of patterns. Considerations of pattern in evolution are biased from the outset if we assume that pattern and directionality will be found. Often our data base is inadequate to demonstrate statistically a trend that we believe that we can see. Both the quantity and the quality of the data base need to be improved, and we especially must make the transition to more quantitative analyses of data. With increased quantification will come opportunities for in-depth examination of patterns, both to test for their reality and to analyze them more specifically.

Table 1. Patterns in the history of life

Patterns at the level of cells and molecules
Accumulation of random changes in evolution (molecular evolutionary clock)
Changes in genome size
Trends in organismal design
Increasing autonomization
Size changes
Changes in organizational complexity
Differentiation and synorganization
Modification of ontogeny and astogeny
Trends in the formation of clades
Unspecialized ancestor, specialized descendants
Early experimentation, later specialization
Irreversibility
Parallel and interactive evolution
Convergence
Trends in diversity
Increase in the occupancy of ecospace
Replacement of taxa through time

Probabilistic and Statistical Approaches

Despite severe problems in applying statistics to paleobiology, attempts should be made to use statistical methods whenever possible. Some statistical approaches are outlined in Table 2. Manipulative experiments are the most powerful, but they may be difficult to design and conduct. Suppose, for example, that a decrease of shell ornamentation were detected that appeared to be correlated with increase in burrowing in bivalves. We hypothe-

Table 2. Statistical approaches to the study of directional trends in evolution

Approach	Amount of observer intervention	Strength of conclusions
Manipulative experiment	High	High
Sample survey	Intermediate	Intermediate
Judgement sample	Low	Low

size that shell ornamentation affects burrowing rate. Ten fossil shells would be selected randomly, plaster casts produced of each, and ornamentation then filed off five randomly selected casts. Burrowing motion of the bivalves could be simulated by burrowing the plaster casts in an appropriate medium using a mechanical arm. Ideally, manipulative experiments of this sort should contain at least two different treatment levels, each of which is separately and independently applied to at least two experimental units for purposes of replication. An example of an approach such as this is the work of Stanley [24].

Nonexperimental approaches can also be used. For example, in a sample survey a researcher specifies a hypothesis concerning Cope's Rule. Ten individuals of a given species of mollusk are selected randomly from each of six (geological) stages and measurements are taken. Subsequent analysis will disclose whether size has increased during the time period selected and if the data are in accord with expectations derived from Cope's Rule.

The least robust approach is a judgment sample test in which, for example, a researcher decides to test Cope's Rule on bivalves and chooses a few families which, in the researcher's opinion, represent the "cleanest" case for the test.

Statisticians have difficulty with concepts such as "natural experiment" or "Gedanken-experiment." The problem is that with repetition the force of the modifier, "natural," is lost – what was originally a correlative study acquires "strength by association" with the concept of the "experiment." In a properly conducted (scientific) experiment, an experimenter applies specific treatments to randomly selected units while others serve as controls. However, natural experiments result from phenomena that are unknown and the "treatments" are applied to the observational units in an uncontrolled way. As a result, there are many possible explanations as to *why* the treatments did or did not cause a difference.

Models of randomness are appropriate in many contexts and often are useful as null hypotheses, but randomness must be judged relative to a defined context. Even if we were to take the philosophical position that no phenomenon is random, all of its causal determinants may never be discovered, leading to the appearance of unpredictable or random behavior. There are many probabilistic models of randomness. The fact that the data in any particular case are consistent with a randomness model does not refute the possibility of a deterministic process, but it might suggest that the forces producing the variation (or presumed pattern) are many, rather small, and operating independently. A number of statistical methods exist which can be applied to the problems of detection of patterns in the fossil record, from relatively simple correlation analysis to relatively complex time series analysis (Connor, this volume).

Generally, observations close in time or space are independent; closeness is measured relative to the temporal or spatial extent of the processes causing the variation. But most statistical techniques assume that observations are independent, contrary to the situation common in paleobiology. The *consequences* of violating the assumption of independence are severe: for example, in many cases standard errors are underestimated. If possible, statistical techniques which do not assume independence of data points should be used.

Time series methods deal with dependence in data in an explicit way. Paleobiologists should encourage statisticians to work on new time series techniques to deal with the problems peculiar to paleobiology. These problems include differing durations of species and the separation of species in evolutionary time by periods of discretely different duration, as well as missing data.

Some statistical methodologies of value include the following [7, 9, 14, 15]:

Survivorship and Reliability Analysis techniques from the fields of engineering and biostatistics for examining, fitting, and testing the distribution of life lengths. These models deal with *censoring* of data, death due to *competing* risks, and nonparametric estimates of life distributions.

Probabilistic Models and Stochastic Processes. The theories of Markov processes, birth and death processes, epidemics, and branching and diffusion are all potentially useful – in most cases, however, the development of theory has outstripped techniques for applying the models to actual data [2, 3].

Simulation. The output of any simulation depends on the specific choices made for the factor levels and the way the factors are functionally related. The design of experiments can be used in simulation studies to expand the usefulness and generality of the output [5, 11].

Hypotheses of patterns and processes can only be tested by evidence collected in an unbiased way after agreement about what sort of evidence constitutes an appropriate test. Concern with standards of evidence (the nature, quantity, types, repeatability, etc., of appropriate evidence) has been a major preoccupation of ecologists. This has enabled ecology to move from a field dominated by anecdotes and unsupported assertions to one in which theory, models, and hypotheses can be corroborated or rejected by acquisition of quantitative observations. Paleobiology might benefit from using similar methodologies and analyses. This was not a focus of this workshop, but we suggest the following as a guideline for the future.

Theory is an important guide for collecting evidence, and data are critical in evaluating theories. However, this interdependency among data, hypothesis generation, and hypothesis testing has led to confusion about that which constitutes a "test" of a hypothesis rather than simply the generation of a new hypothesis. To "test" a hypothesis one must first explicitly state the hypothesis and define a group or population to which test inferences will apply. Data can then be collected from the population, but this must be done in a manner such that the probability of sampling any datum for the group is known and nonzero. Hence, the data could conceivably falsify the hypothesis. Decision rules can be established concerning whether the tested hypothesis should be entertained or rejected, given the sampled data. These decision rules embody statements about the probability of errors that may arise in deciding whether or not the tested hypothesis should be rejected. Failure to reject the tested hypothesis does not necessarily constitute its acceptance but should lead to further, more critical tests of the same hypothesis.

In general, conclusions concerning the reality of trends and directions become convincing only with repetition. Although we urge statistical approaches whenever possible, highly subjective elements enter into most paleobiological considerations of patterns, in part due to the limited data sets available. In general, repetition should be attempted at several different levels – using independent cases, independent methods, independent investigators, and by analyzing different subsets of the data. The key element is congruence (i.e., consilience, consistency, coherency) of results, tested statistically.

Recognized Patterns

Despite difficulties with recognition and identification, most biologists and paleontologists would agree that patterns in evolution exist and that there are trends and vectors (which may have both speed and directional components) over evolutionary time. But which patterns, trends, and vectors exist, how pervasive they are, and how they may be explained remain controversial. For example, at the molecular level there is a good correlation of molecular evolution and time, different molecules evolving at different but relatively constant rates depending on their function and complexity [10, 28]. Patterns at the molecular level appear to be time-dependent only for a given molecule and there is little correlation with change at the organismal level.

There are also putative trends at the level of organismal design (Table 1), but confirmation of these trends is difficult for at least three reasons: the features of interest are often qualitative rather than quantitative traits, the hypothesis of a particular trend has rarely been framed in a manner allowing it to be tested, and the data available usually involve comparisons across clades rather than within clades.

Complexity of structure and autonomization (i.e., the degree of homeostasis or autonomous buffering of environmental variables) are two aspects of organismal design that often are perceived to be closely associated, although they may be quite independent ("complexity" may include such diverse features as shell sculpturing and neuronal connectivity); we will treat them together. Across the spectrum of metazoans and metaphytes, from invertebrates through vertebrates, and algae to seed plants, autonomization and complexity obviously increase, but on lower levels of organization the trends are less clear. The independent evolution of endothermy in vertebrates and insects represents trends towards increasing autonomization and complexity (via the tight integration of nervous and endocrine systems acting on cellular metabolism), presumably driven by the premium on behavioral performance in these animals. All possible combinations of trends of increasing or decreasing complexity and internalization can be found in various clades of metazoans, each presumably driven by the selective factors arising from the ecology of the organisms involved. Note, however, that complexity at the cellular and tissue levels remains approximately constant within phyla, and across most of the metazoan phyla [1].

Although Cope's Rule (a trend of size increase within lineages) generally seems to hold true, there are numerous exceptions ([16, 23]; see LaBarbera, this volume). Aspects of hierarchy and scale often determine the pattern observed. For example, both brachiopods and crinoids reach their maximum size in approximately the middle of their history, and size declines in later times, but this pattern is dictated by size trends within particular classes in each phylum and the true pattern is most likely to be stasis. Stanley [23] has proposed that Cope's Rule arises from an artifact of cladogenesis – most groups arise from small ancestors, and subsequent radiation yields the appearance of a general increase in size. However, data on body sizes of all members of a clade or a series of clades have not been gathered, so whether the pattern of increase in maximum size within a clade results from an increase in mean size or an increase in the variance in body size within a clade remains unknown. A number of selective pressures doubtless underlies size determination of individuals in populations and species and size trends in clades ([16, 23]; see LaBarbera, this volume).

Another aspect of organismal size change during evolution that deserves mention is that of the two distinct quantum changes in the *limits* to size that can be deduced as having occurred. The first, the evolution of multicellularity, represented an escape from the limits on complexity and size imposed by the unicellular condition; the second, the evolution(s) of an internal fluid transport ("circulatory") system, freed metaphytes and metazoans from the limits imposed by diffusion within the organism or between the organism and the environment [12] and permitted the evolution of complexity mentioned above (in aquatic photoautotrophic plants, size and complexity obviously are not so immediately dependent on the evolution of an internal circulatory system).

Patterns of differentiation and synorganization in organisms are similar to patterns of autonomization and complexity. The evolutionary process has produced the diversity of morphologies seen through the Phanerozoic using a surprisingly small diversity of essentially stereotyped material – the cells and tissues. For example, from the Ordovician through the Recent, the number and diversity of tissue types in vertebrates has remained virtually constant. However, although the "bricks" are stereotyped, the architectural variants show clear trends within groups towards increasing differentiation. Classical examples would include the repeated trend from full metamerism to oligometamerism in polychaete annelids and both tagmatization of body segments and differentiation of endo- and exopodites of the appendages of arthropods.

It may be useful to add the concept of vectors to analyses of trends, for then speed and direction naturally follow as parameters to be measured. If variances in speed and direction are high, we will observe no trend or pattern but only "noise". But if speed is slow or has a high variance while directional variance is low, an evolutionary trend will be seen. Vector analysis will also provide a means of analyzing pattern among clades. A clade can show a directional trend if one group speciates rapidly (high speed) with little variance in direction (in respect to features under study) regardless of what has happened to other groups within the clade. A clade can also show a directional trend if all groups speciate (with or without high variance in speed), but with low to modest directional variance (i.e., there is a group trend or a homogeneity in the clade). It is important to determine if trends perceived are real or if they might have resulted from similar vectors in non-monophyletic groups.

Pattern and Process

Pattern and process are closer conceptually than we have thought, and we quickly learned that it was difficult if not impossible to discuss pattern without respect to processes. Pattern can be understood as process in reduced dimensionality. One might argue that patterns are simplified, less continuous characterizations of process. An additional important implication is that concepts of pattern and process are hierarchically slippable ("heterarchical") [8]. Size increase is a *pattern* with respect to the lower-level process that generates it but is itself a *process* to a higher-order history of transformation. A pattern is a pattern because it characterizes a process.

Table 3 contains several modes of explanation for perceived directional trends in evolution. The first, microevolution and its consequences, is probably the most common explanation for evolutionary trends, but other explanations deserve consideration. The second explanation, the topology of phylogenetic trees, relates to the fact that there has been but one history and one genealogy of life. The matrix of intercorrelated characters is very complex, and this historically based framework of homology and descent gives pattern and order by itself. The third mode, differential origination and extinction, includes what has become known as species selection, a somewhat controversial process that relates to the dynamics of patterns of origination and extinction, which can produce trends [25]. The fourth mode involves internal dynamics at the organismal level – the systems of fabricational, functional, and developmental constraints which characterize highly integrated systems. Constraints serve as boundaries that bias the direction of evolution when they are approached (Wagner and Stearns, both this volume; see [17, 20]). The final point ("artifacts") is not a biological explanation, but as a procedural matter we should probably always try to eliminate first element 5 of Table 3 and then element 2 before seeking other explanations.

Table 3. Modes of explanation for patterns

1. Consequences of deterministic processes in natural populations (microevolution)
2. Topology of phylogenetic trees in a random world
3. Differential origin and extinction of taxa (species selection)
4. Fabricational, functional, and developmental constraints (or internal organismal dynamics, evolutionary diffusion in a constraint morphospace, and "ratchets" within Markovian processes)
5. Artifacts

We have chosen one detailed example and a few additional ones which illustrate the interplay between pattern and process in establishment of evolutionary direction. We do this in the full knowledge that presentation of a few examples, even elegant ones, is not sufficient to serve as a general explanation for the appearance of patterns in the evolution of life. We would have liked to discuss relative frequencies of phenomena, but we could not because either the data base is inadequate, the appropriate analyses have not been done, or both. The examples we chose are well studied directional trends which illustrate the interplay of pattern and process.

Evolutionary History of Bivalves: An Example

The evolutionary history of bivalve mollusks outlined by Seilacher [21] illustrates the interplay between two of the possible modes of explanation: microevolution and channeling by a variety of internal and external constraints (Fig. 1). Because of their relatively simple and uniform construc-

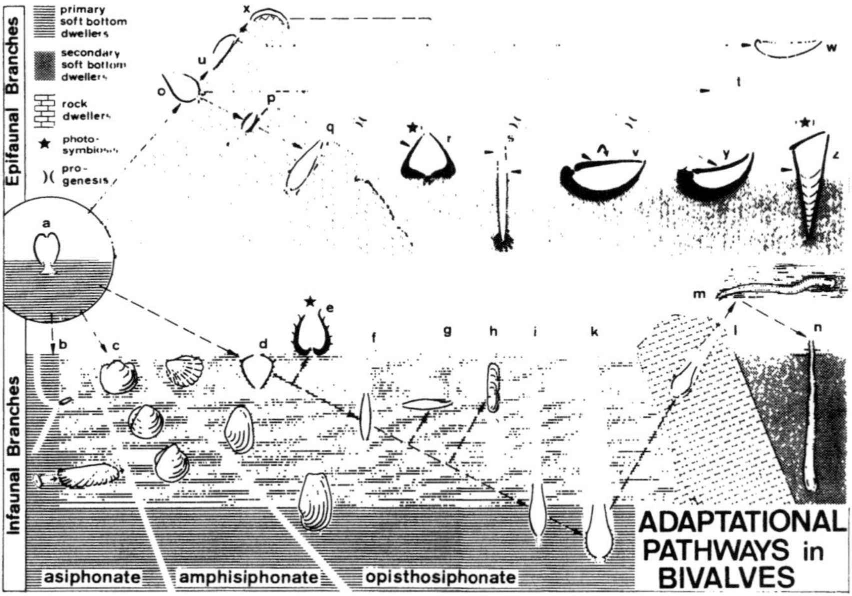

Fig. 1. Evolutionary pathways in the Bivalvia showing several alternative tracks leading from an ancestral, vagile, microphagous soft-bottom dweller (a). See text for explanation

tion, bivalve shells are particularly suited for comparison of evolutionary trends in extant and fossil lineages. As will be seen, "directionality" arises as the outcome of a Markovian progression (Fisher, this volume) in an adaptive landscape. The adaptational pathway is determined by the (a) point of entry into the ecospace (in this case as vagile, microphagous bottom dwellers, Fig. 1 a); (b) nonrandom distribution of adaptive peaks (niches, guilds, "Lebensformtypen") that allow groups to become established along various tracks; (c) constructional and developmental constraints that limit the number of options available at any point; and (d) "racheting" by the irreversibility of most adaptational transformations.

The evolutionary progression can thus be described as sequence of steps. Each step is dependent on the changes that had gone before and in turn provided, retrospectively, the opportunity to improve fitness either by *elaboration* on an adaptive theme or exploitation of a new mode of life made available as an accidental by-product of the immediately preceding changes (*innovation*).

Infaunal branches

Since the transition to filter-feeding in early bivalves removed the necessity for contant mobility, retreat into the sediment was an obvious response to increasing predator pressures. This habitat shift required (a) transformation of the crawling foot into a hydrostatic organ for push-and-pull burrowing, and (b) a ventilation system for continuous flushing of the gills with surface water. Of the three solutions to the ventilation problem (Fig. 1, b: pumping water through a U-shaped dwelling tunnel, *Solemya*; Fig. 1, c: anterior inhalant tube and posterior exhalant siphon, lucinoids; Fig. 1, d: both exhalant and inhalant siphons at posterior and as in all other burrowing bivalves), the third was most successful because it freed the anterior margin for the activity of the foot and allowed shell streamlining without shortening the hinge. Increasingly deeper burrowing proceeded along the following routes:

Fig. 1, d: *Upper tier burrowers*
Elaboration: Establishment of "recovery strategists" with sturdy and highly sculptured shells and the potential to reburrow quickly (cockles, trigoniids, venerids, *Donax*), which allows them to colonize mobile sediments; fusion of ventral mantle edges to keep sediment from intruding into gill chamber.
Innovation: Either return to the surface by the "drag" of photosymbiosis (Fig. 1, e: *Tridacna, Corculum,* and several fossil

venerids) followed by reduction of the foot, or deeper burrowing.

Fig. 1, f: *Middle tier burrowers* streamline the shell and extend the fused mantle into long siphons that are completely retractable but remain separate.

Elaboration: Use of the inhalant siphon to sediment-feed by "vacuum-cleaning" the surface. Burrowing in a horizontal position (Fig. 1, g) increases efficiency of movement to new grazing grounds.

Innovation: Razor-clam shape for quick retreat along a vertical tunnel (*Tagelus*, Fig. 1, h) or deeper burrowing.

Fig. 1, i: *Lower tier burrowers* (gape clams), whose united twin siphon is only partly retractable but can be closed during burrowing to transform the mantle cavity into a hydrostatic skeleton. With this to aid in valve-opening, the ligament can be reduced in spite of increasing sediment pressure.

Elaboration: Well protected suspension feeders.

Innovation: Transformation of the foot into a holdfast and transition to mud boring.

Fig. 1, k: *Mud borers* use the versatility of the new opening mechanism to produce a permanent, club-shaped dwelling tube by scraping with the shell edges.

Elaboration: Invasion of muds that are too stiff for conventional burrowing.

Innovation: Transition to mechanical rock boring.

Fig. 1, l: *Rock borers* concentrate action on the anterior part of the shell edge. Pholads, in which the ligament has completely disappeared, expand the attachment of the anterior adductor muscle beyond the primary hinge axis so that they can open the valves around two axes allowing the rows of marginal teeth successively to "chisel" the rock during a single active stroke.

Elaboration: Establishment in a variety of rock types.

Innovation: Transition to wood.

Fig. 1, m: *Wood borers* reduce valve movement to the secondary (dorsoventral) hinge axis and "file" the wood with a regular grid of teeth formed around an angular pedal gape. They also seal the walls of their boreholes with a calcareous lining.

Elaboration: Use of sawdust as an accessory food source via bacterial symbiosis (*Teredo*).

Innovation: Transition to Fig. 1, n, digesting of small wood fragments.

Fig. 1, n: *Tube-inhabiting secondary soft-bottom dwellers* extend the calcareous tube into the sediment, using the flush for piston-boring to substitute for the lost burrowing ability.
Elaboration: Invasion of soft bottoms.
Innovation: None.

Epifaunal branches

The early commitment to a life attached to rocks led to reduction of the foot and either maintenance of the larval byssus throughout life, cementation, or other means of stabilization.

Fig. 1, c: *Edgewise byssal attachment* leads to reduction of the anterior adductor and migration of the posterior one to a more central position.
Elaboration: Broadening of anterior base (*Mytilus*); escape by swimming (*Lima*, Fig. 1, t).
Innovation: Nestling in rock crevices (Fig. 1, p), pleurothetic attachment (Fig. 1, u) or reclining on soft bottom (Fig. 1, r).

Fig. 1, p: *Nestlers* enjoy increased protection, but their growth is limited by the size of available crevices.
Elaboration: Adaptation by size reduction.
Innovation: Active enlargement of crevice by boring (Fig. 1, q).

Fig. 1, q: *Byssate borers* (e.g., *Lithophaga*), lacking the preadaptations of a siphon and a burrowing mechanism, bore mainly by chemical means and are therefore restricted to carbonate rocks.
Elaboration: Life in wave-exposed cliffs or in symbiosis with corals.
Innovation: None.

Fig. 1, r: *Edgewise recliners* invade soft bottoms, stabilizing themselves by byssus-rooting and differential shell weighting.
Elaboration: Size increase; acquisition of photosymbiosis.
Innovation: Transition to Fig. 1, s.

Fig. 1, s: *Endobyssate mud-stickers* are stabilized by the byssus and by mud deposited around them.
Elaboration: Ability to adjust to changes of the level of the sediment surface by byssus displacement, elongation and development of accessory opening mechanisms (e.g., *Pinna*).

Innovation: Return to surface by the evolution of photosymbiosis.

Fig. 1, u: *Pleurothetic byssate rock dwellers* tend to become unequally valved.
Elaboration: Escape by swimming in flat position (Fig. 1, w).
Innovation: Transition to soft bottoms or to rock cementation (Fig. 1, x).

Fig. 1, v: *Pleurothetic recliners* returned to soft bottoms either via swimmers (pectinids, Fig. 1, w) or via reversal of the original attitude (Fig. 1, u).
Elaboration: Weighting the more convex (now lower) valve in a *Gryphaea*-like fashion.
Innovation: Transition to mud-sticking (Fig. 1, s) with accessory opening mechanisms.

Fig. 1, x: *Cemented rock dwellers* become firmly attached with one valve and consequently lose their rigid morphogenetic programs.
Elaboration: Encrustation of various hard substrates (oysters).
Innovation: Transition to soft bottoms via miniaturized (progenetic) shell encrusters by change in larval substrate preference.

Fig. 1, y: *Pleurothetic cemented recliners,* after having outgrown their initial shell substrates, extend growth programs into shapes that increase stabilization by weighting (differential in gryphid forms, Fig. 1, y), flattening, or outriggers, but always in a pleurothetic mode.

Fig. 1, z: *Cemented mud-stickers* either elongate both valves and develop accessory opening mechanisms (similar to Fig. 1, s), or the attached valve grows into a high cone with the other valve modified as a lid.
Elaboration: Dense growth; obstruction of lower shell cavity by septa; return to surface after acquisition of photosymbiosis (in rudists).
Innovation: Transition to reef-building (rudists).

This flow diagram portrays only a small number of anastomoses; "directionality" is imparted only by the necessarily preceding evolutionary steps. Similar diagrams could be established for other clades and at different taxonomic levels [21].

Additional Examples of Interplay of Pattern and Process

There are other examples of long-term trends involving interplay of pattern and process. One of these has been outlined by Niklas (this volume). Another was presented by Vermeij (in preparation) as the hypothesis of escalation. He argues that within a given environment the competitive and defensive capacities of individuals should increase in expression and incidence under the selective pressure of their enemies. Since the latter in turn increase their capacities with respect to the former, escalation occurs; a pervasive directionality results. This is an argument from principles of natural selection and adaptation ("first principles" to Vermeij) for a particular microevolutionary process underlying a particular pattern in particular environments. The generality of the pattern (i.e., its repetition) is unclear, because available data are insufficient to perform the steps necessary to test the hypothesis for different environments.

One can imagine other patterns, including those which involve major changes in organismic design, as having evolved via an extended microevolutionary process. One example might be the trend towards autonomization in the early history of metazoans, but no data are available nor will they ever be (because it is a unique historical event) for adequate testing of hypotheses of the underlying process for this example; furthermore, microevolutionary theory is sufficiently robust to obviate the need for alternative explanations.

Good example exist to illustrate our four main modes of explanation (Table 1) either by themselves or in concert with each other (e.g., [27] on differential species proliferation and its consequences in African antelope). An example of tree topology is the prevalence of hair and feathers among living vertebrates and the increase of feathered and haired species in the Cenozoic. This trend is a simple outcome of the rapid speciation rates of birds and mammals whose remote ancestors had evolved feathers and hair. While feathers and hair, and associated features, may indeed have direct relevance to the general success of these groups, these structures are probably only indirectly related to events at the level of speciation, except in such non feature-specific aspects as coloration. There are also good examples illustrating uncoordinated trends in a given clade and suggestion of hierarchical levels of processes underlying such patterns [26].

However, there are also directional evolutionary trends for which we have no ready explanation: rapid origin of new body plans early in metazoan history (Valentine, this volume); increase in body size in Cenozoic mammals, correlated with reduced reproductive potential (pressures leading to increase in body size must be strong in order to counter fitness loss; LaBarbera, this volume); increasing complexity of lineages resulting from the construction of complex and highly integrated systems (e.g., the mammalian middle ear); different patterns of symmetry.

Explanations for Patterns

Some members of our group strongly prefer multiple explanations and see patterns as resulting from three or four modes of explanation; others are satisfied that microevolutionary explanations, with natural selection as the exclusive driving force, are sufficient to account for all significant patterns and they see no need to invoke alternative explanations. In general, all members of our group avoid both strict reductionist and unitarian views of natural selection as well as the (antagonistic) dualist view of microevolution versus macroevolution, in favor of hierarchical views which span several levels of biological organization (Wagner, this volume; see [4, 13]). Microevolutionary explanations may fail to satisfy because they are seen as oversimplified models of phenotypic evolution that are too local in their effects. We seek a broader kind of explanation, based in the organism but extending to issues relating to the relative rates of origination and extinction of groups, and patterns of replacement and occupation of ecospace, as well as to issues relating to complex interactions at the level of organisms (e.g., developmental constraints), relative constancy of cell and tissue types over vast periods of time, and the uncertain relation between molecular and morphological evolution.

Trends in Diversity

Our discussions dealt only in passing with trends in diversity, for we felt they were discussed more meaningfully in the groups dealing with extinction and ecology. Succession of major evolutionary faunas (Sepkoski, this volume) requires some explanation, but whether there is a pattern is very uncertain. It may be that each event is unique. The opinions of members of our group vary, but there is a subjective impression that there is resilience in the organic world, and only catastrophies such as major extinction events can produce any substantial resetting of evolutionary direction. The unanswered and probably unanswerable question is whether there is any predictability or directionality to the resetting. Replacements of faunas involve changes that may well include directional evolution of ecological and physiological attributes of the organisms. For example, there may by a tendency towards increased physiological homeostasis and efficiency which might be fostered by processes akin to Vermeij's escalation hypothesis. Perhaps Sepkoski's [22] three faunas (Seilacher would prefer "dynasties") can be interpreted in terms of directed replacements in a world in which both interspecies competition for primary food resources and predation pressure gradually increased. If a Vendian and a Tommotian dynasty are added at the beginning, the pattern may become more visible.

Conclusions and Recommendations for Future Research

Patterns exist in the history of life, as has been known from ancient times, but for the most part they remain intuitional, unquantified, and anecdotal. There are trends, including some fascinating ones, but others (such as Cope's Rule, which was thought to be on a sound footing as a generalization) are not secure (LaBarbera, this volume). We are united in our belief that the fossil record even as presently known contains much more useful information than has been utilized to date. We urge that quantitative, statistically sound, hypothesis-based studies be undertaken of specific trends and of patterns more generally (as in Connor's application of time series statistics to the mass extinction data, this volume). We have failed to get a truly "new" understanding of the issue we addressed in this workshop, possibly because we continue to use the same methods. New methods and new ideas are needed, and extension of such promising approaches as stochastic simulation and statistical analysis will help. Our theories seem to be too general, and the reliable, well analyzed data sets are perhaps too specific (and certainly too few in number). Some believe that new ideas are needed, but if these new ideas include concepts that have not proven to be solidly based or tractable to analysis in the modern living world, much caution is necessary. While there is much to be gained from the fossil record, it is very tempting to go too far and to speculate on unknowable topics.

There is substantial difficulty, given our present knowledge and the state of the fossil record, in recognizing whether or not patterns are present. Close study of the evidence with modern statistical techniques should be pursued. It is possible to fail to detect a pattern that exists if one's procedures lack statistical power. This could occur because the data are highly variable, the alternative hypothesis(es) is indistinguishable from the tested hypothesis (given the data at hand), the sample size is inadequate, or because one slavishly demands significance levels (i.e., $p<0.05$) hallowed by custom rather than by conscious evaluation.

We present a list of suggestions for future training and research of those who wish to study patterns in the history of life.

1. The field is growing more quantitative, and changes in the education of graduate students are necessary. Statistical methods can be borrowed

from other fields to some extent, but we need new kinds of statistical tests. At least some workers in this field must work closely with statisticians and mathematicians to generate new approaches, an endeavor that will require persistence and may entail disappointment.

2. There are differences in approach, and such diversity is good for the field. Some (e.g., Vermeij) argue from first principles of natural selection and adaptation, seeking to pursue specific issues to the point where pattern emerges. Others place their emphasis on patterns being predictable outcomes of processes than can be modeled, and believe that a close and disciplined study will elucidate the underlying processes. Whatever the approach, independent confirmation of findings is essential.

3. Directional patterns in life's history have traditionally been described as secular changes in mean values of the proposed "trait" – size increase, complexity, etc. We need new modes of description as well as explanation to move this subject away from the narrative nineteenth-century "Gesetzmäßigkeiten" to testable ideas. For example, many trends now viewed as only shifts of means may arise as consequences of expansions or contractions of variance about a constant central tendency. Attempts to identify and characterize evolutionary vectors should be undertaken.

4. Attention should be given to questions of relative frequency of events in the fossil record, and documentation is needed both of instances in which there are patterns and trends and of situations in which no pattern emerges. In this context, negative data are just as important as positive data.

5. More ontogenetic information can be gleaned from the fossil record. Careful study of ontogenies within clades can contribute to our understanding of directional trends and the appearance of novel structures. With this foundation, studies of specific problems relating to patterns and trends can be pursued, e.g.: (a) introduction of evolutionary novelty in serially arranged organs; (b) development of polymorphism in colonial organisms, divergent evolution of particular polymorphs and the control of their spatial distribution in the colony; (c) introductions of evolutionary novelties during astogeny of such colonial organisms as bryozoans and graptolites, with special reference to problems of morphogenesis and trophic ecology. What does the succession of ontogenies within a phylectic lineage tell us about evolutionary mechanisms?

6. Careful study of time-controlled, continuous sequences of paleontological data is needed to provide the basis for sound phylogenetic analysis. Parallelism is a heavy constraint on phylogenetic analysis, but a careful study of ontogeny, morphogenesis, fine structure, and functional morphology can in part resolve some of the homeomorphy caused by parallelism.

7. A common, central data base consisting of data gathered with defined standards could be of great importance for the field. It may be impossible to detect and analyze any but the most evident trends without a greatly improved and accessible data base.

8. The search for trends should be made more clade-specific. For example, careful analysis of sizes in all members of a clade is needed in order to determine whether traditional framing of Cope's Rule is appropriate. An analysis of size distributions of all species in a hierarchically nested set of clades would be especially useful.

9. There is need for a microevolutionary model to deal with the evolution of complex functional systems in organisms, because in the absence of such a model the explanatory power of population genetics is unclear.

10. We need operational measures of complexity, preferably quantitative measures. For morphological features, one aspect that might be explored is image processing techniques for computer visualization.

11. We should map Markovian flow diagrams of major taxa as hypotheses for future verification or falsification.

12. It may be that entrenched ideas and cultural biases tied to linguistic issues and intellectual history are at the base of our difficulty in communicating about some of the issues raised. For example, while the English-speaking tradition, dominated by the Neo-Darwinian paradigm, by and large insisted upon external (ecologial) factors and selective pressures upon organisms, the earlier tradition primarily in the continental Western European countries had a different emphasis. Stemming from the "rational morphology" program, it emphasized internal constraints in organisms' evolution. The renewal of interest in functional, constructional, geometrical/topological, and temporal constraints in organism ontogenies together with current progress in developmental genetics may help to reconcile these traditions into a more general synthesis of evolutionary mechanics. More than ever before we need to be clear and precise in our use of words and in developing the concepts associated with these words: we must strive to understand each other.

13. Some phenomena in the history of multicellular organisms may not be understandable in terms of classical and even refined microevolutionary mechanisms; new modes of explanation may be required.

14. We need to explore new and alternative modes of explanation for patterns and trends, especially those which can be detected only with information from the fossil record. The analysis of patterns, trends, directionality, and the like is not of great interest if it involves just the unfolding through time of well understood, classical microevolutionary mechanisms.

We believe that constraints have played important roles in the history of life. We expect that data from large-scale patterns will give new insight into the relationship of order and diversity in the natural world.

References

1. Bonner JT (1965) Size and cycle: An essay of the structure of biology. Princeton, NJ: Princeton University Press
2. Cox DR, Lewis PAW (1968) The statistical analysis of series of events. London: Methuen and Co
3. Cox DR, Miller HD (1965) The theory of stochastic processes. New York: J Wiley and Sons
4. de Ricqlès A (1984) Unité et diversité du vivant; L'évolution des organismes. In: Encyclopaedia Universalis, 2nd ed, pp 388–393. Paris: Encyclopaedia Universalis
5. Fagerstrom R, Meeter D (1983) The design and analysis of a simulation: tolerance quantiles and random walk designs. Commun Stat: Sim Comp 24:541–558
6. Gould SJ (1982) In praise of Charles Darwin. Discover 3:20–25
7. Gross AJ, Clark VA (1975) Survival distributions; applications in the biomedical sciences. New York: J Wiley and Sons
8. Hofstader DR (1979) Gödel, Escher, Bach: An eternal golden braid. Brighton: Harvester Press
9. Kalbfleisch JD, Prentice RL (1980) The statistical analysis of failure time data. New York: J Wiley and Sons
10. Kimura M (1983) The neutral theory of molecular evolution. Cambridge: Cambridge University Press
11. Kleijnen JPC (1974) Statistical techniques in simulation, part 2. New York: Marcel Dekker
12. LaBarbera M, Vogel S (1982) The design of fluid transport systems in organisms. Am Sci 70:54–60
13. Larson A (1984) Neontological inferences of evolutionary pattern and process in the salamander family *Plethodontidae*. In: Evolutionary Biology, eds Hecht MK, Wallace B, Prance GT, pp 119–217. New York: Plenum Publishing Co
14. Mann N, Schafer R, Singpurwalla N (1974) Methods for statistical analysis of reliability and life data. New York: J Wiley and Sons
15. Nelson W (1983) Applied life data analysis. New York: J Wiley and Sons
16. Newell ND (1949) Phyletic size increase, an important trend illustrated by fossil invertebrates. Evolution 3:103–124
17. Reif W-E, Thomas RDK, Fischer MS (1985) Constructional morphology: the analysis of constraints in evolution. Acta Biotheor 34:233–248
18. Rensch B (1960) The laws of evolution. In: The evolution of life, ed Tax S, pp 95–116. Chicago: University of Chicago Press
19. Riedl R (1978) Order in living organisms. Brisbane, Chichester, New York, Toronto: J Wiley and Sons
20. Roth G, Wake DB (1985) Trends in the functional morphology and sensorimotor control of feeding behavior in salamanders: an example of the role of internal dynamics in evolution. Acta Biother 34:175–192

21. Seilacher A (1985) Bivalve morphology and function. In: Mollusca, notes for a short course, ed Broadhead T. Knoxville, TN: University of Tennessee. Stud Geol 9:88–101
22. Sepkoski JJ Jr (1981) A factor analytic description of the Phanerozoic marine fossil record. Paleobiology 7:36–53
23. Stanley SM (1973) An explanation for Cope's rule. Evolution 27:1–25
24. Stanley SM (1975) Why clams have the shape they have: an experimental analysis of burrowing. Paleobiology 1:49–58
25. Stanley SM (1979) Macroevolution. San Francisco: WH Freeman and Co
26. Van Valen LM (1985) A theory of origination and extinction. Evol Theory 7:133–142
27. Vrba ES (1984) Evolutionary pattern and process in the sister group Alcelaphini-Aepycerotini (Mammalia: Bovidae). In: Living Fossils, eds Eldredge N, Stanley SM, pp 62–79. Berlin Heidelberg New York Tokyo: Springer Verlag
28. Wilson AC, Carlson SS, White TJ (1977) Biochemical evolution. Ann Rev Biochem 46:573–639

Patterns and Processes in the History of Life,
eds. D.M. Raup and D. Jablonski, pp. 69–98. Dahlem Konferenzen 1986
Springer-Verlag Berlin, Heidelberg

The Evolution and Ecology of Body Size

M. LaBarbera

Dept. of Anatomy
The University of Chicago
Chicago, IL 60637, USA

Abstract. An organism's size affects virtually all aspects of its physiology and ecology. There are presently no theoretical models which can explain the broad patterns of shape and functional changes observed; empirical descriptions of these patterns have suffered from a lack of rigor in choice and analysis of data.

The trend of persistent size increase in lineages of animals may be an artifact; the frequency of dwarfing may be hidden by taphonomic (preservational) and observational biases. The selective factors underlying persistent size changes are likely to differ between terrestrial vertebrates and marine invertebrates. The genetics of size change is poorly known and cannot be deduced from strictly allometric studies.

Body size is likely to have profound effects on the probabilities of speciation and extinction, and these effects would probably be amplified during periods of mass extinction.

Introduction

"You now see how, from the things demonstrated thus far, there clearly follows the impossibility (not only for art, but for nature herself) of increasing machines to immense size. Thus it is impossible to build enormous ships, palaces, or temples, for which oars, masts, beamwork, iron chains, and in sum all parts shall hold together; nor could nature make trees of immeasurable size, because their branches would eventually fail of their own weight; and likewise it would be impossible to fashion skeletons for men, horses, or other animals which could exist and carry out their functions proportionably when such animals were increased to immense height – unless the bones were made of much harder and more resistant material than the usual, or where deformed by disproportionate thickening, so that the shape and appearance of the animal would become monstrously gross."

Galileo Galilei, *Two New Sciences*

An organism's size is perhaps its most apparent characteristic. However, despite early attempts to establish the limitations that size per se brings to animal design, size was largely ignored as a variable in biology. The works of Hemmingsen [74, 75] and Schmidt-Neilsen [152] on the physiology of body size, Gould [60] on body size in evolution, the symposium volume edited by Pedley [123] on size and functional morphology, and a general theoretical model by McMahon [112] have collectively served to bring the importance of body size as a biological variable forcibly to the attention of biologists. The recent explosive growth of both theoretical and empirical studies on the influence of size on all aspects of biology and evolution and the extension of the focus of such studies from physiology and functional morphology to ecological characteristics has resulted in a spate of books on the subject [33, 124, 153]. As a result, the field of allometry, the study of the influence of size on form and function, has rather suddenly become a prominant focus in ecology and evolutionary biology.

The Kinds of Allometry

Cock [44], in an exemplary presentation, distinguishes three main categories of primary data in allometric studies – static, cross-sectional, and longitudinal – and four different levels of allometric analysis – ontogenetic, intraspecific, interspecific, and evolutionary allometry. Different combinations of primary data types and levels of analysis will produce different results. The mixture of different data types in allometric studies is not uncommon and may be difficult to detect without close reading; the results of such studies are, of course, of dubious validity for any purpose. It is equally important to realize that the results of allometric studies at any of the four levels of analysis are rarely transferable to another level [39]. Many authors have taken ontogenetic or interspecific trends as indicative of evolutionary trends; such conclusions should be viewed with extreme skepticism.

The Analysis of Allometric Data

The "simple law of allometry" (or, more tersely, the allometric equation), $Y = a\ M^b$, dominates the field; its use is usually justified by either a geometrical argument (see below) or the argument that this form is the most appropriate representation of relative growth rates [60]. Before discussing techniques for fitting the allometric equation to data and deriving estimates of the scaling coefficients and exponents, the sole rival in the modern literature to the allometric equation should be discussed.

Jolicoeur's Principal Components Model

Jolicoeur [84] suggested that the first principle component in a principal components analysis represents a measure of size and that the scaling exponents of the n dimensions in the multivariate analysis were equal to the loading of each dimension divided by $n^{-1/2}$. This multivariate approach is tempting in its simplicity, and the technique has been extensively used. The only critical test of this suggestion was performed by Jungers and German [87], who compared Jolicoeur's principal components method to conventional regression techniques. Although principal components analysis correctly rank-ordered the scaling exponents of each variable, the exponents derived were not equal to those derived from bivariate analyses. By definition, the squares of the exponents derived from a principal components analysis must average to one [44, 87]. The largest exponent will thus appear to be positively allometric and smaller exponents will be more negative; inclusion of another dimension *must* change the exponents for all other variables [87]. The absolute values and relative magnitudes of scaling exponents derived from a principle components analysis must be viewed with extreme suspicion – only by happenstance will they match exponents derived from more conventional bivariate analysis of the same data.

Variables in Allometric Analyses

Cock's comment ([44], p 181) that "The standard of reporting and analyzing metrical growth data is still often deplorably low" unfortunately remains true. If the range of sizes included in an analysis is small (less than an order of magnitude), the probability that the scaling exponent and coefficient will be distorted by sampling error is high. If a conflict exists in the literature on the value for a particular exponent, in general the study involving the larger size range should be given greater weight.

Fitting Data to the Allometric Equation

The Log Transform. To simplify the estimation of scaling coefficients and exponents, the simple equation of allometry is usually transformed by taking logarithms:

$$\log Y = \log a + b \log M.$$

In this form, least squares linear regression techniques can be used to estimate the coefficient and exponent. A log transformation also normalizes the distribution of the Y variable at any given value of M. (Y is assumed to be lognormally distributed [12, 13] as is true of a broad variety of data [12, 13, 73, 165, 189].) However, a log transform introduces a systematic bias into estimates of the allometric coefficient [12, 16, 167, 185]. Least squares linear regression techniques find the best fit line to *mean* values of the y variables, but the mean of a log transformed variable is the *median* of the original (lognormal) distribution. The correction for this bias is conveniently estimated from the standard error of log a [167]. Since the correction depends on the variance of the log-transformed data and thus will be unique to each study, all published values for scaling coefficients must be viewed as suspect (a particularly unfortunate circumstance for studies such as [61, 62, 65, 184], where ratios of scaling coefficients were the core of the analyses).

Regression Models. Statisticians distinguish two regression models on the relationship between the variables, the variance structure of the data, and associated error [99]. Model I regressions such as the common least squares linear regression assume that the error term ($\log \varepsilon$) is normally distributed with a mean of zero and constant variance, the distribution of log Y is normal at each value of log M, the variance of log Y is constant across the range of log M, and log M is an independent variable. This last point is often misunderstood. In biometrics, an independent variable is one whose value is both *known without error* and *set by the investigator*. The value of log M is never determined without error; although the error may be minor, in many allometric studies values of M are determined indirectly (from other allometric equations or from secondary sources) and are likely to have a large error term (see [44], p 144). In addition, values of M are not set by the investigator; allometric data thus violates the statistical assumptions underlying Model I regression techniques.

In Modell II regressions such as major axis or reduced major axis (geometric mean) regressions, both variables are assumed to have an associated error term; there is no "independent" variable and a line is fit by minimizing the sum of products of residuals of both log Y and log M. Major axis and reduced major axis assume different error structures and variance relations of the two variables. See [159] for an excellent discussion of all three regressions in the context of allometry.

The choice of regression model is not trivial, for the three techniques yield different results. Least squares linear regression consistently yields the lowest values for the scaling exponent, while either reduced major axis or major axis may yield the highest value depending on the variance structure of the data; the differences between the three techniques decrease as the correlation coefficient (r) increases. The scaling exponent determined by reduced major axis is equal to the exponent determined by least squares linear regression divided by the correlation coefficient [143]. Some [4, 85, 164] object to the use of Model II regressions in allometric studies as sacrificing biological relevance to a statistical straightjacket; all seem to misunderstand the techniques and the non-intuitive definitions of dependent and independent variables in biometrics. Indeed, in evolutionary studies a Model II regression such as reduced major axis would be particularly advantageous. As Lande [94, 96] shows, the slope of the evolutionary trajectory under pure selection for body size should be equal to the additive genetic correlation between the variable of interest and body size times $h_y\sigma_y/h_x\sigma_x$, where h_y and h_x are the square roots of the heritabilities of the variable and body size, respectively; σ_y/σ_x is the slope of the reduced major axis regression of the log-transformed variable vs. log body size. In general, Model II regressions are to be preferred [63, 90, 93, 99, 143, 159, 166]. The easiest option would be the use of least squares linear regression, reporting both the regression slope and that slope divided by the correlation coefficient (= reduced major axis slope). Appropriate tests for significance of differences between slopes determined by reduced major axis are available [42]; standard errors will be numerically equal to the standard errors for least squares regressions [44].

Allometric Models

Constant Geometry

For any series of *geometrically similar* objects (i.e., objects whose linear dimensions differ only by a constant multiplier), surface areas are proportional to some characteristic length squared and volumes are proportional to some characteristic length cubed. Any pair of functionally related variables which are dependent on different aspects of the geometry must change as size changes, since length, area, and volume of geometrically similar objects change at different rates.

Geometric similarity or its equivalent, *isometric growth,* is usually the null hypothesis in allometric studies and usually implies a change in functional relations with change in size. To compare empirical results with null hypotheses, it is necessary to compare the exponents relating one parameter (geometric or other) to another. Most allometric studies assume that whole-body densities of animals are the same and substitute mass for volume, yielding the "simple law of allometry":

$$Y = a\ M^b.$$

where Y is any variable of interest, a is a coefficient of proportionality (the scaling coefficient), b is the scaling exponent, and M is the organism's mass.

Functional Constancy

Most allometric models are based on the maintenance of some functionally relevant variable constant over a range of sizes; they often predict a regular distortion of the organism's geometry to counteract functional changes. The oldest of such models is the "surface law of metabolism," where body temperature is the constant parameter. On the (implicit) assumptions of geometric similarity and uniform and constant body temperatures and thermal conductances, metabolic rates would be set by the rate of heat loss and thus by the surface area of the body, i.e., metabolic rates (MR) should scale as body mass to the two-thirds power.

The history of the "surface law" is an object lesson in the seductive potential of allometric "laws." Although none of the assumptions were ever critically tested and the vast majority of the data were at variance with the prediction, the "surface law" was considered valid until Kleiber [92] demonstrated that mammalian metabolic rates scaled as $M^{0.75}$ [153]. Discrepancies between the "surface law" and relevant data were glossed over, in large measure because of the satisfyingly "physical" nature of the theory. Indeed, despite 50 years of additional data, the "surface law" persists in the literature. Theoretical models which purport to predict scaling relations should be carefully tested; empirical scaling relations are "noisy" and the desire of biologists for laws of the same power as those in physics is strong.

The most important theoretical models in modern allometry consider environmentally imposed or self-generated physical forces; they are "mechanical" models which maintain a constant safety factor. The assumption that physical properties of materials are scale-independent appears to be valid for muscle [59], mammalian bone [18, 45], and tendon [178]. Since mechanical models of allometry predict changes in shape associated with changes in size, they are usually tested against a null hypothesis of geometric similarity.

Static Stress Similarity

Static stress similarity assumes that force per unit area in a supporting structure is maintained constant by a regular increase in relative cross-sectional area as organisms' sizes increase. The predictions of a static stress similarity model differ depending on the loading regime. Static stress similarity predicts that for self-loaded columns, cross-sectional areas must remain proportional to body mass and thus $d \propto M^{1/2}$ and l is constant; for selfloaded beams, $d \propto M^{2/5}$ and $l \propto M^{1/5}$.

Elastic Similarity

An elastic similarity model assumes that the likelihood of failure by buckling or bending is kept constant. Elastic similarity was first proposed by Rashevsky [137, 138] and was revived and extended by McMahon [112]. To maintain proportional deflections constant as size changes, McMahon has shown that regular distortions of shape must occur such that $l \propto M^{1/4}$, $d \propto M^{3/8}$, and thus $l \propto d^{2/3}$. Elastic similarity assumes that forces are proportional to the weight of the organism. Since McMahon's elastic similarity model presently dominates the field of allometry, I will attempt to evaluate the evidence relating to its validity.

Trees. The elastic similarity model has been tested using data on the dimensions of trees [112, 114, 118]. Trees would seem to be a particularly robust test since the mechanical situation is straightforward – loading is almost exclusively by gravitational and drag forces, trees do not move or exhibit complex behavior, and it can be argued that there is an evolutionary premium on achieving large size (height) with a minimum investment of biomass.

McMahon [114, 118] models trees as fractals (i.e., self-similar structures) following an elastic similarity model where, at any point in the tree

$d \propto l^{3/2}$; the mean for all five values (four species) in the data was 1.50 More straightforward data is available on scaling exponents for tree height vs. basal diameter for trees from the Brookhaven Forest [186]; the mean for these three values (three species) was 1.67. The model is even more poorly fit by shrubs; exponents for diameter vs. length for three species of shrubs are 1.05, 1.23, and 0.77 [186].

Under an elastic similarity model, $M \propto d^{8/3}$ (i.e., $d^{2.67}$); a geometric similarity model would predict $M \propto d^3$, while static stress similarity would predict $M \propto d^{5/2}$ [115, 116]. Murray [121] found a scaling exponent of 2.49 ($r = 0.99$) for weight vs. circumference (the latter is directly proportional to diameter) of nine species of trees. The mean exponent for total tree weight vs. basal diameter for 34 species of trees is 2.44 (s.d. $= 0.23$) [183]. Both data sets support a static stress similarity model; the exponent predicted by elastic similarity only barely lies within one standard deviation of the mean exponent for the 34 species.

Limb Bones of Mammals. McMahon [113] claimed good support for his model from data on the scaling of length to diameter of limb bones of all ungulates, all artiodactyls, and bovids, but data for cervids and suids were far from the predictions. I compared data from [2, 9, 113, 115, 134]: the data are in excellent agreement on bovids for the single point where the two data sets overlap, but data for larger groups of mammals [1, 18] are far from elastic similarity's predicted exponent of 0.66. Correcting the data as reduced major axis (RMA) slopes amplifies the disparity. Present data support the elastic similarity model *only* for the tibia in ungulates, the humerus, ulna, tibia, and entire fore-and hind limbs in bovids, and the femur in ceratomorphs.

Elastic similarity predicts that $l \propto M^{1/4}$ and $d \propto M^{3/8}$ [112]. Taking data from [2, 19, 86, 105], 39 of the 48 values agree more closely with a model of geometric similarity; only bone lengths in bovids and ceratomorphs support elastic similarity. Virtually all data on lengths and diameters of bones for a broad assemblage of birds and mammals show exponents that support a geometric similarity model [1].

Body Lengths. Data on the scaling of head plus body length and total mass for 240 land mammals and 24 marine mammals [50] yields mass exponents of 0.314 (95% C.I. $= \pm 0.018$) and 0.339 (± 0.009), respectively, significantly different from elastic similarity's prediction of 0.25 and statistically indistinguishable from geometric similarity. This conclusion is not changed using an RMA regression model. Economos [50] claims that the mass exponent for length in terrestrial mammals larger than 20 kg (0.266 ± 0.022) confirms elastic similarity, but an RMA regression model increases the exponent to 0.292, significantly different from both elastic and geometric similarity. The exponent of body mass as a function of length for insects is 2.62 (RMA; 2.70) [146] to 2.36 (RMA; 2.54) [67]; both are significantly different from the predictions of geometric similarity (3.0) and elastic similarity (4.0).

Total Skeletal Mass. Elastic similarity predicts that skeletal mass should be directly proportional to body mass; McMahon ([116], p 141) explained the observed exponents of 1.03 to 1.18 in vertebrates [3, 132, 142] as a result of "the diversity of species compared," but the exponents for rattlesnakes (1.18) and whales (1.14) are in no better agreement than those for more diverse groupings [3]. Available data indicate that the relative dimensions and relation between bone dimensions and body mass for terrestrial animals other than bovides are best predicted using a model of geometric similarity.

Other evidence cited in support of the elastic similarity model is of highly variable quality. The relationship between chest circumference and body mass in primates [112, 115, 116] agrees well with elastic similarity, although only five species are represented by the 35 data points in this mix of intra- and interspecific allometry. The relationship between body surfacee area and body mass [112, 115, 116] is claimed to be "reasonably well fitted" ([112], p 1203) or "a plausible fit" ([116], p 152) to the prediction of an elastic similarity model, but the line predicted by elastic similarity *visibly* underestimates the slope of the data. McMahon ([116], p 152) cites Stahl's [168] finding that body surface area varies as body mass to the 0.65 power as being "in support of elastic similarity," but this exponent cannot, of course, distinguish elastic similarity's 0.63 from geometric similarity's 0.67. The "solid foundations" ([51], p 465) of the elastic similarity model are nonexistant and this model threatens to become this century's "surface 'law' of metabolism" through inappropriate application; only the dimensions of some (not all) long bones in bovids and ceratomorphs and limb excursion angles during locomotion in terrestrial mammals are well predicted by an elastic similarity model.

Mechanical Similarity. Prange's [131] scaling model (termed "mechanical similarity" [86]) postulates that, if $M \propto l^x$ and $d \propto l^y$, then to maintain functional equivalence for bending, buckling, and torsion, $y = (x+2)/4$. Elastic similarity is a special case of the mechanical similarity model [86]. A mechanical similarity model implies that dynamic rather than static stresses are limiting [101].

Jungers [86] suggests that mechanical similarity is more generally applicable to the bones of terrestrial mammals than is elastic similarity. If $d \propto l^x$, an elastic similarity model predicts $x = 1.50$, and a geometric similarity model predicts $x = 1.00$. I calculated the scaling of diameter to length from data in [2, 19, 86, 105, 134]; in no case do the data agree with the elastic similarity model. While a mechanical similarity model predicts the observed exponent more closely than does elastic similarity, in 19 of the 24 cases mechanical similarity predicts exponents higher than those observed. The diameter of the vertebral centrum in rattlesnakes scales as centrum length to the 1.301 power [133], again significantly lower than the prediction of elastic similarity; mechanical similarity would predict an exponent of 1.277.

Tensile Loading. The concentration of attention on bending and buckling of skeletal elements is unfortunate. Peterson et al. [126, 127] have studied the scaling of skeletal structures exclusively exposed to tensile loadings. In the stalk bearing the fruit of the sausage tree *Kigelia pinnata,* for example, stalk length is independent of both stalk diameter ($r = 0.025$) and fruit mass ($r = -0.097$), but log stalk diameter is significantly correlated ($r = 0.89$) with log fruit mass; the mass exponent is 0.361 ± 0.069 (RMA slope = 0.406), in accord with a static stress model [126]. Elastic similarity would predict stalk length $\propto d^{2/3}$ and diameter $\propto M^{3/8}$. In the Achilles tendon of mammals [127], tendon length and diameter have mass exponents of 0.342 ± 0.028 and 0.361 ± 0.029, respectively (RMA; 0.347 and 0.368). Tendon length scales as tendon diameter to the 0.931 ± 0.069 power (RMA; 0.947). These results are consistent with geometric similarity but may more reasonably be interpreted as reflecting the energy storage function of the Achilles tendon [37], assuming a size-independent tendon toughness (energy storage per unit volume before failure). Results such as these should not be considered exceptional; scaling can only be understood in the context of mechanical and physiological reality.

Metabolic Rate

The literature on the scaling of metabolic rate with body size is huge and unwieldy; see [124] for a useful compilation of the literature up to 1982. In general, in large groups of organisms such as homiotherms, invertebrate and vertebrate poikilotherms, trees, and prokaryotic and eukaryotic unicells, metabolic rates scale as body mass to approximately the 0.75 power. Differences between these different groups of organisms are primarily expressed in the scaling coefficients; i.e., for a given body mass, the metabolic rates of poikilotherms and homiotherms are approximately 8 [9, 75] and 230 [75] times higher than those of unicells, respectively. Differences between groups even greater than these have been claimed [145]. At least in vertebrates, the decrease in specific metabolic rate with increasing size cannot be attributed to increases in non-metabolizing tissues such as bone [174].

Data for invertebrates are generally more variable than that for vertebrates, in large measure because of the difficulty of establishing comparable conditions (see [14, 15]). The exponent for mammals has been claimed to be somewhat high due to differences in the normal activity patterns of the animals involved [135]. None of these problems, however, seems likely to change the general relationship to any great degree, and some of this scrutiny has resulted in the conclusion that the pattern is even more robust than appreciated (e.g., despite earlier reports to the contrary, marsupial and placental mammals show identical scaling of metabolic rates [78]). For general discussions of metabolic rate scaling, see [124] and [153].

Heusner [77] has questioned the reality of the 0.75 exponent for metabolic rate, terming it a "statistical artifact" of inappropriate grouping of data. However, Heusner confuses *intra*specific (exponent approximately 0.66) and *inter*specific (exponent approximately 0.75) metabolic rate allometry – both exponents are "real," but they refer to different phenomena [53, 180]. One implication of the difference between intra- and interspecific metabolic rate scaling exponents is that the scaling coefficient itself must be proportional to body mass to the 0.09 power [180].

Numerous attempts have been made to explain the basis for the 3/4 exponent of metabolic rate on body mass. Blum [21] has suggested that this exponent is determined by the surface-to-volume ratio of a *four*-dimensional geometry; what the fourth spatial dimension might be in animals is obscure. Several authors [49, 51, 112, 115–117] have attempted to derive the 3/4 exponent from an elastic similarity model. They argue that metabolic power should be proportional to muscle cross-sectional area; elastic similarity predicts that muscle cross-sectional area and thus metabolic rates should be proportional to $(M^{3/8})^2 = M^{3/4}$. This rationale fails to explain the common slope among invertebrates or the common slope exhibited by unicells, bacteria, or trees, all of which lack muscles. All other derivations [49, 56, 68, 129, 160] fail to cover the huge diversity of organisms that exhibit a 3/4 exponent for metabolic rate scaling. Indeed, at present there *is* no general explanation for the 3/4 mass exponent for metabolic rate; the most all-encompassing of design generalities in biology must, at present, be treated simply as an empirical fact.

Demographics

The intrinsic rate of increase (r_m) of a population is a fundamental parameter in ecology and evolutionary biology. "The intrinsic rate of increase of a population... can therefore be viewed as an adaptive topography for life history characters..." ([95], p 611). Values for the exponent of r_m on body mass are presented in Table 1. The exponents imply that small organisms devote a larger fraction of their available energy to reproduction [20, 54], which has been interpreted as implying that stable environments should favor large body sizes [181].

Broad generalities that lump major taxonomic groups may be inappropriate; the relationship between r_m and body mass is probably unique to each major taxon [9]. Stearns [170] found that family and order level effects (i.e., lineage-dependent traits) accounted for 29–36% of the species-level variation in life-history traits, while size accounted for the remainder; "size is a trait so important that its consequences retain their force regardless of lineage" ([170], p. 186).

Caughly and Krebs [36] argue that, at least for mammals, populations are regulated by either intrinsic (behavioral or physiological changes in r_m) or extrinsic (changing food supply) mechanisms. This dichotomy in regula-

Table 1. The exponent of intrinsic rate of increase (r_m) on body mass for plants and animals. The numbers in square brackets in the exponents column are the mass exponents recalculated as RMA slopes

Taxon	Exponent	CI	r	n	Reference
Unicellular animals	−0.28	–	–	–	(54)
Diatoms	−0.13 [−0.16]	−0.21/−0.05	−0.81	10	(8)
Dinoflagellates	−0.15 [−0.17]	−0.28/−0.05	−0.87	6	(8)
Heterotherm metazoa	−0.27	–	–	–	(54)
Herbivorous mammals	−0.36 [−0.38]	–	−0.96	8	(36)
Terrestrial mammals	−0.26 [−0.38]	–	−0.69	44	(76)
Mammals	−0.26 [−0.35]	−0.23/−0.29	−0.75	81	(154)
Virus-mammals	−0.26 [−0.27]	–	−0.95	–	(20)

tory mechanisms should be reflected in a bimodality in r_m values; given the relationship between r_m and body mass, this implies a bimodality in body masses which is indeed observed in terrestrial mammals [36]. Given the strong relationships between body mass, r_m, and other life-history parameters [33, 34, 120, 124], similar bimodal distributions would be expected for other life-history traits of higher vertebrates. Analyses of the relationships between life-history traits, ecology, and body size for animals with indeterminate growth are given in [91, 156, 157].

Ecology

In a stable population, the secondary production to standing biomass ratio, P/B, is equal to the birth rate, turnover rate, or specific mortality rate of the population [10, 182]. All animals investigated to date have an exponent for the mass dependence of P/B in the range of about 0.3–0.4 with the conspicuous exception of insects, where mass dependence is either absent or weakly positive [10, 52, 182]. The scaling coefficients for fish and mammals are 4–5 and 20–25 times that for invertebrates, respectively [10]; thus fish and mammals should show much higher specific mortality rates than invertebrates. Some groups of invertebrates are even more insulated from mortality; small size per se may be a refuge from predation [8, 10].

The scaling of P/B is independent of population density [10]; size of an individual at sexual maturity appears to be the determining factor. Productivity could be expected to scale as metabolic rates [7, 10, 98], but the observed P/B ratio of field populations differs from laboratory determinations of the mass exponent (-0.25) of specific metabolic rate (but see [98]) apparently due to the differences in size structure of populations of small and large animals [7, 10].

Hutchinson [79, 80] proposed that there was a limiting degree of similarity between coexisting species necessary to insure minimal niche overlap. Such a principle would be of obvious relevance to the study of the evolution of body size, but to date the only careful test of this hypothesis is Roth's [148]; she concludes that the probability of divergence is constant regardless of the value of the ratio. A similar analysis on a number of mammalian data sets is given in [147]. The assumed competitive advantage of larger forms can be enhanced, negated, or reversed depending on the details of the size-frequency distribution of the prey [187]. Neither the concept of character displacement nor limiting similarity are applicable over the full size range of animals [187].

Case [35] suggested that the evolution of larger body sizes is favored in situations where low quality foods are abundant or where food availability fluctuates; the commonly observed correlation between large size and low food quality for terrestrial mammals is thus a result of differential abundance rather than food quality per se. This model was derived for mammals but should be applicable to aquatic detrivores; it does not seem to be relevant to invertebrate benthic suspension feeders, given the strong overlap in preferred particle sizes.

Harestad and Bunnell [72] found that the home range of herbivores scaled as $M^{1.02}$ (RMA; 1.18), the home range of omnivores scaled as $M^{0.92}$. (RMA; 0.97), and the home range of carnivores scaled as $M^{1.36}$ (RMA; 1.51). The home range of carnivorous birds scaled as $M^{1.31}$, while that of carnivorous lizards scaled as $M^{0.95}$. These results imply that home range increases more rapidly with size than does metabolic rate and suggest a declining ability of the habitat (through increasing patchiness or grain) to support a population as home range size increases [72].

Damuth [46] found a strong allometry in population density of terrestrial herbivorous mammals, with the number of animals per square kilometer proportional to $M^{-0.75}$ ($n=307$, $r=-0.86$; RMA slope$=0.87$). Peters and Raelson [125] expanded Damuth's data base and found that a single relationship for all geographic regions was inappropriate; between tropical and temperate regions, herbivore density was proportional to body mass to the -0.30 to -0.66 power, while density was proportional to body mass to the -0.48 to -1.14 power. No single value for the ratio of herbivore to carvivore biomass was appropriate for all body sizes and geographic areas. On a global basis, herbivore density was proportional to $M^{-0.882}$ (RMA; -1.082) and carnivore (including omnivores) density was proportional to $M^{-1.15}$ (RMA; -1.32); for all mammals on a global basis, density was proportional to $M^{-0.859}$ (RMA; -1.086). The average biomass per species of herbivore is proportional to $M^{0.12}$, while that of carnivores is proportional to $M^{-0.15}$ [125]; using RMA slopes, these values would be $M^{-0.082}$ and $M^{-0.32}$, respectively. The minimum sustainable population size (independent of trophic mode) is proportional to $M^{-0.851}$ (RMA; -0.882) [125].

Different exponents for population densities and home range sizes imply that the number of mammals sharing a home range is proportional to $M^{0.27}$, implying increasing sociality with increasing body size [30–32, 47]. Using the RMA corrected exponents of Harestead and Bunnell [72] and Peters and Raelson [125], the number of herbivorous mammals sharing a home range would be proportional to $M^{0.098}$, while the number of carnivorous mammals sharing a home range would be proportional to $M^{0.36}$. Damuth [47] estimated the exponent for herbivores as 0.35. Which of these estimates is most accurate is of some interest, since Damuth concluded that the productivity of the environment is independent of body size, while the present analysis would imply that the perceived habitat for herbivores deteriorates as body size increases.

The allometry of a number of other ecological parameters have been investigated. Examples include the ecological cost of transport, daily movement distances, daily food consumption [57], foraging time, efficiency, growth rates, litter size, and r and K selection [30–32] (but see [140, 141] for exceptions to these generalizations), and the allometry of population cycles [128]. Convenient summaries can be found in [33, 124].

Size Change in the Fossil Record

Cope's Rule, the axiom that species within a lineage tend to get larger with evolutionary time, has been generally accepted in paleontology and evolutionary biology [22, 23, 66, 70, 122, 161, 169]. Newell [122] presented evidence that Cope's Rule was applicable to some groups of forams, corals, bryozoans, echinoderms, brachipods, molluscs, and arthropods, but size increase in arthropods was rare and most bryozoans, brachiopods, and graptolites showed little size increase throughout their fossil record. Boucot [24] noted that size *decrease* is a common phenomenon in brachiopod clades. Newell viewed size decrease in lineages as less common but noted that "With any given method of collecting, small fossils are more likely to be overlooked than are large ones; hence, the discovery record is relatively less complete for the former" ([122], p 122).

Stanley [169] suggested that Cope's Rule should better be viewed as evolution *from* small body size rather than *towards* large body size. Small forms tend to be unspecialized, while large forms (through allometric constraints) tend to be more specialized [23, 169]; most higher taxa have arisen from relatively smaller ancestors (see [169], Table 1). "It would almost seem as though some sort of exclusion principle were operating and that major structural change could take place only in relatively small organisms" ([23], p 13).

Plots of species diversity vs. body size for a variety of taxa are given in [23, 36, 80, 110, 169, 175]; only about 9% of the clades of living terrestrial mammals are large (=over about 30 cm head length) [177]. The generality of right-skewed diversity/size plots is presumed to imply that ecosystems have more niches for small organisms than for large organisms and that, if the probability of an adaptive breakthrough is constant for any species, then the probability is high that the novelty will occur in a relatively small form [169]. Bonner [22, 23] suggests that *empty* niches are more likely for organisms of large body size, Van Valen [176] suggests that empty niches are transitory and occur at random in ecospace, while Walker and Valentine [179] claim that between 10–50% of all niches are always vacant.

Stanley [169] suggested that a "random walk" in changing body sizes would give the appearance of Cope's Rule. Boucot [24] claims that rates of size change are equal for both increases and decreases in body size, but that abrupt decreases in size during evolution are more likely than abrupt increases because the minimum viable size for any species is closer to the median body size than the maximum body size of the clade. Boucot [24] argues that there is a correlation between the rate of evolution of body size and small population sizes and implies that, among marine invertebrates, size change is more likely among species with nonplanktotrophic larvae (which have poor dispersal ability, thus yielding small effective population sizes) than among species with planktotrophic larvae.

Bonner [23] suggested that there were actually three modes of evolution: a) rapid evolution (1–30 darwins) of species over periods of 10^3–10^4 years, where the probabilities of size increase and decrease were equal (e.g., evolution of body size in island faunas); b) medium evolutionary rates (0.01–0.1 darwins) involving families over periods of 5–45 million years, where size increase followed by structural innovation was the more common mode (e.g., the evolution of modern horses from *Eohippus*); and c) slow evolutionary rates (less than 0.01 darwins) involving entire phyla or kingdoms, where structural innovation precedes size increase (e.g., graptolites to horses). Boucot [24] correlates high rates of size change with high rates of evolution (speciation?); it is unclear whether this is equivalent to Bonner's high evolutionary rates.

The generality of Cope's Rule and, particularly, the evolutionary importance of dwarfing (evolutionary decrease in body size) is a subject of contention. Van Valen ([177], p. 89) claims that "a large mammal never becomes a small one," but Bonner [23] notes that the evolution of body size of faunas on island tends not to follow Cope's Rule and that on islands size increases or decreases seem equally likely, but it is about fifteen times more likely that size will change than that it will remain the same. Lomolino [103] found that body size on islands was proportional to the body size on the mainland to the 0.73–0.95 power, depending on the taxonomic group. Carnivores, artiodactyls, and lagomorphs tended to be dwarfed, rodents tended towards gigantism, and insectivores and marsupials showed no change or no clear trend. For all species, island body size was proportional to mainland body size to the 0.95 power (s.d. = ± 0.007). Thus, small mammals tend to increase in size on islands while large mammals tend to decrease, a result which could be taken as support for either Stanley's [169] and Van Valen's [177] contention that for each niche there is an optimal body size or for Van Valen's [176] contention that all niches are normally full and that one species' gain is another's loss.

Dwarfing is not restricted to islands, however. Sweet [173] documents a case of dwarfing within a clade of the living salamander genus *Desmognathus*. Marshall and Corruccini [108] present data on dwarfing in fossil lineages of Australian marsupials; they attribute dwarfing in these marsupials to resource limitation. Some [63, 108] have claimed an asymmetry in the evolution of body size in mammals, with dwarfed mammals having a relatively larger trophic apparatus than other mammals of the same body size, but this conclusion probably is an artifact of a principal components analysis; Prothero and Sereno [135] found that tooth area was isometric with body weight in a lineage of dwarfed rhinoceroses. They attribute dwarfing in their study to a dietary switch to a high quality but patchy resource.

Dwarfing has only rarely been noted in marine invertebrates, but the phenomenon of stunting (environmental suppression of growth) has probably been confounded with dwarfing, particularly in fossils (see [69]). Criteria for differentiating the possible causes of small body size in marine invertebrate fossils have been presented [106]; Mancini [107] gives an example of such dwarfing which, following Gould [64], he attributes to selection for early reproduction in an unstable habitat. Much of the record of size decrease among marine invertebrates may be hidden; Hallam [70] notes that younger taxa in both the bivalves and ammonites he studied are often smaller than their ancestors but that transitional stages are invariably absent. There is evidence for a size decrease in lineages of the radiolarian genera *Pterocanium* [100] and *Eucystidium* [88].

Hallam [70] documents persistent but irregular trends for size increase in Jurassic bivalves and ammonites; 70% of these lineages double in size and some increase by as much as four times. Hallam finds that the rate of size increase (in darwins) is inversely proportional to the species longevity and concludes that extinction probability is coupled to population size (which decreases with increasing body size; see [70], Fig. 3). Van Valen [177] finds a similar result for foraminiferans, where large body size (over 5 mm diameter) genera have an average half-life of about 9 myr while small genera have a half-life of about 25 myr. However, Van Valen reports the opposite for genera of large mammals, which persist longer than genera of small mammals. Hallam [71] reports an increase in size within both species and lineages of the oyster *Gryphaea*; here the largest species had the longest species duration.

Slower evolutionary rates and higher probabilities of extinction in large animals might result from size-related trends in r_m, generation time, or population size. Smaller population sizes and higher mobilities of large mammals might yield a decreased genetic variability; there is a weak negative correlation ($r = -0.20$) between body size and heterozygosity in mammals [188]. However, the giant clam *Tridacna* shows higher levels of heterozygosity than any other bivalve [6]. This discrepancy in pattern between mammals and bivalves might be due to the high dispersal ability of invertebrates with planktotrophic larvae (although *Tridacna* larvae have a rather short larval period with respect to other bivalves); gene flow rather than body size may be the most important variable. The relationship between body size and larval developmental mode in marine invertebrates and fish is complex [82, 171, 172], although planktotrophic species always show greater dispersal abilities (see [11, 82]). Patterns of geographic distribution, speciation rates, and extinction rates in planktotrophs and nonplanktotrophs are distinct and follow the expected pattern if dispersal and gene flow are the relevant factors [81, 82]. Brown [29] predicted, on the basis of ecological energetics, that large mammal species should show higher extinction rates on islands than small species and that carnivores should show higher extinction rates than herbivores; an argument based on relative population sizes would yield the same predictions. These patterns are indeed observed [28].

There are a few other well documented cases of persistent size increase in the fossil record. There is an exponential increase in maximum axis diameter in Late Silurian-Devonian vascular plants [38] and a similar exponential increase in size for all classes of molluscs throughout the Phanerozoic [151]. Hindiid demosponges show a linear increase in diameter from the Middle Ordovician-Permian [55]; the foram chronospecies/lineage *Globorotalia plesiotumida – G. tumida* doubles in size across the Miocene-Pliocene boundary [104]. Kellogg [88] reports a size increase of about 20% in the radiolarian *Eucystidium matuyamai.* Raup and Crick [139], in a reanalysis of Brinkmann's data on ammonites, find a size increase of 48% in *Zugokosmosceras* but note that the interpretation of this result is difficult; whether a random walk or sustained selective pressure best explains this data depends on the sampling interval chosen, and even the 48% size increase is less than two thirds of one standard deviation from the zero change that would be predicted by a random walk in body size.

A number of authors have used allometry to help interpret the fossil record; space does not permit a full survey, but some selected references will give a flavor of the variety of approaches. Runnegar [149] uses growth allometry in modern annelids to attempt to establish the affinities of the Precambrian worm *Dickinsonia*; Jerison [83] and Radinsky [136] analyze the evolutin of brain size in mammalian carnivores and ungulates. There have been attempts [48, 52, 181] to use empirical allometric relations to understand the ecology and taphonomy of fossil mammal and reptile communities, to reconstruct the ecology of pterosaurs [27], and to predict the body weights of fossil artiodactyls [155]. Powell and Stanton [130] use empirical allometric relations to estimate life expectancy, biomass, demographics, and trophic relations in five species of Eocene molluscs and estimate total energy flow in this community.

If large-bodied organisms tend to be stenotypic, then one might expect differential survival of large- and small-bodied groups in times of mass extinctions. This certainly seems to be the case for the terminal Cretaceous extinction event where virtually all large-bodied vertebrates were eliminated, but to my knowledge there have been no investigations on the influence of body size on survival of clades in the far more severe Permian mass extinction. The Precambrian-Cambrian boundary has been variously interpreted as marking an extensive adaptive radiation [26, 43] or an extinction event followed by an adaptive radiation [158]. If the former is accepted, then the Precambrian-Cambrian transition is marked by a pervasive size increase in the faunas (see [25], Fig. 7); if the latter interpretation is followed, then a selective extinction of large forms such as "sea pens" and worms over a meter in maximum dimension [43, 149] must have occurred. The idea that adaptive breakthroughs occur primarily at small body sizes [169] appears to be vindicated by the observed Cambrian faunas; the first bivalves [150] are less than about a millimeter in maximum dimension, and the most striking characteristic of the Tommotian shelly fauna [17, 109] is its small size.

Evolutionary Allometry

The term "evolutionary allometry" has been applied to situations as diverse as the study of a clade of fossil species or genera or living descendents of a common ancestor [44]; many of the examples mentioned above thus fall between the concepts of interspecific and evolutionary allometry. It has often been assumed that evolutionary allometry can be used to illuminate directly the mechanisms of evolution or the constraints under which evolution operated; as will be seen, the situation is not nearly so straightforward, and most of these conceptions are based on incorrect or unjustified assumptions.

For example, it is sometimes assumed [64, 66, 119] that a close correspondence between scaling exponents for evolutionary and ontogenetic allometries implies that evolution proceeds by changing the timing of development. However, the correlation between the types of allometries is low [39, 44]. It is unsafe to assume that selection has acted on size alone even if the evolutionary lineage falls on the line of intraspecific phenotypic allometry [40]; such an assumption would be equivalent to resurrecting Haeckel's Biogenetic Law (see [58, 64]). Even if evolution acted on body size alone, the exponent of evolutionary allometry would be determined by the static *genetic* exponent, not the static *phenotypic* exponent [94, 96]. Although the latter two exponents tend to be similar for linear skeletal measurements in mammals [96] or for developmentally or functionally related characters [95], the two exponents need not be equal or even have the same sign [4]. There is no necessary correspondence between the genetic and phenotypic covariation patterns [94]; since the genetic allometry exponent determines the evolutionary trajectory, the phenotypic allometry exponent should not be used to infer the nature of past selective forces.

It is true, at least for mammals, that the genetics of ontogeny is in accord with the time-dependent changes predicted by von Baer's law [41, 96], but the situation may be more complex for invertebrates. McEdward [111] reports, not unexpectedly, that ontogenetic allometries in four species of sea urchins are strongly dependent on egg size. Remarkably, for two of these species separation of the blastomeres after either the first or second cleavage results in ontogenetic allometries more similar to those of species with small egg sizes than the allometries of unmanipulated eggs ([162] and McEdward, personal communication) – the stem cell in development (egg or separated blastomere) senses and compensates for its size!

Body size per se is highly heritable (h^2 over 50% [4]), but it is a highly polygenic trait (over 100 loci in mice) and thus can be expected to have a large number of associated genetic correlations, both positive and negative [4, 94]. Lande and Arnold [97] present examples of the non-intuitive results that can occur under selection on correlated characters. In primates, the genetic correlation between brain and body size is positive, but it may be necessary to postulate two selective pressures – a positive selection for brain size and a negative selection on body size – to account for the observed trend [96].

If developmental constraints or canalizations are assumed to underlie an evolutionary trend, the relevant allometries are again genetic. The correlation between the ontogenetic phenotypic and genetic allometry exponents (equal to the square root of the heritability of the exponents) is highly variable [39]; in mice it varies from 0.18–0.63. Brain size may be altered by selection on body size in mice, but the effects of selection for body size are highly dependent on when selection acts during ontogeny [5]. In contrast, brain size in domestic quail proved unresponsive to selection on body size [144].

The discussion of evolutionary allometry need not be restricted to metric phenotypic characters; life-history traits can be treated as another allometric character. Given some rather permissive assumptions, Lande [95] has shown that the mean relative fitness of an individual is equal to the reciprocal of the mean age of reproduction (the generation time), while the mean absolute fitness is equal to the antilog (base e) of r_m, the intrinsic rate of increase. The selection differential is proportional to the slope of a regression of relative fitness on the character of interest [95]. As was shown above, the interspecific allometry of both generation time and intrinsic rate of increase in vertebrates is approximately proportional to $M^{-0.25}$. It is, of course, dangerous to assume that intra- and interspecific allometry exponents are similar, but in this case it seems reasonable to assume that, even intraspecifically, smaller body sizes imply that generation time should decrease (given less growth necessary before first reproduction) and r_m should increase. This implies that, in vertebrates, selection should strongly act to decrease body size unless a countering selective force is acting; Cope's Rule is thus something of a mystery. The situation in many invertebrates, where time of first reproduction is determined by time to reach a critical body size rather than age per se, may be very different. For example, time of first reproduction is positively correlated with growth rates, genetic heterozygosity, and "adult" body size (a poorly defined trait in animals with indeterminate growth) in oysters [163], and a similar situation is seen in snails [89]; if this pattern is common among marine invertebrates, Cope's Rule could result directly from selective pressure to increase r_m. The results of selection in all cases will be the sum of the response to direct selection on that character and the responses of genetically correlated traits; in a large population, major components of fitness will tend to have negative genetic correlations [95]. Lande and Arnold [97] present a multivariate technique for measuring the intensity of selection and the selection gradient for both quantitative characters and life-history traits.

Lemen and Freeman [102] analyzed the variance of size and shape (measured using multivariate statistics) in 72 genera and three families of bats (a total of 250 species). Lemen and Freeman found that variances in size within genera were nearly equivalent to those within families, while shape variances were much lower. They interpret this pattern to mean that evolution proceeds at two different levels within families of mammals. Within a genus, shape is highly conservative and diversification occurs primarily in size. The evolution of new genera (with new shapes) in a family occurs by decoupling those features (previously highly correlated) that defined the genus; modification of these shape features forms a new group with a distinct average shape which, in turn, diversifies in size.

Acknowledgements. I thank L. Radinsky, S. Emerson, K. Niklas, R. Chazdon, W. Jungers, and K. Prestwich for helpful discussions. M. Handcock and R. Chappell freely gave advice on statistical questions and J. Hives did yeoman's service in a library organized for the convenience of physicians rather than scholars.

References

1. Alexander RM (1982) Size, shape, and structure for running and flight. In: A Companion to animal physiology, eds Taylor CR, Johansen K, Bolis L, pp 309–324. New York: Cambridge University Press
2. Alexander RM, Jayes AS, Maloiy GMO, Wathuta EM (1979) Allometry of the limb bones of mammals from shrews (*Sorex*) to elephant (*Loxodonta*). J Zoll 189:305–314
3. Anderson JF, Rahn H, Prange HD (1979) Scaling of supportive tissue mass. Q Rev Biol 54:139–148
4. Atchley WR (1983) Some genetic aspects of morphometric variation. In: Numerical taxonomy, ed Felsenstein J, pp 346–363. Berlin: Springer-Verlag
5. Atchley WR, Riska B, Kohn LAP, Plummer AA, Rutledge JJ (1984) A quantitative genetic analysis of brain and body size associations, their origin and ontogeny: data from mice. Evolution 38:1165–1179
6. Ayala FJ, Hedgecock D, Zumwalt GS, Valentine JW (1973) Genetic variation in *Tridacna maxima*, an ecological analog of some unsuccessful evolutionary lineages. Evolution 27:177–191
7. Banse K (1979) On weight dependence of net growth efficiency and specific respiration rates among field populations of invertebrates. Oecologia 38:111–126
8. Banse K (1982) Cell volumes, maximal growth rates of unicellular algae and ciliates, and the role of ciliates in the marine pelagial. Limnol Oceanogr 27:1059–1071

9. Banse K (1982) Mass-scaled rates of respiration and intrinsic growth in very small invertebrates. Mar Ecol Prog Ser 9:281–297
10. Banse K, Mosher S (1980) Adult body mass and annual production/biomass relationships of field populations. Ecol Monogr 50:355–379
11. Barlow GW (1981) Patterns of parental investment, dispersal and size among coral reef fish. Env Biol Fish 6:65–85
12. Baskerville GL (1970) Testing the uniformity of variance in arithmatic and logarithmic units of a Y-variable for classes of an X-variable. Oak Ridge National Laboratory Publ ORNL-IBP-70-1. Oak Ridge, TN: Oak Ridge National Laboratory
13. Baskerville GL (1972) Use of logarithmic regression in estimation of plant biomass. Can J Forest Res 2:49–53
14. Bayne BL, Thompson RJ, Widdows J (1976) Physiology: I. In: Marine mussels: their ecology and physiology, ed Bayne BL, pp 121–206. Intl Biol Prog Monogr 10. Cambridge: Cambridge University Press
15. Bayne BL, Widdows J (1978) The physiological ecology of two populations of *Mytilus edulis L.* Oecologia 37:137–162
16. Beauchamp JJ, Olson JS (1973) Corrections for bias in regression estimates after logarithmic transformation. Ecology 54:1403–1407
17. Bengston S, Fletcher TP (1983) The oldest sequence of skeletal fossils in the Lower Cambrian of southeastern Newfoundland. Can J Earth Sci 20:525–536
18. Biewener AA (1982) Bone strength in small mammals and bipedal birds: do safety factors change with body size? J Exp Biol 98:289–301
19. Biewener AA (1983) Allometry of quadrupedal locomotion: the scaling of duty factor, bone curvature and limb orientation to body size. J Exp Biol 105:147–171
20. Blueweiss L, Fox H, Kudzma V, Nakashima D, Peters R, Sams S (1978) Relationships between body size and some life history parameters. Oecologia 37:257–272
21. Blum JJ (1977) On the geometry of four-dimensions and the relationship between metabolism and body mass. J Theoret Biol 64:599–601
22. Bonner JT (1965) Size and cycle: An essay on the structure of biology. Princeton: Princeton University Press
23. Bonner JT (1968) Size change in development and evolution. In: Paleobiological aspects of growth and development, ed Macurda DB, pp 1–15. The Paleontological Society, Memoir 2 [J Paleontol 42 (5)]. Menosha, WI: George Banta
24. Boucot AJ (1976) Rates of size increase and of phyletic evolution. Nature 261:694–696
25. Brasier MD (1979) The Cambrian radiation event. In: The origin of major invertebrate groups, ed House MR, pp 103–159. Systematics Association Spec vol 12. New York: Academic Press
26. Brasier MD (1982) Sea-level changes, facies changes and the Late Precambrian-Early Cambrian evolutionary explosion. Precambrian Res 17:105–123
27. Brower JC, Veinus J (1981) Allometry in pterosaurs. Univ Kansas Paleont Contr 105:1–32
28. Brown JH (1971) Mammals on mountaintops: nonequilibrium insular biogeography. Am Nat 105:467–478
29. Brown JH (1981) Two decades of homage to Santa Rosalia: toward a general theory of diversity. Am Zool 21:877–888
30. Calder WA III (1982) A tradeoff between space and time: dimensional constraints in mammalian ecology. J Theoret Biol 98:393–400

31. Calder WA III (1982) The pace of growth: an allometric approach to comparative embryonic and post-embryonic growth. J Zool 198:215–225
32. Calder WA III (1983) Ecological scaling: mammals and birds. Ann Rev Ecol Syst 14:213–230
33. Calder WA III (1984) Size, function, and life history. Cambridge: Harvard University Press
34. Case TJ (1978) On the evolution and adaptive significance of postnatal growth rates in the terrestrial vertebrates. Q Rev Biol 53:243–282
35. Case TJ (1979) Optimal body size and an animal's diet. Acta Biother 28:54–69
36. Caughley G, Krebs CJ (1983) Are big mammals simply little mammals writ large? Oecologia 59:7–17
37. Cavagna GA, Heglund NC, Taylor CR (1977) Walking, running, and galloping: mechanical similarities between different animals. In: Scale effects in animal locomotion, ed Pedley TJ, pp 111–126. New York: Academic Press
38. Chaloner WG, Sheerin A (1979) Devonian macrofloras. Spec Paper Palaeontol 23:145–161
39. Cheverud JM (1982) Relationships among ontogenetic, static, and evolutionary allometry. Am J Phys Anthropol 59:139–149
40. Cheverud JM (1984) Quantitative genetics and developmental constraints on evolution by selection. J Theoret Biol 110:155–171
41. Cheverud JM, Rutledge JJ, Atchley WR (1983) Quantitative genetics of development: genetic correlations among age-specific trait values and the evolution of ontogeny. Evolution 37:895–905
42. Clarke MRB (1980) The reduced major axis of a bivariate sample. Biometrika 67:441–446
43. Cloud P, Glaessner MF (1982) The Ediacarian period and system: metazoa inherit the earth. Science 217:783–792
44. Cock AG (1966) Genetical aspects of metrical growth and form in animals. Q Rev Biol 41:131–190
45. Currey J (1984) The mechanical adaptations of bones. Princeton: Princeton University Press
46. Damuth J (1981) Population density and body size in mammals. Nature 290:699–700
47. Damuth J (1981) Home range, home range overlap, and species energy use among herbivorous mammals. Biol J Linn Soc 15:185–193
48. Damuth J (1982) Analysis of the preservation of community structure of assemblages of fossil mammals. Paleobiology 8:434–446
49. Economos AC (1982) On the origin of biological similarity. J Theoret Biol 94:25–60
50. Economos AC (1983) Elastic and/or geometric similarity in mammalian design? J Theoret Biol 103:167–172
51. Economos AC (1984) The surface illusion and the elastic/geometric similarity paradox, encore. J Theoret Biol 109:463–470
52. Farlow JO (1976) A consideration of the trophic dynamics of a late Cretaceous large-dinosaur community (Oldman Formation). Ecology 57:841–857
53. Feldman HA, McMahon TA (1983) The 3/4 mass exponent for energy metabolism is not a statistical artifact. Resp Physiol 52:149–163
54. Fenchel T (1974) The intrinsic rate of natural increase: the relationship with body size. Oecologia 14:317–326

55. Finks RM (1971) A new Permian eutaxicladine demosponge, mosaic evolution, and the origin of the Dicranocladina. J Paleontol 45:977–997
56. Frasier CC (1984) An explanation of the relationships between mass, metabolic rate and characteristic skeletal length for birds and mammals. J Theoret Biol 109:331–371
57. Garland T Jr (1983) Scaling the ecological cost of transport to body mass in terrestrial mammals. Am Nat 121:571–587
58. Garstang W (1922) The theory of recapitulation: a critical restatement of the biogenetic law. J Linn Soc Lond 35:81–101
59. Goldspink G (1977) Mechanics and energetics of muscle in animals of different sizes, with particular reference to muscle fibre composition of vertebrate muscle. In: Scale effects in animal locomotion, ed Pedley TJ, pp 37–56. New York: Academic Press
60. Gould SJ (1966) Allometry and size in ontogeny and phylogeny. Biol Rev 41:587–640
61. Gould SJ (1971) Geometric similarity in allometric growth: a contribution to the problem of scaling in the evolution of size. Am Nat 105:113–136
62. Gould SJ (1972) Allometric fallacies and the evolution of *Gryphaea*: a new interpretation based on White's criterion of geometric similarity. In: Evolutionary biology, eds Dobzhansky T, Hecht MK, Steere WC, vol 6, pp 91–119
63. Gould SJ (1975) On the scaling of tooth size in mammals. Am Zool 15:351–362
64. Gould SJ (1977) Ontogeny and Phylogeny. Cambridge: Harvard University Press
65. Gould SJ (1979) An allometric interpretation of species-area curves: the meaning of the coefficient. Am Nat 114:335–343
66. Gould SJ (1982) Change in developmental timing as a mechanism of macroevolution. In: Evolution and development, ed Bonner JT, pp 333–346. Dahlem Konferenzen. Berlin, Heidelberg, New York: Springer-Verlag
67. Gowing G, Recher HF (1984) Length-weight relationships for invertebrates from forests in south-eastern New South Wales. Aust J Ecol 9:5–8
68. Gray BF (1981) On the "surface law" and basal metabolic rate. J Theoret Biol 93:757–767
69. Hallam A (1965) Environmental causes of stunting in living and fossil marine benthonic invertebrates. Palaeontology 8:132–155
70. Hallam A (1975) Evolutionary size increase and longevity in Jurassic bivalves and ammonites. Nature 258:493–496
71. Hallam A (1982) Patterns of speciation in Jurassic *Gryphaea*. Paleobiology 8:354–366
72. Harestad AS, Bunnell FL (1979) Home range and body weight – a reevaluation. Ecology 60:389–402
73. Harvey PH (1982) On rethinking allometry. J Theoret Biol 95:37–41
74. Hemmingsen AM (1950) The relation of standard (basal) energy metabolism to total fresh weight of living organisms. Rep Steno Mem Hosp 4:1–58
75. Hemmingsen AM (1960) Energy metabolism as related to body size and respiratory surfaces, and its evolution. Rep Steno Mem Hosp 9:7–110
76. Hennemann WW III (1983) Relationships among body mass, metabolic rate and the intrinsic rate of natural increase in mammals. Oecologia 56:104–108
77. Heusner AA (1982) Energy metabolism and body size. I. Is the 0.75 mass exponent of Kleiber's equation a statistical artifact? Resp Physiol 48:1–12
78. Hinds DS, McMillen RE (1984) Energy scaling in marsupials and eutherians. Science 225:335–337

79. Hutchinson GE (1959) Homage to Santa Rosalia or Why are there so many kinds of animals? Am Nat 93:145–159
80. Hutchinson GE, McArthur RH (1959) A theoretical ecological model of size distributions among species of animals. Am Nat 93:117–125
81. Jablonski D (1980) Apparent versus real biotic effects of transgressions and regressions. Paleobiology 6:397–407
82. Jablonski D, Lutz RA (1983) Larval ecology of marine benthic invertebrates: paleobiological implications. Biol Rev 58:21–89
83. Jerison HJ (1973) Evolution of the brain and intelligence. New York: Academic Press
84. Jolicoeur P (1963) The multivariate generalization of the allometry equation. Biometrics 19:497–499
85. Jungers WL (1984) Aspects of size and scaling in primate biology with special reference to the locomotor skeleton. Yrbk Phys Anthropol 27:73–97
86. Jungers WL (1985) Body size and scaling of limb proportions in primates. In: Size and scaling in primate biology, ed Jungers WL, pp 345–381. New York: Plenum
87. Jungers WL, German RZ (1981) Ontogenetic and interspecific skeletal allometry in nonhuman primates: bivariate versus multivariate analysis. Am J Phys Anthropol 55:195–202
88. Kellogg DE (1983) Phenology of morphologic changes in radiolarian lineages from deep-sea cores: implications for macroevolution. Paleobiology 9:355–362
89. Kemp P, Bertness MD (1984) Snail shape and growth rates: evidence for plastic shell allometry in *Littorina littorea*. Proc Natl Acad Sci USA 81:811–813
90. Kidwell JF, Chase HB (1967) Fitting the allometric equation – a comparison of ten methods by computer simulation. Growth 31:165–179
91. Kirkpatrick M (1984) Demographic models based on size, not age, for organisms with indeterminate growth. Ecology 65:1874–1884
92. Kleiber M (1932) Body size and metabolism. Hilgardia 6:315–353
93. Kuhry B, Marcus LF (1977) Bivariate linear models in biometry. Syst Zool 26:201–209
94. Lande R (1979) Quantitative genetic analysis of multivariate evolution, applied to brain: body size allometry. Evolution 33:402–416
95. Lande R (1982) A quantitative genetic theory of life history evolution. Ecology 63:607–615
96. Lande R (1985) Genetic and evolutionary aspects of allometry. In: Size and scaling in primate biology, ed Jungers WL, pp 21–32. New York: Plenum Press
97. Lande R, Arnold SJ (1983) The measurement of selection on correlated characters. Evolution 37:1210–1226
98. Lavigne DM (1982) Similarity in energy budgets of animal populations. J Anim Ecol 51:195–206
99. Laws EA, Archie JW (1981) Appropriate use of regression analysis in marine biology. Mar Biol 65:13–16
100. Lazarus D (1983) Speciation in pelagic Protista and its study in the planktonic microfossil record: a review. Paleobiology 9:327–340
101. Lighthill J (1977) Comments on Dr. Prange's paper. In: Scale effects in animal locomotion, ed Pedley TJ, pp 182–183. New York: Academic Press
102. Lemen CA, Frreeman PW (1984) The genus: a macroevolutionary problem. Evolution 38:1219–1237

103. Lomolino MV (1985) Body size of mammals on islands: the island rule reexamined. Am Nat 125:310–316
104. Malmgren BA, Berggren WA, Lohmann GP (1984) Species formation through punctuated gradualism in planktonic foraminifera. Science 225:317–319
105. Maloiy GMO, Alexander RM, Njau R, Jayes AS (1979) Allometry of the legs of running birds. J Zool 187:161–167
106. Mancini EA (1978) Origin of micromorph faunas in the geologic record. J Paleontol 52:321–333
107. Mancini EA (1978) Origin of the Grayson micromorph fauna (Upper Cretaceous) of north-central Texas. J Paleontol 52:1294–1314
108. Marshall LG, Corruccini RS (1978) Variability, evolutionary rates, and allometry in dwarfing lineages. Paleobiology 4:101–119
109. Matthews SC, Missarzhevsky VV (1975) Small shelly fossils of late Precambrian-Cambrian age. J Geol Soc Lond 131:289–304
110. May RM (1978) The dynamics and diversity of insect faunas. In: Diversity of insect faunas, eds Mound LA, Waloff N, pp 188–204. Oxford: Blackwell Scientific Publications
111. McEdward LR (1984) Morphometric and metabolic analysis of the growth and form of an echinopluteus. J Exp Mar Biol Ecol 82:259–287
112. McMahon TA (1973) Size and shape in biology. Science 179:1201–1204
113. McMahon TA (1975) Allometry and biomechanics: limb bones in adult ungulates. Am Nat 109:547–563
114. McMahon TA (1975) The mechanical design of trees. Sci Am 233:93–102
115. McMahon TA (1975) Using body size to understand the structural design of animals: quadrupedal locomotion. J Appl Physiol 39:619–627
116. McMahon TA (1980) Scaling physiological time. Lect Math Life Sci 13:131–163
117. McMahon TA, Feldman HA (1983) The 3/4 mass exponent for energy metabolism is not a statistical artifact. Resp Physiol 52:149–163
118. McMahon TA, Kronauer RE (1976) Tree structures: deducing the principle of mechanical design. J Theoret Biol 59:443–466
119. McNamara KJ (1982) Heterochrony and phylogenetic trends. Paleobiology 8:130–142
120. Millar JS, Zammuto RM (1983) Life histories of mammals: an analysis of life tables. Ecology 64:631–635
121. Murray CD (1927) A relationship between circumference and weight in trees and its bearing on branching angles. J Gen Physiol 10:725–728
122. Newell ND (1949) Phyletic size increase, an important trend illustrated by fossil invertebrates. Evolution 3:103–124
123. Pedley TJ (1977) Scale effects in animal locomotion. New York: Academic Press
124. Peters RH (1983) The ecological implications of body size. Cambridge: Cambridge University Press
125. Peters RH, Raelson JV (1984) Relations between individual size and mammalian population density. Am Nat 124:498–517
126. Peterson JA, Benson JA, Ngai M, Morin J, Ow C (1982) Scaling in tensile "skeletons": structures with scale-independent length dimensions. Science 217:1267–1270
127. Peterson JA, Benson JA, Morin JG, McFall-Ngai MJ (1984) Scaling in tensile "skeletons": scale dependent length of Achilles tendon in mammals. J Zool 202:361–372

128. Peterson RO, Page RE, Dodge KM (1984) Wolves, moose, and the allometry of population cycles. Science 224:1350–1352
129. Platt T, Silvert W (1981) Ecology, physiology, allometry and dimensionality. J Theoret Biol 93:855–860
130. Powell EN, Stanton RJ Jr (1985) Estimating biomass and energy flow of molluscs in palaeo-communities. Palaeontology 28:1–34
131. Prange HD (1977) The scaling and mechanics of arthropod exoskeletons. In: Scale effects in animal locomotion, ed Pedley TJ, pp 169–181. New York: Academic Press
132. Prange HD, Anderson JF, Rahn H (1979) Scaling of skeletal mass to body mass in birds and mammals. Am Nat 113:103–122
133. Prange HD, Christman SP (1976) The allometrics of rattlesnake skeletons. Copeia 1976:542–545
134. Prothero DR, Sereno PC (1982) Allometry and paleoecology of medial Miocene dwarf rhinoceroses from the Texas Gulf Coastal Plain. Paleobiology 8:16–30
135. Prothero J (1984) Scaling of standard energy metabolism in mammals: I. Neglect of circadian rhythms. J Theoret Biol 106:1–8
136. Radinsky L (1978) Evolution of brain size in carnivores and ungulates. Am Nat 112:815–831
137. Rashevsky N (1944) Studies in the physicomathematical theory of organic form. Bull Math Biophys 6:1–59
138. Rashevsky N (1961) Mathematical principles in biology and their applications. Springfield, OH: Thomas
139. Raup DM, Crick RE (1981) Evolution of single characters in the Jurassic ammonite *Kosmoceras*. Paleobiology 7:200–215
140. Reaka ML (1979) The evolutionary ecology of life history patterns in stomatopod Crustacea. In: Reproductive ecology of marine invertebrates, ed Stancyk SE, pp 235–260. Columbia: University of South Carolina Press
141. Reaka ML (1980) Geographic range, life history patterns, and body size in a guild of coral-dwelling mantis shrimps. Evolution 34:1019–1030
142. Reynolds WW, Karlotski WJ (1977) The allometric relationship of skeleton weight to body weight in teleost fishes: a preliminary comparison with birds and mammals. Copeia 1977:160–163
143. Ricker WE (1973) Linear regressions in fisheries research. J Fish Res Bd Can 30:409–434
144. Ricklefs RE, Marks HL (1984) Insensitivity of brain growth to selection of four-week body mass in Japanese quail. Evolution 38:1180–1185
145. Robinson WR, Peters RH, Zimmermann J (1983) The effects of body size and temperature on metabolic rate of organisms. Can J Zool 61:281–288
146. Rogers LE, Hinds WT, Buschbom RL (1976) A general weight vs. length relationship for insects. Ann Entomol Soc Am 69:387–389
147. Roth VL (1979) Can quantum leaps in body size be recognized among mammalian species? Paleobiology 5:318–336
148. Roth ML (1981) Constancy in size ratios of sympatric species. Am Nat 118:394–404
149. Runnegar B (1982) Oxygen requirements, biology and phylogenetic significance of the late Precambrian worm *Dickinsonia,* and the evolution of the burrowing habit. Alcheringa 6:223–239

150. Runnegar B, Bentley C (1983) Anatomy, ecology and affinities of the Australian Early Cambrian bivalve *Pojetaia runnegari Jell.* J Paleontol 57:73–92
151. Runnegar B, Jell PA (1976) Australian Middle Cambrian molluscs and their bearing on early molluscan evolution. Alcheringa 1:109–138
152. Schmidt-Nielsen K (1975) Scaling in biology: the consequences of size. J Exp Zool 194:287–308
153. Schmidt-Nielsen K (1984) Scaling: Why is animal size so important? Cambridge: Cambridge University Press
154. Schmitz OJ, Lavigne DM (1984) Intrinsic rate of increase, body size, and specific metabolic rate in marine mammals. Oecologia 62:305–309
155. Scott KM (1983) Prediction of body weight of fossil Artiodactyla. Zool J Linn Soc 77:199–215
156. Sebens KB (1979) The energetics of asexual reproduction and colony formation in benthic marine invertebrates. Am Zool 19:683–697
157. Sebens KP (1982) The limits to indeterminate growth: an optimal size model applied to passive suspension feeders. Ecology 63:209–222
158. Seilacher A (1984) Late Precambrian and Early Cambrian metazoa: preservational or real extinctions? In: Patterns of change in earth evolution, eds Holland HD, Trendall AF, pp 159–168. Dahlem Konferenzen. Berlin, Heidelberg, New York, Tokyo: Springer-Verlag
159. Seim E, Saether B-E (1983) On rethinking allometry: which regression model to use? J Theoret Biol 104:161–168
160. Sernetz M, Rufeger H, Kindt R (1982) Interpretation of the reduction law of metabolism. Expl Biol Med 7:21–29
161. Simpson GG (1953) The major features of evolution. New York: Columbia University Press
162. Sinervo BR, McEdward LR, Strathmann RR (1984) The effect of experimentally reduced egg size on form, function and rate of development of planktotrophic larval echinoids. Am Zool 24:131A
163. Singh SM, Zouros E (1978) Genetic variation associated with growth rate in the American oyster (*Crassostrea virginica*). Evolution 32:342–353
164. Smith RJ (1981) Interspecific scaling of maxillary canine size and shape in female primates: relationships to social structure and diet. J Hum Evol 10:165–173
165. Smith RJ (1984) Determination of relative size: the "criterion of subtraction" problem in allometry. J Theoret Biol 108:131–142
166. Sokal RR, Rohlf FJ (1981) Biometry. San Francisco: WH Freeman and Company
167. Sprugel DG (1983) Correcting for bias in log-transformed allometric equations. Ecology 64:209–210
168. Stahl WR (1967) Scaling of respiratory variables in mammals. J Appl Physiol 22:453–460
169. Stanley SM (1973) An explanation for Cope's rule. Evolution 27:1–26
170. Stearns SC (1983) The impact of size and phylogeny on patterns of covariation in the life-history traits of mammals. Oikos 41:173–187
171. Strathmann RR, Strathmann MF (1982) The relationship between adult size and brooding in marine invertebrates. Am Nat 119:91–101
172. Strathmann RR, Strathmann MF, Emson RH (1984) Does limited brood capacity link adult size, brooding, and simultaneous hermaphroditism? A test with the starfish *Asterina phylactica*. Am Nat 123:796–818

173. Sweet SS (1980) Allometric inference in morphology. Am Zool 20:643–652
174. Ultsch GR (1974) The allometric relationship between metabolic rate and body size: role of the skeleton. Am Midl Nat 92:500–504
175. Van Valen L (1973) Body size and numbers of plants and animals. Evolution 27:27–35
176. Van Valen L (1974) Two modes of evolution. Nature 252:298–300
177. Van Valen L (1975) Group selection, sex, and fossils. Evolution 29:87–94
178. Wainwright SA, Biggs WD, Currey JD, Gosline JM (1976) Mechanical design in organisms. London: Edward Arnold
179. Walker TD, Valentine JW (1984) Equilibrium models of evolutionary species diversity and the number of empty niches. Am Nat 124:887–899
180. Weiser W (1984) A distinction must be made between the ontogeny and the phylogeny of metabolism in order to understand the mass exponent of energy metabolism. Resp Physiol 55:1–9
181. Western D (1980) Linking the ecology of past and present mammal communities. In: Fossils in the making, eds Behrensmeyer AK, Hill AP, pp 41–54. Chicago: The University of Chicago Press
182. Western D (1983) Production, reproduction and size in mammals. Oecologia 59:269–271
183. White J (1981) The allometric interpretation of the self-thinning rule. J Theoret Biol 89:475–500
184. White JF, Gould SJ (1965) Interpretation of the coefficient in the allometric equation. Am Nat 99:5–18
185. Whittaker RH, Marks PL (1975) Methods of assessing terrestrial productivity. In: Primary productivity of the biosphere, eds Leith H, Whittaker RH, pp 55–118. New York: Springer Verlag
186. Whittaker RH, Woodwell GM (1968) Dimension and production relations of trees and shrubs in the Brookhaven Forest, New York. J Ecol 56:1–25
187. Wilson DS (1975) The adequacy of body size as a niche difference. Am Nat 109:769–784
188. Wooten MC, Smith MH (1985) Large mammals are genetically less variable? Evolution 39:210–212
189. Wright S (1968) Evolution and the genetics of populations, vol 1. Genetic and biometric foundations. Chicago: University of Chicago Press

Patterns and Processes in the History of Life,
eds. D. M. Raup and D. Jablonski, pp. 99–117. Dahlem Konferenzen 1986
Springer-Verlag Berlin, Heidelberg

Progress in Organismal Design

D. C. Fisher

Museum of Paleontology
University of Michigan
Ann Arbor, MI 48109, USA

Abstract. The question of whether or not large-scale historical trends in the design of organisms represent progress (i.e., improvement of design) has attracted attention since the beginning of evolutionary thought but has proved extremely resistant to objective analysis. Some of this difficulty reflects the range of time scales and phylogenetic contexts for which the question might be posed, but a more fundamental problem involves the definition of improvement itself. When improvement of design is discussed within the context of current evolutionary thought, it is frequently portrayed as the expected outcome of the sustained operation of natural selection on variation within populations. Such an interpretation of large-scale historical trends in morphology frequently involves some degree of orthoselection and/or adaptive replacement. An alternative interpretation is that many such trends reflect little more than the Markovian aspect of the evolutionary process. Trends do have a finite, and not always small, probability of occurrence in systems whose underlying causal structure behaves in a pseudorandom fashion. According to this second interpretation, the appropriate level for causal analysis may be well below that at which the trend is manifested. This is equivalent to suggesting that the causal basis of the trend may be extremely heterogeneous. Both selectionist and Markovian models of morphologic change can be tested, but it is always a *particular* model, defined by its own assumptions and boundary conditions, that is corroborated or refuted, not the whole class of selectionist or Markovian models. Certain large-scale morphologic trends documented in the fossil record (e.g., increases in brain size within mammals) indeed appear to represent the intermittent or sustained operation of directional selection. Other trends (e.g., changes in morphologic complexity or in amount of genetic information within most phyla or classes) show patterns that are best interpreted as a simple random walk or a branching process (depending on the phylogenetic structure of the problem). However, a large number of trends (e.g., body size within certain higher taxa) can be explained by "diffusion" models or models of branching processes in which directionality is imposed by the position of the initial state of the system relative to the total range of accessible states. Because of the frequency of environmental change, the multiplicity of factors underlying fitness, the possibility of frequency-dependent and epistatic interactions among features, and the consequent possibility of nontransitive fitness relations between phenotypes, selection acting within populations frequently, though not inevitably, fails to produce unidirectional trends. The extent to which unidirectional trends dominate, or fail to dominate, the fossil record is therefore not a measure of the adequacy of neo-Darwinian mechanisms as causes of large-scale patterns in evolution.

Introduction

The recent resurgence of interest in macroevolutionary patterns, although distinctive in the quantitative methodologies being employed, is an extension of a much older inquiry into the major features of the history of life. One of the principal issues in this inquiry is whether or not there is directionality – and in particular, progress, or improvement – in the history of life. Gould [13] argues that this issue has been one of three prime focal points in the expression of contrasting views on evolutionary process. For instance, improvement was a central element of Osborn's [26] "aristogenesis" but was denied, at least as a long-term trend, by Hyatt's [21] theory of "racial life cycles." One of the reasons this issue remains complex is that quite apart from the different answers associated with divergent views on evolutionary mechanisms, we expect the appearance and interpretation of directionality to vary depending on both the scale at which it is observed and the historical interval over which it is followed. Neontological data and certain segments of the fossil record favored by excellent preservation have long been sufficient to document change over relatively short time spans, and a simple comparison of organisms with abiotic systems says something about change at the largest scale. The more problematic issue is thus whether the history of life shows directed change at intermediate scales and whether any such change qualifies as improvement. In addition, this problem takes on a different character within different phylogenetic contexts.

Any interpretation of history will depend on how we reconstruct the relationships and sequence of appearance of organisms, but beyond this, both the mechanisms for explaining change and the associated patterns of change depend on whether the change is conceived of as taking place within a single lineage, within a single clade (monophyletic group), or between/among clades. Few, if any, evolutionary biologists/paleontologists have a single, "one-size-fits-all" answer to the question of whether trends in the history of life represent vectorial tendencies or whether any such tendencies could be characterized as improvement [8, 39, 41, 46].

The particular version of the venerable issue of directionality that is the assigned focus of this paper is whether trends in morphology, observed at the clade level and higher, reflect improvement in the design of organisms or simply the outcome of evolution acting as a Markovian process. Most of the trends in question represent "net" change across a time series, but in some cases trends may approach monotonicity. Previous studies have documented an enormous variety of evolutionary trends and have given many of these at least tentative explanations, but relatively few trends have attracted wide attention as cases of improvement. Candidates at a large temporal and taxonomic scale include patterns of increase in a) anatomical complexity [37], b) perception of the environment [1], c) control over the environment, d) amount of genetic information [22], e) correspondence between functional and epigenetic patterns of interaction [35], and f) rate of energy utilization. On a finer temporal and taxonomic scale, possible examples include increased a) leg length in cursorial tetrapods and arthropods, b) brain size in mammals [27], c) hydrodynamic efficiency in fishes and marine mammals, d) effectiveness of pollination mechanisms in plants, and e) effectiveness of water retention in desert-adapted organisms.

Within current evolutionary thought the tendency to interpret large-scale trends as improvement in design is frequently associated with the neo-Darwinian paradigm. In this context trends may be explained as the long-term result of the intermittent or sustained action of directional selection and/or adaptive replacement [33, 39] through competition. However, I will argue that the validity of this expectation of long-term improvement is severely compromised by complications that operate *within* the neo-Darwinian paradigm. The interpretation of trends as Markovian in nature is labeled "steady-statist" by Gould [13], but this is not so much intended to deny change, even at a very general level, as it is to correct the impression that a sustained trend requires a unitary causal mechanism. It demonstrates that trends are frequent products of systems whose causal basis is heterogeneous and localized at a hierarchical level well below that at which the trend is manifested. The Markovian interpretation is not necessarily meant to deny the efficacy of microevolutionary processes as causes of macroevolutionary patterns, but rather to evaluate critically the level of generality at which single causes are thought to operate [28]. This insight is occasionally misrepresented by those who would interpret criticism of orthoselectionist explanations of large-scale trends as corroboration for non-Darwinian mechanisms of evolution.

The first step in evaluating "progressivist" and "Markovian" interpretations of large-scale morphologic trends is to treat some of the problems associated with a definition of progress, or improvement. Markovian processes and their application to evolutionary problems have been well reviewed by Raup [28] and will need only scant treatment here, although some slightly more "structured" models will be considered. I will then discuss some of the difficulties of progressivist extrapolations of neo-Darwinian processes, arguing that long-term improvement is in many cases not a valid prediction. Finally, I will consider possible tests of alternative interpretations of large-scale trends and the theoretical implications of these studies.

Definition of Progress

Progress is a notoriously slippery concept, but it has played an important role in western philosophical traditions [13, 24]. Its general meaning, to the extent of specifying some increase in "quality" (i.e., a "change for the better"), is straightforward, but objective analysis begins to founder at the point of developing adequate criteria by which quality might be judged. Depending on the problem at hand, progress might be seen as increased complexity, increased simplicity, increased efficiency, increased control over limiting resources, increased longevity, and so on. The list could be extended almost indefinitely, but its most striking feature is that despite some tendency for association among certain criteria, different criteria are ultimately noncoincident. No one criterion or set of criteria seems adequate for all situations, and application of different criteria leads to inconsistencies in what is judged to represent progress. The assessment of progress therefore has an inescapable "axiological" [2] or evaluative aspect.

In search of an operational alternative, it may be tempting to forsake the notion of improvement and measure progress simply as displacement from an earlier condition. However, this allows any observed change to count as progressive. Such a concept may have local significance (i.e., with respect to a previously identified transformation) and may be useful in contexts such as the distinction between plesiomorphic and apomorphic conditions. However, this usage represents an a posteriori sense of "progress" quite distinct from its more general, a priori sense.

Definitions of progress in evolutionary terms have been attempted previously by a great many writers (e.g., [2, 12, 19, 20, 34, 39, 44, 46]). Their approach has in most cases been to admit the difficulties involved but nevertheless attempt to select one or more "least objectionable" criteria of progress. However, such an approach seems unlikely to escape charges of inconsistency and/or subjectivity. As an alternative, let us consider possible *properties* of a notion of progress. I have been unable to specify a set of *necessary* and *sufficient* properties, but it may at least illumine the problem to specify two that appear to be *necessary*. One of these is transitivity. If the transition from *A* to *B* is considered an improvement, and likewise for the transition from *B* to *C*, then *C* must be considered an improvement relative to *A*. The necessity for transitivity arises because progress is essentially a univariate concept, even though it might be accomplished by innumerable styles and directions of change [24]. It thus behaves as a single axis in a multivariate state-space. Part of the difficulty of reconciling any one definition of progress with the diverse phenomenology of the natural world is a consequence of the information loss that inevitably accompanies projection of conspicuously multidimensional data onto a single axis. As it happens, all of the criteria of progress mentioned in the first paragraph of this section show transitivity (assuming an appropriate measure of each; e.g., if *B* is more complex than *A*, and *C* is more complex than *B*, then *C* is more complex than *A*). Therefore, transitivity does not allow us to distinguish among these alternatives. However, it will help evaluate additional definitions and will be useful in judging putative cases of progress.

A second necessary characteristic of progressive trends is that they show some degree of emergence relative to their lower-level causal structure. Trends can usually be described as resulting from interactions at any of a series of levels, and the processes conceived as operating at different levels are usually different in character [1]. However, progressive trends tend to be most clearly interpreted as results of interactions at the level at which the trend itself is perceived. Progressive trends depend in part on processes acting at lower levels, but they typically develop out of complex interactions among multiple lower-level processes, modulated by interlevel feedback as well. In contrast, trends that are directly reducible to processes acting at a lower level are usually considered mere manifestations of inherent – and unchanging – properties of the systems that display them. Such a realization is generally incompatible with the type of qualitative historical change that the notion of progress requires. For instance, "progressive" decrease in the temperature of an object surrounded by a cooler medium involves a series of states whose mutual relationships are transitive, but the higher-level process of cooling is reducible to a process involving redistribution of kinetic energy at molecular and submolecular levels. When the boundaries of the system are properly recognized and its dynamics understood, we see that there has been no change in essential properties of the system. We are thus not inclined to think of cooling as representing progress.

While emergence is an important attribute of progressive trends, it is not absolute in the sense that transitivity can be. Our perception of emergence and reducibility is dependent on how we have characterized system boundaries and the degree to which we understand system dynamics [23]. This emphasizes, in a different way, the subjectivity of judgments regarding progress. In addition, all that is needed to threaten a given criterion of progress is that it be reducible *in principle* to lower-level processes. As with cooling, it is not required that we have precise knowledge of all details of lower-level interactions. The degree to which the criteria of progress mentioned above show emergence is bound to be interpreted differently by different workers; fortunately, we need not solve this problem at this time.

I am not convinced that any single criterion of progress is objective, operational, and appropriate in all contexts. At the same time, I do not wish to retreat from the major question posed here (regarding the relative merits of progressivist and Markovian interpretations of large-scale morphologic trends) with the lame excuse that it is unresolvable because one of the alternatives is undefined. In my opinion, the best approach to defining progress in an evolutionary sense is to relate it to the replacement of one phenotype by another. Mechanisms for accomplishing this include selection and competition, but other processes, operating at a variety of levels, may be involved as well. I hasten to add that I do *not* consider all or even most such replacement as representing progress. I only suggest that if there are cases where a pattern of replacement demonstrates transitivity and emergence, it is at least to that extent a plausible candidate for recognition as a progressive trend. The advantage of this approach is that it relies on organisms interacting with other organisms within a specified physical context (or range of contexts) to develop "for themselves" an ordering scheme. This ordering scheme consists of a statement of the actual and potential replacement relations between pairs of phenotypes (or pairs of distributions of phenotypes within populations). This at least avoids the imposition of an externally defined ordering scheme such as is implicit in other criteria of progress. For a historical sequence to be transitive, a population representing any younger stage in the transformation must be capable of replacing a population representing any older stage. It should be clear immediately that not all transformation series in evolution behave in this fashion. For a sequence to be emergent it must, for instance, fail to reduce simply to a case of environmental tracking. Whether any evolutionary sequence is emergent in this sense is probably best regarded as an open question.

Markovian Processes

Markovian processes and their application to evolutionary patterns are discussed in some detail by Raup [28]. Markovian models are in part path-dependent, in that the state of the system at time $= t_n$ constrains the allowable states at t_{n+1}. However, the actual transition in state between t_n and t_{n+1} always involves some probabilistic element. In this sense, the system behaves in a random fashion, but with at least a short-term memory [28]. The simplest Markovian model of fairly general applicability is a random walk in which step size is constant, the probability of moving in each of two directions is equal, and the direction of movement is determined independently at each step. This model is a clear and analytically tractable analog of a time series in which some descriptive variable of a system (e.g., some morphologic attribute of the members of a lineage) increases or decreases in a pseudorandom fashion. A slightly more complex random walk model might, for instance, let step size be variable, determined (independently at each step) by a probability distribution. By suitable manipulation of such a probability distribution, Markovian models can be produced that show a systematic bias, or directionality, in their outcomes. However, these are implicitly excluded in the framing of the question on which this paper focuses and are therefore avoided in the following discussion. The applicability of random walk models is not restricted to systems (if there are any) that are acausal or random in some pure sense. They are explicitly intended for systems in which the causality is sufficiently complex that it produces statistically random behavior. This generally means that the causality is best represented as operating at a lower level. Translating to the case of organisms showing morphologic change, the application of a random walk model does not necessarily mean that the observed morphologic changes are considered adaptively neutral (though they may be).

Another important type of Markovian model involves branching processes. These show path-dependent and probabilistic properties similar to those demonstrated by random walks. However, the allowable state changes include addition and loss of lineages from the system [10, 28]. Branching processes can be treated as models of evolution at the clade level and above. They have been used very effectively in research on patterns of clade diversity and morphologic change [29–31, 43].

An important lesson of Markovian models is that some degree of apparent order can arise by chance (or its observational equivalent – the concatenation of lower-level causes [31]). Even when the directions of change are equiprobable, directional trends are commonly produced. Their likelihood is of course dependent on their length and uniformity, but as an example, Raup [28] mentions that in 20% of reasonably long random walks the path taken by the system will remain either higher or lower than the starting point 97.6% of the time [10]. As such, directional trends do not necessarily require explanations framed in terms of single causes acting throughout the duration of the trend. Raup [28] adds the important caution that in using probabilistic methods to study evolutionary events we should avoid the temptation to develop more and more elaborate models in an attempt to reproduce real-world data. By suitable juggling of parameters, a model can be devised to reproduce almost any pattern. However, unless the values of parameters and the structure of the model as a whole can be defended independently of the patterns produced by the model, the "explanation" embodied in the model must be considered entirely ad hoc. At the risk of committing exactly this error, I wish to examine one elaboration on the theme of branching models. In most branching models, lineages propagate within a relatively unconstrained space. If the vertical dimension is taken to represent time – the series of stages through which the branching system develops – the horizontal dimension (or plane) either is given no metric significance at all (if only the number of branches existing during a given interval is important) or is used to carry information about some other attribute of branches – usually the morphology typical of members of a lineage. In the latter case, the occupation of horizontal space (i.e., different morphologies) is usually assumed to be controlled by factors related to branch dynamics. These include the number or frequency of branching and extinction events and the magnitude and direction of morphologic change at and between such events. Directional morphologic change can occur within a single lineage or averaged across all lineages of a clade. However, in a sufficiently large collection of such clade histories we do not expect to find consistently preferred directions of morphologic change, assuming that there are none at the level of branch dynamics.

In contrast, I wish to consider models of branching systems that have symmetrical branch dynamics, as above, but which differ in that they operate within a system of constraints on allowable morphologies. These constraints, or limits, represent specifiable interactions or relationships (biomechanical, developmental, or otherwise) whose own dynamics can be studied empirically or theoretically. While the joint effects of some of these constraints may be measured by variance/covariance matrices [4, 14], this latter approach does not offer the direct characterization of individual constraints needed in the present models. The effects of constraints can be treated as separate from normal branch dynamics, for the constraints are postulated here to come into play only in the context of extremal phenotypes. In these cases, and only in these, the limiting relationships are superimposed on normal branch dynamics. I do not envision such constraints as affecting the "motor of change" [13] for nonextremal phenotypes, nor any other process as acting on nonextremal phenotypes to bias their directions of change. If the initial state (i.e., the ancestral morphology) of such a system is located symmetrically relative to the constraints (or vice versa), we would still predict no overall bias in direction of change. However, if the constraints are disposed asymmetrically (nonradially) relative to the initial state, particularly if the initial state is located near one extreme of the range of allowable states, trends in some directions will encounter the constraints more frequently (or sooner) than trends in other directions. Thus, on average, the most conspicuous, sustained trends will be in the direction of least morphologic constraint. Systems of this sort have been analyzed using diffusion models [38, 40], and evolution within such systems has been described metaphorically as "diffusion within a structured design-space" [11]. In the simplest cases the extremal phenotypes occupy literally peripheral positions in the morphologic space accessible to a clade, but in more complex cases constraints may prohibit certain intermediate phenotypes or even directions of movement (as opposed to positions) within morphologic space.

The class of models described above has numerous predecessors (e.g., [32]). In large part, however, it is patterned after Stanley's [42] analysis of Cope's Rule – the generalization that body size tends to increase during the evolutionary history of many groups. Although the actual history of size change in most groups is complex (LaBarbera, this volume), increase in mean body size over time has been documented within many clades [42]. Stanley explains this in part by arguing that there is a finite, biomechanically determined size range consistent with many Baupläne and by demonstrating that in many cases the originators of new clades were located near the lower end of the allowable size range for their Bauplan. The reasons for this are not germane at the moment, but the important consequence is that the subsequent history of such clades records significant intervals of increase in mean size, as lineages "diffuse" to fill out the spectrum of allowable sizes. In fact, Stanley's model does not represent pure diffusion, since he also postulates that there is an optimum body size for each Bauplan, frequently located, as it happens, at a somewhat larger size than that typifying the clade founders. This latter factor introduces a selectional component that would act on all members of a clade and would thus comprise part of their branch dynamics. For the sake of discussion, the diffusion models described here explicitly exclude this type of factor. Many instances of directional change might profitably be considered in terms of such models.

Does Natural Selection Lead to Improvement?

As noted above, improvement is an explicit prediction of some evolutionary theories and not of others – but what of the neo-Darwinian paradigm? Darwin himself [6] was in most cases quite clear that natural selection does not lead inevitably to improvement. In fact, Gould ([13], pp 13–14) maintains

> ... that an explicit denial of innate progression is the most characteristic feature separating Darwin's theory of natural selection from other nineteenth century evolutionary theories. Natural selection speaks only of adaptation to local environments, not of directed trends or inherent improvement. Darwin wrestled with the issue of progression and finally concluded that his theory provided no rationale for belief in evolutionary directions – adaptation to local environments means just that and nothing more.

At the same time, neither Gould nor Darwin denies "the possibility of direction as an outcome of natural selection" ([13], p 14). Indeed, "... one pathway to adaptation lies in structural 'improvement' conferring a more general success upon its bearer..." ([13], p 14). Most dicussions of evolutionary theory from the perspective of the neo-Darwinian paradigm adopt a very similar stance, arguing against the idea that natural selection leads, as a general rule, to cumulative progression (e.g., [8, 41, 46]) and yet maintaining the possibility of some directionality, plausibly interpreted as improvement, on intermediate scales.

Nevertheless, some recent critics of the sufficiency of the neo-Darwinian paradigm suggest that it does indeed entail the expectation of long-term improvement (e.g., [15, 16]). Gould [15], despite recognition of several complicating factors, has even made this expectation fundamental to his recent description of the "paradox of the first tier" – the evident conflict between a theory that (he argues) leads us to expect progress and a record in which progress is less than abundantly manifest. In order to sort out these divergent views it is necessary to examine the context and nature of selection more closely.

In all but exceptional cases, a population's exposure to natural selection takes place within a changing environment. Therefore, comparison of phenotypes representing a lineage at two different stages in its history requires judging one phenotype in one context against another phenotype in a different context. Since we expect the outcome of ecological interactions to be context-dependent, any such judgment is immediately suspect. In addition, if we are to assess improvement on the basis of potential replacement relations, we must somehow evaluate the two phenotypes interacting within the *same* context. This may be relatively straightforward for "consecutive" phenotypes (one of which did in fact interact with and replace the other), but it presents a problem for phenotypes that are separated by one or more transformation stages (as is generally the case in an evaluation of transitivity). The only nonarbitrary solution to this problem seems to be to require a case for improvement to be based on a positive replacement potential in *any* of the contexts likely to be encountered by the two phenotypes in question. I submit that it is difficult to predict that selection acting in *one* context will routinely produce a phenotype with such far-reaching potential. This makes it very difficult both to predict and to demonstrate improvement. This view may be somewhat exaggerated, since both environmental change and context dependency may be more or less pronounced depending on the details of each case, but we at least have an indication that there may be problems.

If "morphology" and "design" are interpreted in the narrow sense of "anatomical configuration", we encounter yet another problem with the idea that natural selection predicts improvement of design. Natural selection operates with respect to values of relative fitness, and design in this narrow sense is only one component of fitness. Other components include rates of development, timing of maturation, energetic requirements ... in other words, most aspects of physiology, behavior, and ecology. It is entirely plausible that in a given case a modification in one or more of these other areas could greatly enhance fitness, but at some "cost" to morphology. This problem could be circumvented by defining "morphology" and "design" much more broadly so that they denote the configuration of the entire life history. However, this accommodation only emphasizes the complexity of the problem.

To explore this issue further it is useful to have in mind a model of the interactions that comprise and/or determine the development, life history, and probability of reproductive success of individuals of a species. Such a model, referred to as a "causal plexus" has been introduced elsewhere [11] as part of a discussion of adaptation. The causal plexus is a schematic, algorithmic representation of interactions that characterize the biology of organisms (a genotype, a species, etc.). It consists of a network of links that represent interactions or relationships that are either mutual or polarized (cause-effect). These links connect nodes that represent features, attributes, or consequences of interactions. The apparent structure of the causal plexus depends somewhat on the scale at which it is analyzed. The causal plexus of a genotype is a probabilistic characterization of interactions that comprise the lives of individuals bearing that genotype, from the initiation of development through their complete contribution to succeeding generations. It begins with heritable information (both chromosomal and extrachromosomal) and subsequently expands as a mapping of epigenetic interactions. Features and attributes of the organism interact with each other and with their environment, and some (probably many) of these interactions have consequences that ultimately bear on fitness, defined as probability or propensity for reproductive success [25]. Interactions and attributes that are related to fitness in this way constitute a subset of all the links and nodes that make up the causal plexus. In schematic terms, these links and nodes, traced through the life history of the organism, would be seen as converging on fitness (or relative fitness) itself, considered as a context-dependent property of the genotype. This description is essentially a distillation of a very widely appreciated view of the complexity inherent in the structure of organisms and their relationship to their environment (e.g., [9, 17, 45]).

One of the most important aspects of the causal plexus of any organism is its extremely recursive, complexly interconnected structure (manifested by such phenomena as polygenic effects, pleiotropy, and the failure of any one-to-one mapping between structure and function). The causal plexus is hierarchical in many respects, but different hierarchies overlap in such a complicated fashion that no one of them can be said to characterize adequately the organization of the system. Systems with these structural properties are referred to as "structurally and interactionally complex" [47] or "heterarchical" [18] and are an important focus of research on artificial intelligence. In this context, and potentially in evolutionary biology as well, what is intriguing about these systems is not only their complexity as static structures, but also their response to alteration of their individual components. In some cases they can show surprising resistance to overall change (canalization?) even when considerable numbers of components are perturbed. In other cases they can show a surprising susceptibility to "self-re-organization" when even minor changes are imposed. Additional work is necessary to clarify the extent to which both the structure and the content of such systems contribute to the variety of possible responses to change.

The complexity of the causal plexus and its sensitivity to change are directly responsible for phenomena such as epistatic interactions and frequency-dependent selection. As argued by Wimsatt [48] and Sober [41], the effects of these phenomena may lead to nontransitivity of fitness relations among phenotypes. In other words, even in a regime dominated by directional selection acting on individuals within a population, the fact that *B* is fitter than (and replaces) *A* and that *C* is fitter than (and replaces) *B* does not guarantee that *C* is fitter than (or would replace) *A*. This clearly challenges any expectation that natural selection leads to improvement. Improvement is not something we can predict simply from knowing that selection operates; rather, it is a consequence of the structure, and particularly the stability (in the face of variation in context), of the causal plexus on which selection acts. Improvement is by no means prohibited, but neither is it inevitable. The idea that selection does lead to improvement may gain some support within population genetics from the statement that selection increases the average fitness within a population. However, the validity of this claim is dependent on the conventions used to characterize relative fitness, the avoidance of multilocus genetic models, and the explicit exclusion of complicating factors such as frequency-dependent selection.

If there is reason to be skeptical of improvement on the microevolutionary scale of selection within populations, even greater challenges are encountered at higher levels [15]. For example, competition has attracted much criticism as a mechanism for replacement between species [5], and even in cases where competition may be a factor, it does not always lead to sustained trends that could be interpreted as improvement [7]. In addition, the importance of "incumbency" in resisting replacement [3] and of mass extinction in forcing turnover (possibly on the basis of species properties that are not the primary target of selection during more "normal" times; see Jablonski, this volume) may cut short many trends that would otherwise constitute improvement [15]. At this level too, if progress obtains, it may be as much a consequence of *patterns of organization* of interactions as it is a direct reflection of underlying processes.

Testing Interpretations of Trends

Are the alternative interpretations dealt with here mutually exclusive? In discussing a single, observed trend a decision as to whether it *shows* improvement rests simply on comparison of its endpoints, and it is possible that a transformation compatible with a Markovian model could result in a change that would be evaluated as improvement. In this case these two interpretations seem compatible. However, if we ask whether a single, observed trend was *produced* by a process of (i.e., a process dominated by a mechanism for) improvement, the issue is whether or not a biasing factor, acting more or less throughout the transformation, is responsible for the outcome. This possibility is indeed incompatible with Markovian models as they are construed for purposes of this discussion. If the focus of attention is multiple, observed trends (either "replicates" of a similar trend or a collection of trends that differ in detail), it is much less likely that they would all even *show* improvement without the action of some biasing factor. The diffusion models discussed above differ from Markovian models through the deterministic effects of their postulated (and independently testable) constraints. They differ likewise from improvement models in that they interact deterministically only with extremal phenotypes. Improvement models, or indeed any model involving adaptation as an important factor, will also involve analysis of constraints, but in this context constraints are essentially special cases of a much more broadly acting selectional regime.

If a trend is interpreted as representing improvement, produced by natural selection operating with respect to fitness differences between alternative phenotypes, then these phenotypes – or more explicitly, the features showing the trend – must have been integrated into the causal plexus of their species in such a way as to have a net positive effect on fitness (relative to the fitness of more primitive conditions). To test a model of neo-Darwinian improvement we must evaluate the hypothesis that the features in question are adaptations and that natural selection would have tended to favor the sequence of changes that constitute the trend. In order to be tested at all, such a hypothesis must be specific enough to be refuted. One route to testing involves assessing the physical properties of the feature(s) being studied to determine whether they would have the postulated effect on fitness. Other tests may involve different morphologic features (or aspects of the physical environment or associated biota) that are, according to the hypothesis of adaptation, causally related to the feature on the basis of which the trend was recognized. If these associated features do not show a pattern consistent with the hypothesis of adaptation, it may be refuted. The problem of testing hypotheses of adaptation is controversial [16] but is discussed extensively elsewhere [11, 36].

A specific case of alleged improvement may also be tested by determining whether it demonstrates properties required by this interpretation. Assessing a case for transitivity applies technically only to histories of transformation in which at least three reference points have been distinguished – rather than the minimum of two required by the notion of improvement itself. However, most segments of history can be subdivided in some fashion, and this does not seem to be an inordinately stringent demand to make in the analysis of "major patterns" in the history of life. Although most assessments of transitivity are likely to be theoretical, an empirical basis for recognizing transitivity might consist of a relatively derived daughter species invading the geographic ranges of, and replacing, a succession of increasingly plesiomorphic parent species. This pattern of replacement would be much more likely if improvement had in fact occurred than if it had not. There is always a danger that confounding variables (e.g., environmental change or morphologic differences other than those recognized in the documentation of the trend) might obscure the pattern, but this must be dealt with in any complex analysis. The property of emergence is not as useful empirically as is transitivity. This is because it consists of a relation between an observed pattern and its putative cause(s). If a trend is interpreted as improvement, some degree of emergence has already been assumed or argued.

Interpretations of trends as reflecting only the Markovian aspects of evolution are also subject to test. Raup [28] describes the two principal approaches to testing based on analytical and Monte Carlo methods. Analytical methods offer more precise predictions of the likelihood of various outcomes but are insufficiently developed to handle some of the more mathematically complex Markovian models. Monte Carlo methods have a much broader application but have greater computational requirements and less precise results. In either case, it turns out that the range of behavior allowed by most Markovian models (and therefore consistent with the neutral, random, or pseudorandom/causally heterogeneous interpretation of the trend) is quite broad [28]. This means that on the basis of pattern alone, only what we would intuitively judge to be a rather striking trend would be recognized by the test as requiring a special, higher-level explanation. This has the obvious advantage of providing excellent insurance against Type 1 errors (rejecting the null hypothesis when it is in fact true). Speaking from this perspective, Raup ([28], p 70) offers this summary: "If an array of real-world data does not depart significantly from chance expectations, it is folly to attempt a specific biological interpretation of the data." However, this position is relatively vulnerable to Type 2 errors (accepting the null hypothesis when it is in fact false). If we are concerned about this side of the coin as well, the degree of "folly" involved in considering "biological interpretations of the data" will really depend on whether there is supporting evidence beyond the pattern-data that the Markovian model can digest.

Markovian models of morphologic change may also be tested when we have cases that can be interpreted as replicate "runs" of a similar "natural experiment." If improvement is interpreted broadly enough, these may consist of almost any sample of lineage or clade histories, but on a smaller taxonomic scale examples of parallel or iterative evolution and/or ecological interaction should be analyzed. Significant replication of outcomes (as determined by analytical or Monte Carlo methods) is a potential falsifier of (unbiased) Markovian models, because it implicates a systematic biasing factor.

The diffusion/constraint models discussed above are in a sense hybrids of Markovian and selectionist/deterministic models and therefore require elements of both testing procedures. Within the central area of the branching system, well away from extremal phenotypes, we would not expect patterns to depart from random expectations. The most critical aspect of testing such models will involve the location and nature of constraints. Since these will be manifested in a conventional selectionist mode, they can be tested as indicated above, but only in the context of extremal phenotypes themselves. It is particularly obvious in this case (though it is equally true for all types of models considered above) that these tests bear only on a particular version of a given type of model and not, for instance, on diffusion models generally.

Conclusion

Our interest in the interpretation of trends is due to the insight they might provide into the process of evolution. Our initial approach to interpreting trends is often no more than an expression of prior views on evolution, but by testing a range of interpretations we expect to obtain feedback that will be instrumental in remodeling those views. Recent criticism of the neo-Darwinian paradigm has focused on its supposed expectation of sustained improvement through the operation of natural selection and has contrasted this with the paucity of well documented cases of improvement. However,

I have argued that in many cases the expectation of improvement is itself flawed. To the extent that this is true, our choice of interpretations for trends, at least within the range of alternatives considered here, does not really bear on the question of whether they are caused ultimately by Darwinian processes. Nevertheless, this choice has enormous implications for our perceptions of the level and the time scale on which those processes operate.

Acknowledgements. This paper has been improved significantly by discussion with colleagues and friends during the Dahlem Workshop and at the University of Michigan. I wish particularly to thank D. B. Wake, D. A. Meeter, S. J. Gould, G. P. Wagner, and R. S. Cox.

References

1. Arnold AJ, Fristrup K (1982) The theory of evolution by natural selection: a hierarchical expansion. Paleobiology 8:113–129
2. Ayala FJ (1974) The concept of biological progress. In: Studies in the philosophy of biology, eds Ayala FJ, Dobzhansky T, pp 339–355. London: Macmillan
3. Bakker RT (1977) Tetrapod mass extinctions – a model of the regulation of speciation rates and immigration by cycles of topographic diversity. In: Patterns of evolution, ed Hallam A, pp 439–468. Amsterdam: Elsevier
4. Cheverud J (1984) Quantitative genetics and developmental constraints on evolution by selection. J Theor Biol 110:155–171
5. Connor EF, Simberloff DS (1979) The assembly of species communities: chance or competition? Ecology 60:1132–1140
6. Darwin C (1859) On the origin of species by means of natural selection. London: Murray
7. Dawkins R, Krebs JR (1979) Arms races between and within species. Proc Roy Soc Lond B 205:489–511
8. Dobzhansky T, Ayala FJ, Stebbins GL, Valentine JW (1977) Evolution. San Francisco: Freeman
9. Dullemeijer P (1980) Functional morphology and evolutionary biology. Acta Biotheor 29:151–250
10. Feller W (1968) An introduction to probability theory and its applications. New York: Wiley
11. Fisher DC (1985) Evolutionary morphology: beyond the analogous, the anecdotal, and the ad hoc. Paleobiology 11:120–138
12. Goudge TA (1961) The ascent of life. Toronto: University Press
13. Gould SJ (1977) Eternal metaphors of paleontology. In: Patterns of evolution, ed Hallam A, pp 1–26. Amsterdam: Elsevier
14. Gould SJ (1984) Covariance sets and ordered geographic variation in *Cerion* from Aruba, Bonaire and Curacao: a way of studying nonadaptation. Syst Zool 33:217–237

15. Gould SJ (1985) The paradox of the first tier: an agenda for paleobiology. Paleobiology 11:2–12
16. Gould SJ, Lewontin RC (1979) The spandrels of San Marco and the Panglossian paradigm: A critique of the adaptationist programme. Proc Roy Soc Lond B 205:581–598
17. Hildebrand M, Bramble DM, Liem KF, Wake DB (eds) (1985) Functional vertebrate morphology. Cambridge: Harvard University Press
18. Hofstadter DR (1979) Gödel, Escher, Bach: An eternal golden braid. New York: Random House
19. Huxley JS (1942) Evolution: The modern synthesis. New York: Harper
20. Huxley JS (1953) Evolution in action. New York: Harper
21. Hyatt A (1880) The genesis of the Tertiary species of *Planorbis* at Steinheim. Anniv Mem Boston Soc Nat Hist (1830–1880): 1–114
22. Kimura M (1961) Natural selection as the process of accumulating genetic information in adaptive evolution. Genet Res 2:127–140
23. Klee RL (1984) Micro-determinism and concepts of emergence. Phil Sci 51:44–63
24. Mayr E (1982) The growth of biological thought. Cambridge: Harvard University Press
25. Mills S, Beatty J (1979) The propensity definition of fitness. Phil Sci 46:263–286
26. Osborn HF (1922) Orthogenesis as observed from paleontological evidence beginning in the year 1889. Am Nat 56:134–143
27. Radinsky L (1978) Evolution of brain size in carnivores and ungulates. Am Nat 112:815–831
28. Raup DM (1977) Stochastic models in evolutionary paleontology. In: Patterns of evolution, ed Hallam A, pp 59–78. Amsterdam: Elsevier
29. Raup DM (1985) Mathematical models of cladogenesis. Paleobiology 11:42–52
30. Raup DM, Gould SJ (1974) Stochastic simulation and the evolution of morphology – towards a nomothetic paleontology. Syst Zool 23:305–322
31. Raup DM, Gould SJ, Schopf TJM, Simberloff DS (1973) Stochastic models of phylogeny and the evolution of diversity. J Geol 81:525–542
32. Raup DM, Michelson A (1965) Theoretical morphology of the coiled shell. Science 147:1294–1295
33. Raup DM, Stanley SM (1978) Principles of paleontology, 2nd ed. San Francisco: Freeman
34. Rensch B (1947) Evolution above the species level. New York: Columbia University Press
35. Riedl R (1978) Order in living organisms. New York: Wiley & Sons
36. Rudwick MJS (1964) The inference of function from structure in fossils. Brit J Phil Sci 15:27–40
37. Saunders PT, Ho MW (1981) On the increase in complexity in evolution. II. The relativity of complexity and the principle of minimum increase. J Theor Biol 90:515–530
38. Sawyer S (1976) Branching diffusion processes in population genetics. Adv Appl Prob 8:659–689
39. Simpson GG (1953) The major features of evolution. New York: Columbia University Press
40. Slatkin M (1981) A diffusion model of species selection. Paleobiology 7:421–425
41. Sober E (1984) The nature of selection. Cambridge, MA: MIT Press
42. Stanley SM (1973) An explanation for Cope's rule. Evolution 26:1–26

43. Stanley SM, Signor PW III, Lidgard S, Karr AF (1981) Natural clades differ from "random" clades: simulations and analyses. Paleobiology 7:115–127
44. Stebbins GL (1969) The basis of progressive evolution. Chapel Hill: University of North Carolina Press
45. Wake DB, Roth G, Wake MH (1983) On the problem of stasis in organismal evolution. J Theor Biol 101:211–224
46. Williams GC (1966) Adaptation and natural selection. Princeton: Princeton University Press
47. Wimsatt W (1974) Complexity and organization. In: Boston studies in the philosophy of science, eds Schaffner KF, Cohen RS, pp 67–86. Dordrecht: Reidel
48. Wimsatt W (1980) Reductionistic research strategies and their biases in the units of selection controversy. In: Scientific discovery, ed Nickles T, pp 213–259. Dordrecht: Reidel

Patterns and Processes in the History of Life,
eds. D. M. Raup and D. Jablonski, pp. 119–147. Dahlem Konferenzen 1986
Springer-Verlag Berlin, Heidelberg

Time Series Analysis of the Fossil Record

E. F. Connor

Dept. of Environmental Sciences
University of Virginia
Charlottesville, VA 22903, USA

Abstract. When treated as a statistical time series, the fossil record can be examined for temporal trends and the presence of periodic patterns. The relationship between two or more time series can also be examined. Ordinal techniques for the analysis of temporal trends and frequency domain techniques for the examination of periodicity and the relationship between two time series are recommended. The lack of point estimates of the temporal occurrence of fossils and the irregular spacing and sparseness of the temporal record are the major obstacles to applying time series analysis to the fossil record.

Introduction

The fossil record is unrivaled as a temporal record in its length and the diversity of biological phenomena it chronicles. While is spans over 600 million years of the history of our planet, it also chronicles the evolution of life in form, distribution, and abundance and serves as a barometer of past environments and geophysical events. As a statistical record of temproal pattern and variation in the evolution of life it has received considerable attention over the last 100 years. Much of the classical work involves inferences of secular trends in the morphology of organisms within particular phyletic lineages, or general claims for secular morphological trends across taxa [36, 42, 43]. However, inferences of secular trends have also been the subject of recent examinations of the fossil record. For example, the notion that organisms have evolved to be more physically complex has recently been examined by Flessa et al. [22], and the question of whether the number of kinds of organisms has increased over the span of the fossil record has been intensively examined [3, 33, 34, 40, 44].

Most recently inferences that aspects of the fossil record exhibit periodicity [19, 20, 35] have led to attempts to test for the presence of periodicity in the rates of extinction of families of marine organisms, to estimate this periodicity, and to relate these periodic events to potentially causal geophysical or astrophysical processes. In the course of examining these inferences, the analysis of temporal variation in the fossil record has been expanded to include all the major areas of inquiry that embody the field of statistical time series analysis: the examination of a) largely monotonic temporal trends, b) periodicity, and c) the relationships between two or more statistical time series.

The paper explores the issues surrounding statistical inference as it relates to analysis of the fossil record as a statistical time series. It is not an attempt to review all of time series analysis, but rather to deal with the unique problems that the time series analyst will confront when examining the fossil record.

A Brief Overview of Time Series Analysis

Time series analysis emcompasses a broad array of statistical techniques that can be applied to observations generated under both experimental and non-experimental conditions. In the former instance inferential statistical procedures are well developed, but in the latter only a limited repertoire of inferential techniques are available. For this reason, and because of the difficulty of collecting sufficiently long time series of data, time series analysis has seldom been used in scientific hypothesis testing. Time series techniques usually are applied to explore data in order to generate new hypotheses about, or models of the processes that underlie, an observed series of data. In engineering and economics, time series techniques are often used to predict future values of an observed time series.

In normal theory applications of statistical inference involving the general linear model, observations are assumed to arise independently so that the probability distributions associated with the observations can be characterized by their means and variances. However, in instances where observations arise as part of a temporal sequence, neighboring observations in the sequence will be related. That is, the value of an observation x_t will depend to some extent on the values of $x_{t-1}, x_{t-2}, \ldots$ This will be true unless the observations are generated by a purely random process. Because of this dependence, in addition to specifiying the mean and variance of temporally related observations, it is also necessary to specify the temporal covariance or correlation structure of the data. A large part of time series analysis revolves around the study of the temporal covariance or correlation of data.

Although techniques are available for determining if time series of data display temporal trends in the average level of the series or its variance (see below), most time series techniques require that the investigator assume or insure that such trends are not present. In other words, it is necessary that these characteristics of the time series remain "stationary" over time. Data are often transformed or filtered to insure that this assumption is met. This allows the time series analyst to focus on the temproal correlation structure of the data after removing temporal trends in the mean and variance of the series.

The temproal covariance and correlation of a single time sequence of data are characterized by the autocovariance function and the autocorrelation function, respectively. For a time series of observations arising from a stationary normal process, the autocovariance function $\gamma(k)$ is

$$\gamma(k) = E[(x(t) - \mu)(x(t+k) - \mu)], \tag{1}$$

where E is the expectation operator, μ is the population mean of the underlying process, and k, which takes on values $1, 2, \ldots N-1$, indicates the distance between observations in the series. The autocorrelation function $\varrho(k)$ is simply the autocovariances divided by the variance of the series

$$\varrho(k) = \frac{\gamma(k)}{\sigma^2}. \tag{2}$$

$\varrho(k)$ is estimated by the sample autocovariance function,

$$c(k) = \frac{1}{N} \sum_{t=1}^{N-k} (x_t - \bar{x})(x_{t+k} - \bar{x}), \tag{3}$$

where $\bar{x}$ is the sample mean of the series. The sample autocorrelation function is again the ratio of the sample autocovariances to the sample variance,

$$r(k) = \frac{c(k)}{s^2}. \tag{4}$$

Both of these functions depict the temporal covariance or correlation structure of the data as a function of the spacing (k) between data points in the time series. c(1) and r(1) are estimates of the autocovariance and autocorrelation of all adjacent observations. c(2) and r(2) are estimates of the autocovariance and autocorrelation between data points two observations apart. These functions are usually plotted against k, the lag between observations, and their shape can reveal much about the nature of the observed time series. Since these functions are computed from time series of data that are either continuous or discretely spaced at equal intervals, the lag k autocovariances and autocorrelations describe the covariance in the data as a function of time. The autocovariance and autocorrelation functions are termed "time domain" descriptions of the temporal structure of the observed series.

The autocovariation of an observed series of data can also be depicted in the "frequency domain." This describes the covariance of observations occurring at particular temporal frequencies, say daily, yearly, or every 10 million years. This can be achieved by performing the Fourier transform of the autocovariance function

$$\bar{\Gamma}(f) = \Delta \sum_{k=-\infty}^{\infty} r(k) \cos 2 fk, \quad -\frac{1}{2\Delta} < f < \frac{1}{2\Delta}. \tag{5}$$

$\Gamma(f)$ is known as the power spectrum of the underlying process and represents a decomposition of the variation in the series into the contributions attributed to particular frequencies for a range of frequencies.

The expectations of the autocorrelation function and the power spectrum for a white noise random process provide bases for hypothesis tests to evaluate the significance of observed autocorrelations and power. However, these functions can provide much more information about the structure of a time series than simply whether or not it differs from a white noise random process. Both can be used as diagnostic tools for formulating process models of the observed time series or for selecting parametric models to fit the observed time series. The selection, formulation, and evaluation of parametric statistical models to account for an observed time series, employing these diagnostic tools, comprises the bulk of parametric time series analysis.

Time series analysis can be expanded to the bivariate and multivariate case in order to deal with the relationship between two or more time series. Cross-correlation functions and cross-spectra can be computed to estimate the correlation between time series at particular lags or frequencies, respectively. The expectations of these functions then serve as tests of the null hyphothesis that two series are uncorrelated, and they serve as diagnostic tools in formulating models to relate two time series. For a very accessible introduction of time series analysis the reader should refer to Gottman [24], while for in-depth coverage of frequency domain analysis and time domain time series analysis, Jenkins and Watts [27] and Koopmans [31], and Box and Jenkins [8] are recommended, respectively.

The Fossil Record as a Statistical Time Series

The fossil record is a composite of many time series of morphologic, phyletic, diversity, and biogeographic information. Components of the fossil record defined taxonomically, spatially, or temporally can be examined as individual time series or as suites of time series. The composite record as it relates to a particular criterion variable can be examined as well. Using spatially aggregated data introduces all the uncertainties associated with spatially correlating the stratigraphic relationships of fossils, but it is the only way to obtain temporal sequences spanning long time periods. Aggregating data across taxa may be equivalent to lumping Equids with Cardiid mollusks if taxonomic categories are not equivalent in different phyletic lines. Of course the quality and quantity of fossil data varies temporally so that the temporal biases illuminated by Raup [33] also plague the interpretation of long time series of fossil data. However, these are substantive paleontological and biological issues that each investigator must face when examining any fossil evidence.

The aspect of the fossil record that renders its examination as a statistical time series very difficult is the tremendous uncertainty with which we actually know the temporal distributions of fossils. This is partly because the fossil record is a biased sample of the history of life, but largely because even for those organisms that have a good fossil record we can make only interval estimates of their temporal occurrence. Radiometric dating techniques allow us only to bracket estimates of the temporal occurrence of a particular fossil, and even this interval estimate often has an error of several million years [25]. When attempting to examine long sections of the fossil record we may only have interval estimates of the boundaries of geologic series or stages and not of particular fossil deposits. This broadens the estimates of the temporal occurrence of fossils even further.

Because of our inability to develop point estimates of the temporal occurrence of each fossil, we are confronted with examining a time series that has been deposited largely continuously, but its temporal dependence is known only at a small number of descretely and irregularly spaced times. Furthermore, even at these times we do not have instantaneous estimates of our criterion variables, but rather estimates based on the accumulation of data between dated events. In practice these cumulative estimates are treated as if they were instantaneous by assigning their temporal dependence based on some arbitrary convention. For example, in their analysis of periodicity in extinction rates in the fossil record of marine organisms, Raup and Sepkoski [35] treated all extinctions within an interval as occurring at its end. One could just as well have selected the midpoint of each interval or spread extinctions evenly across the width of each interval. The outcome of time series analyses performed on data tabulated under differing conventions could be quite different, and results based on arbitrary convention may bear no resemblence to the actual time sequence of events.

There are also substantial statistical problems associated with estimating the autocorrelations and power spectra of time series with irregularly spaced observations. The problem of irregularly spaced observations is usually treated in an ad hoc fashion. Sometimes it is solved by interpolating between adjacent data points. This approach is useful when a few observations are missing from an otherwise equally spaced record but for the fossil record would require substituting many interpolated values. Dunsmuir [14] and Barham and Dunstan [4] review a growing statistical literature concerning the estimation of parametric time series models and power spectra when data are irregularly spaced or missing. However, when many observations are missing or if they are highly irregularly spaced these techniques are not particularly reliable. At present, the unique challenge of the fossil record to the time series analyst is to develop techniques to deal with unequally and irregularly spaced observations from a sparse temporal record.

Temporal Trends

The paleontologist's motivation for examining fossil data for the presence of temporal trends lies either in specific theories concerning the course of evolution or in assessing empirical patterns suggested by inspection of fossil data. The time series analyst will more often be motivated by the need to insure that a particular time series is stationary before proceeding to examine its temporal autocorrelation. Regardless of the motivation, the same series of tests can be applied, although their interpretation will depend on the specific motivation.

A series of parametric and nonparametric tests for temporal trend in the average level of a time series and its dispersion can be applied in a straightforward manner. Probably the most obvious approach to test for temporal trends in the average level of a time series is to regress the criterion variable on its temporal occurrence. The advantage of this approach is that if a trend is detected and the investigator intends to perform further analyses, then the parametric regression equation can be used to generate a new criterion variable, the residual, which will no longer show temporal trends. However, the application of least-square regression to autocorrelated data requires that estimates of the regression model account for this autocorrelation. The Cochrane-Orcutt [10] procedure is one commonly used approach that provides unbiased estimates of the regression coefficients accounting for autocorrelation. Weaker tests based on partitioning the series into two or more sections and examining the null hypothesis of no difference in location (means or medians) between sections of the time series versus the alternative hypothesis of temporally ordered increases or decreases in the level of the series are also possible. The parametric t-test, Cox and Stuart's [12] modifications of the nonparametric sign test, and the median test [9, 11] can all be used in this way. Under the assumption of normality the regression and t-test approaches will be more efficient than Cox and Stuart's tests, but for non-normal data the nonparametric approach may be considerably more efficient. Tests based on Spearman's or Kendall's rank order correlation coefficients, Daniels' test [9, 13] and the Mann-Kendall test [9, 32] can also be applied to test for temporal trends. Even when the observations are normally distributed, both of these tests are nearly as efficient as the test based on the parametric regression coefficient. If the examination of trend is motivated by an hypothesis of trend or by an inspection of the data, and not by the desire to make a nonstationary series stationary, then Daniels' or the Mann-Kendall will be the most powerful and robust tests. If the goal of the investigation of trends is their removal, the regression approach will be most useful.

Inferences concerning the dispersion of some criterion variable derived from the fossil record are very uncommon. Ashton and Rowell [2] examined morphological variability in trilobite faunas as a function of species richness, but I am unaware of specific theories concerning temporal trends in variability that have been examined using fossil data. Coupled with one of the above-described tests on the average level of a time series, a test for temporal trends in the variance (or some other measure of dispersion) of a time series would constitute a test for strong stationarity. The parametric variance ratio test or Box and Anderson's modification [7], could be applied to compare the variances of the first and second halves of the time series. However, this test is sensitive to violations of the assumption of normality, so that the jackknife test or the Moses rank-like tests, also applied to a split time series, may be more reliable [26]. If the time series is divided into c groups of k consecutive observations and some measure of dispersion is computed for each group (range, variance, etc.), then the Cox and Stuart test may be applied to these estimates. However, when applied to test for temporal trends in dispersion the Cox and Stuart test requires the assumption of no trend or a linear trend in the mean of the series [9]. The removal of a temporal trend to insure a stationary series can be accomplished by transforming the data, fitting a parametric model of the temporal trend, or applying various filtering techniques. Transformations of the data may be useful in removing trends in the variance of a series but will not be helpful for trends in the average level of a series. Parametric models, on the other hand, are useful in removing linear and quadratic temporal trends in the mean of a series. The use of filters such as substituting the first differences of a series ($x_t - x_{t+1}, x_{t+1} - x_{t+2}, \ldots$) for the raw observations can remove long-term temporal trends, but also attenuate the variation in the series that is attributable to low-frequency phenomena [27]. If filters are applied to remove temporal trends, then the effect of the filter on the temporal autocorrelation structure of the data should be investigated.

In order to illustrate these tests and other aspects of the analysis of the fossil record as a statistical time series, I have compiled data on the numbers of extinctions, originations, and the standing diversity of families of marine organisms from Sepkoski [39] with his revisions current through July 1984. The stratigraphic categories and the estimates of the temporal boundaries of stages are as presented by Sepkoski [38, 39], Harland et al. [25], and Raup and Sepkoski [35]. For families with poorly defined stratigraphic limits, the bottom of the stage or series of first occurrence and the top of the stage or series of last occurrence were assumed to delimit a family's range in the fossil record. Data are presented for 80 stratigraphic stages, deleting the Riphean and the lower and middle Vendian, because of their low diversity, and the Recent, because it has not yet terminated. The entire record as tabulated here is comprised of 3708 families of marine organisms and spans 599.9 million years. All events are by convention assumed to have occurred instantaneously at the end of each geologic stage.

Figure 1 presents the raw numbers of extinction and originations as tabulated from Sepkoski [39], and Fig. 2 illustrates the standing diversity in

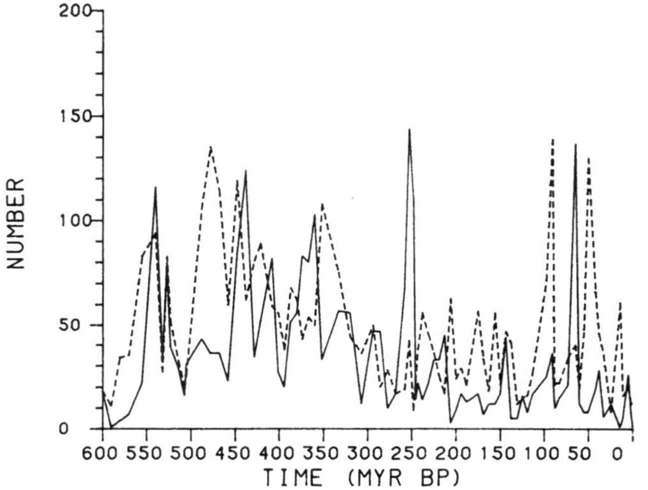

Fig. 1. Number of extinctions (*solid line*) and originations (*dashed line*) occurring in each stratigraphic stage for 3708 families of marine organisms

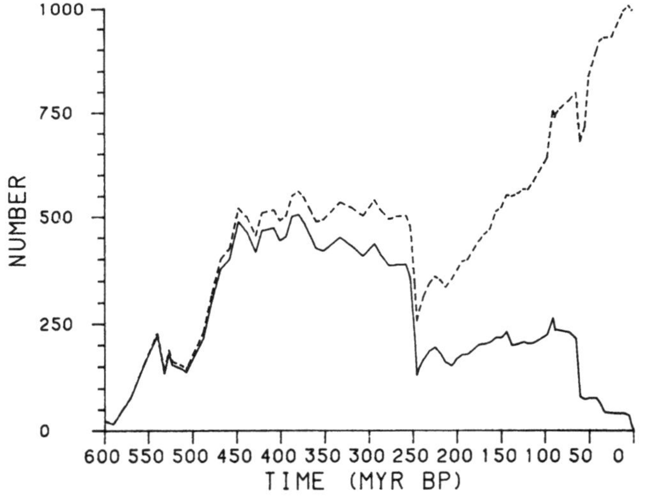

Fig. 2. Standing diversity of families of marine organisms for all taxa (*dashed line*) and with extant taxa removed (*solid line*)

each stage with and without removing extant taxa (Raup and Sepkoski's Pull of the Recent). Neither the number of extinctions or originations appear to exhibit long-term trends, although the number or originations has declined. However, standing diversity appears to have increased from the Vendian to the Recent. Table 1 summarizes the results of applying several tests for temporal trend in location and dispersion to each of these series. Those tests that require splitting the series were performed on the data from the first and last forty stratigraphic stages. Cox and Stuart's test for trend in location was applied using Conover's [11] suggestion that differences between observations paired starting from the ends of the two halves of the series be used as a test statistic. Cox and Stuart's test for dispersion was applied to variances computed from groups containing data from four adjacent geologic stages. The results of these tests are not always in agreement. Of the tests on trend in location, the median test and the sign test are not very powerful, and the sign test is sensitive to nonlinear trends. Figure 3 shows the temporal distribution of the residual $(D_i - \hat{D}_i)$ estimated from the parametric linear regression of diversity on time, and a plot of the first differences of total diversity.

A more interesting time series might be some estimate of the rate of extinction and origination of families over a similar time course. Figure 4 illustrates the temporal pattern in extinction rates and origination rates computed as;

$$\text{E rate(i)} = \frac{\text{Ei}}{D_{i-1} + O_i} \times 100, \quad \text{and} \tag{6}$$

$$\text{O rate(i)} = \frac{\text{Oi}}{D_{i-1} + O_i} \times 100 \tag{7}$$

for the entire data set including extant taxa, where E_i, O_i, and D_i are the numbers of extinctions, originations, and the standing diversity in the ith geologic stage. These series display a temporal decrease that arises because of the temporal increase in diversity illustrated in Fig. 2. Figure 5 presents the residuals $(O_i - \hat{O}_i)$ and $(E_i - \hat{E}_i)$ derived from parametric linear regressions of the temporal dependence of the extinction rates and origination rates and from these rates estimated by applying the "Pull of the Recent" filter suggested by Raup and Sepkoski [35]. The Pull of the Recent filter generates a stationary series for extinction rates but not for origination rates, since for origination rates both the numerator and denominator of the expression are affected. Furthermore, the Pull of the Recent filter dis-

Table 1. Test for temporal trend in location and dispersion

Time Series of Data	Tests on Location						Tests on Dispersion			
	Median test (x_1)	Cox and Stuart sign test num(+'s)	t-test (t)	Linear regression (r)	Daniels' test (r)	Mann-Kendall test (r)	Variance ratio $F_{39, 39}$	Jackknife test Q	Moses rank-like W	Cox and Stuart sign test num(+'s)
All Taxa										
Number of Originations	3.2 $p>0.1$	27 $p<0.05$	2.43 $p<0.02$	0.29 $p<0.0048$	0.32 $p<0.004$	0.24 $p<0.002$	1.3589 $p>0.1$	0.3791 $p>0.35$	104 $p>0.48$	5 $p>0.5$
Number of Extinctions	20 $p<0.001$	33 $p<0.001$	4.46 $p<0.001$	0.30 $p<0.0035$	0.43 $p<0.001$	0.30 $p<0.001$	2.76 $p<0.002$	0.27 $p>0.5$	83 $p>0.1$	7 $p>0.5$
Standing Diversity	7.2 $p<0.02$	24 $p>0.05$	5.75 $p<0.001$	−0.83 $p<0.0001$	−0.80 $p<0.001$	−0.66 $p<0.001$	1.6705 $p>0.05$	2.4635 $p<0.01$	102 $p>0.5$	6 $p>0.5$
Origination Rate	20 $p<0.001$	31 $p<0.001$	4.77 $p<0.001$	0.73 $p<0.0001$	0.78 $p<0.001$	0.61 $p<0.001$	21.8 $p<0.002$	7.14 $p<0.001$	115 $p>0.48$	7 $p>0.5$
Extinction Rate	33.8 $p<0.001$	33 $p<0.001$	5.27 $p<0.001$	0.59 $p<0.0001$	0.77 $p<0.001$	0.58 $p<0.001$	6.43 $p<0.002$	3.5 $p<0.001$	67 $p<0.002$	8 $p>0.5$
Extant Taxa Removed										
Number of Originations	24.2 $p<0.001$	34 $p<0.001$	6.47 $p<0.001$	0.64 $p<0.0001$	0.74 $p<0.001$	0.59 $p<0.001$	6.38 $p<0.002$	3.7 $p<0.001$	65 $p<0.002$	8 $p>0.5$
Standing Diversity	12.8 $p<0.001$	38 $p<0.001$	5.97 $p<0.001$	0.34 $p<0.0009$	0.32 $p<0.004$	0.24 $p<0.002$	4.21 $p<0.002$	6.28 $p<0.001$	67 $p<0.002$	10 $p<0.05$
Origination Rate	12.8 $p<0.001$	29 $p<0.01$	3.83 $p<0.001$	0.67 $p<0.0001$	0.64 $p<0.001$	0.49 $p<0.001$	27.03 $p<0.002$	5.28 $p<0.001$	84 $p>0.1$	7 $p>0.5$
Extinction Rate	1.8 $p>0.2$	29 $p<0.01$	0.29 $p>0.6$	−0.013 $p-0.45$	0.10 $p>0.37$	0.07 $p>0.33$	1.44 $p>0.2$	0.46 $p>0.5$	100 $p>0.7$	5 $p>0.5$

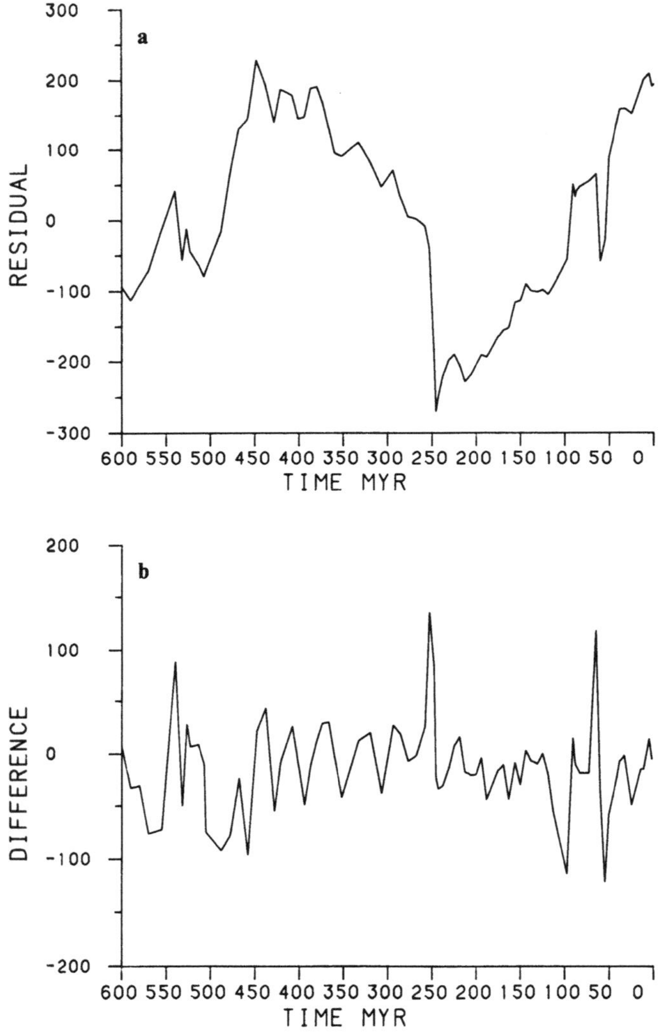

Fig. 3 a, b. Trend removal from diversity data for all taxa. **a** diversity residual obtained from linear regression of standing diversity on time, **b** first differences of diversity time series

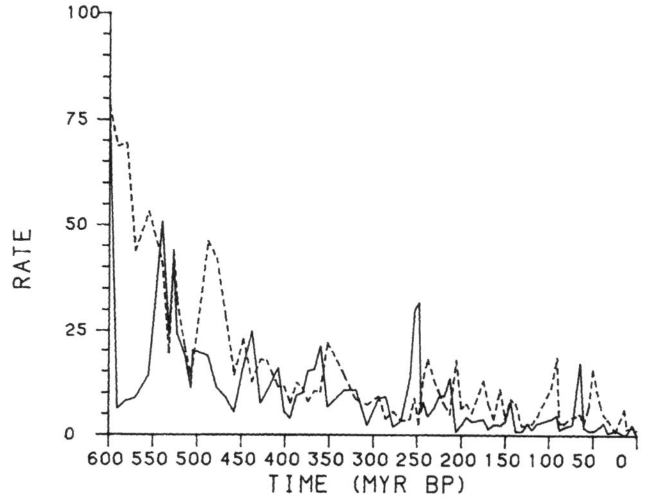

Fig. 4. Extinction rates (*solid line*) and origination rates (*dashed line*) for all taxa

torts the relative magnitude of the extinction rate at the end of the time series when extant taxa have essentially been removed.

Nonrandomness and Periodicity in Time Series

The questions of whether a particular sequence of observations constitutes a random sequence and of whether data display periodic behavior are most often addressed in the frequency domain. The periodogram and the power spectrum provide bases for tests regarding these questions. Kitchell and Pena [30] illustrate the use of time domain techniques in analyses of the fossil record.

The periodogram is an asymptotically unbiased but inconsistent estimator of the power spectrum and is computed from the Fourier transform of the sample data. This transform approximates the observed data set as the sum of a series of sine and cosine functions.

$$x_t = \alpha_0 + \sum_{i=1}^{q} (\alpha_i \cos(2\pi f_i t) + \beta_i \sin(2\pi f_i t)), \tag{8}$$

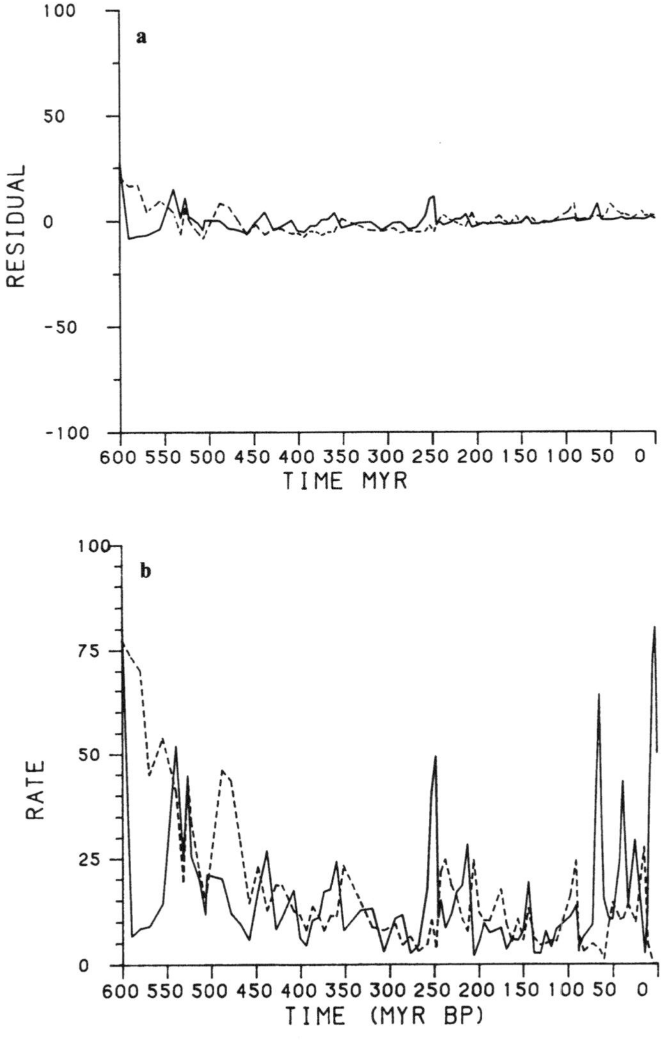

Fig. 5 a, b. Trend removal from extinction rates. **a** extinction rate residual obtained from linear regression, **b** extinction rate (*solid line*) and origination rate (*dashed line*) for data with extant taxa removed (Raup and Sepkoski's [35] Pull of the Recent filter)

where q is odd, $f_i = i/N$ is the ith harmonic of the fundamental frequency $1/N$, N is the length of the temporal record, and t is the temporal occurrence of an observation. The Fourier coefficients, the α_i's and β_i's, are estimated as

$$a_i = \frac{2}{N} \sum_{t=1}^{N} x_t \cos(2\pi f_i t), \quad \text{and} \tag{9}$$

$$b_i = \frac{2}{N} \sum_{t=1}^{N} x_t \sin(2\pi f_i t) \tag{10}$$

for $i = 1, 2, \ldots q$. The periodogram is simply

$$I(f_i) = \frac{N}{2} (a_i^2 + b_i^2)$$

for $f_i = 1/N, 2/N, \ldots 1/(2\Delta)$, and the $I(f_i)$, or the intensity at frequency f_i, are plotted against f_i (8). The Nyquist frequency $1/(2\Delta)$ is the highest frequency that can be detected and it corresponds to a period of twice the width (Δ) of the spacing between data points. Jenkins and Watts (27) refer to a modified version of the periodogram as the sample spectrum

$$I(f) = \frac{2}{N} (a_f^2 + b_f^2), \qquad 0 < f < 1/(2\Delta). \tag{12}$$

The sample spectrum is also an asymptotically unbiased but inconsistent estimate of the power spectrum. The sample spectrum is related to the autocovariance function by the inverse Fourier transform of Eq. 5:

$$I(f) = 2(c(0)) + 2 \sum_{k=1}^{N-1} c(k) \cos(2\pi f k). \tag{13}$$

Both the periodogram and the sample spectrum are designed to detect periodicities at fixed frequencies embedded in noisy data. These fixed frequencies are the harmonics or multiples of the fundamental frequency ($1/N$) which has a period equal to the length of the series.

Consistent estimates of the power spectrum are usually obtained by smoothing the sample spectrum or the fourier cosine transform of the autocovariance function [27]. This introduces some bias to the estimated power spectra, but the marked decrease in the variance of the estimates far outweighs this small bias. More recently the method of maximum entropy has been used to provide consistent and unbiased estimates of the power spectrum [5].

These procedures are usually applied to a data record with equally spaced observations, and their properties are well-known only in this instance. Dunsmuir [14], Barham and Dunstan [4], Jones [28], Scheinok [37], Gaster and Roberts [23], and Koopmans [31] discuss alternative procedures for estimating the periodogram and spectrum from unequally spaced data. However, the properties of these alternative procedures are not known when most of the data are missing and the temporal occurrence of missing observations is very irregular as in the case of the fossil record. For example, the procedure outlined by Dunsmuir [14] requires that the data record be treated as regularly spaced, but with missing observations. For the 600 million year section of the fossil record presented in Figs. 1–5, this would require that the sampling interval be every one million years (to insure that all data are included), which means that 520 out of 600 data points would be treated as missing.

Another approach to estimating the periodogram for irregularly spaced data is to account explicitly for the irregular spacing of the data in the Fourier transform. This is equivalent to employing Simpson's or the trapezoid rule to compute the Fourier integral of the area under the curve of the data for discrete and irregularly spaced observations. Under this procedure the Fourier coefficients are computed by

$$a_i = \frac{2}{L} \sum_{k=1}^{N} \frac{1}{2} (x_k \cos(2\pi f_i t_k) + x_{k+1} \cos(2\pi f_i t_{k+1}))(t_{k+1} - t_k), \tag{14}$$

and

$$b_i = \frac{2}{L} \sum_{k=1}^{N} \frac{1}{2} (x_k \sin(2\pi f_i t_k) + x_{t+k} \sin(2\pi f_i t_{k+1}))(t_{k+1} - t_k), \tag{15}$$

where L is the length of the record, $f_i = i/L$, $1/L \leqq f_i \leqq 1/(2\Delta)$, is the average spacing between ovbservations, and N is the number of observed data points. The periodogram is simply the sum of the squares of these coefficients:

$$I(f_i) = (a_i^2 + b_i^2). \tag{16}$$

Because only the relative magnitudes of these intensities are needed, the normalizing factor $L/2$ can be omitted. Using this approach the computed

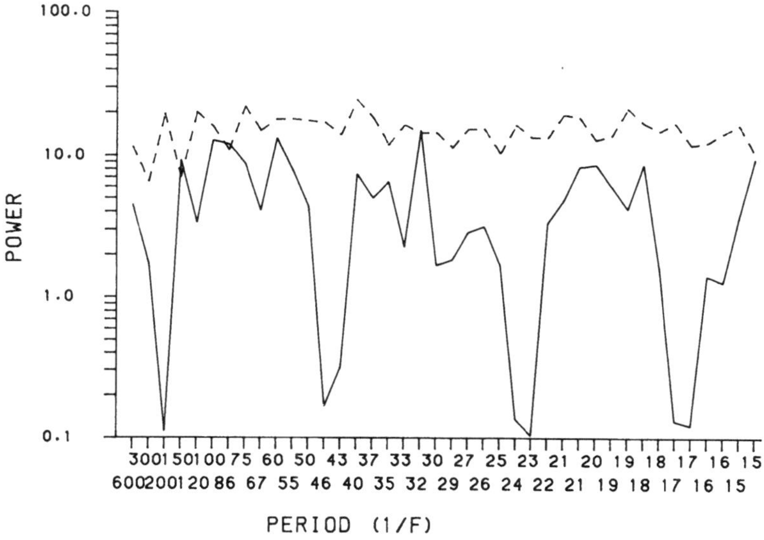

Fig. 6. Periodogram of detrended extinction rate for all taxa over 599.9 myr record (*solid line*). Dashed line indicates the upper 95th percentile of the bootstrapped distribution of power at each frequency

periodicities refer to periodicities in units of the actual time scale, rather than as integral multiples of the average time between observations.

Figure 6 illustrates the periodogram for the extinction rate residuals of Fig. 5 computed by accounting for the irregularly spaced data. Notice that the highest intensity occurs at a period of approximately 31 million years or a frequency of 0.03 cycles per million years. This is the value for periodicity in extinction rate proposed by Fischer and Arthur [20].

Limits Imposed on the Determination of Periodicity by Finite Length of Record

The fact that a temporal record is of finite length places certain lower limits on our ability to determine periodicity. These limits are fundamental in the sense that they would apply to a system even if we had perfect knowledge of the criterion variable and the geological time scale. They arise solely from the properties of the Fourier transform and hence will apply to any analysis of periodicity.

If we note the length of the fossil record by L, then the frequencies that will appear in the Fourier series will be given by $f_n = n/L$, where n is integer. This statement follows from the fact that the Fourier transform expands a series such as the extinction record in terms of sines and cosines whose frequencies are such that they fit exactly onto the length L. The periods that appear on the horizontal axis of the periodogram, therefore, will be given by $T_n = L/n$, from which it follows that the intervals between successive periods will be

$$\Delta T_n = T_n - T_{n+1} = L\left(\frac{1}{n} - \frac{1}{n+1}\right) = \frac{L}{n(n+1)} \approx \frac{L}{n^2}. \tag{17}$$

The limit on the accuracy of determining periodicity follows from the fact that we cannot determine the frequency of a peak in the Fourier transform to an accuracy higher than that given by the spacing between periods at which the transform is calculated. This is equivalent to saying that you cannot determine a length to an accuracy much greater than the smallest division on your ruler. To understand the implications of this limit for an analysis of the type carried out by Raup and Sepkoski [35], note that if we let T_p denote the periodicity of the extinctions, then the Fourier harmonic corresponding to T_p will be given by $T_p = 1/n_p$. This means that the intervals between Fourier periods around the crucial value of T_p will be

$$\Delta T_p = \frac{L}{n_p^2} = \frac{T_p^2}{L}. \tag{18}$$

Two numerical examples will show the importance of this result:

1) Suppose that we wish to determine the periodicity of extinctions to within one million years (myr). Taking T = 26 myr, we find that the length of record must be

$$L = \frac{T_p^2}{\Delta T_p} \approx 676 \text{ myr}. \tag{19}$$

In other words, to determine T_p to within 1 myr, we need to have an extinction record extending over the entire Phanerozoic.

2) Suppose that we have L = 250 myr, T = 26, a situation corresponding to Raup and Sepkoski's (35) analysis. Then

$$\Delta T_p = \frac{T_p^2}{L} = 2.5. \tag{20}$$

This means that we cannot determine the extinction periodicity to better than about 10% at best. Any errors associated with problems in the fossil record or the geological time scale will increase this uncertainty (J. Trefil, submitted).

A Test for Nonrandomness

For each of the three estimates of the power spectrum presented above, the intensities at the harmonic frequencies are independent and represent the amount of variation in the series attributable to each frequency. If the data represent a white noise process, then all frequencies will be equally represented and the periodogram will be a horizontal line. The periodogram can be viewed as the empirical distribution function of intensity and the sum of the intensities divided by the total variation in the record as the empirical cumulative distribution function. A test for deviation from the null hypothesis of equal representation of all frequencies is based on the Kolmorgorov-Smirnov test [11, 27]. The distribution function for a random sequence should be uniform so that its cumulative distribution will be a straight line with slope equal to 1. For a one-sided test of the null hypothesis stated

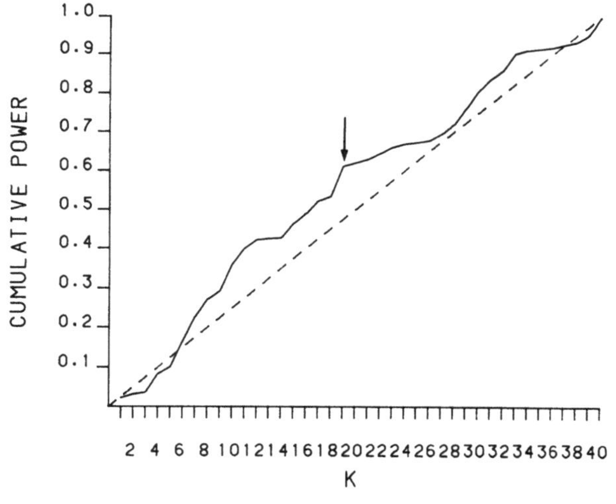

Fig. 7. Cumulative distribution of power (*solid line*) and the expected cumulative distribution of power for a white noise random process (*dashed line*). The maximum vertical distance between these two distributions (T) is indicated by the *arrow*

above versus the alternative that at least one value of I (f_i) is greater than the intensity expected for a white noise process, the maximum vertical distance of the observed above the hypothesized cumulative distribution function (T) serves as the test statistic. The critical values of T for specified α are tabulated by Conover [11] for sample sizes of forty or less and can be approximated for larger samples. Jenkins and Watts [27] suggest that the sample size be taken as the maximum number of frequencies calculated. Figure 7 illustrates this test applied to the periodogram of the extinction rate residual plotted in Fig. 6. The T value obtained from inspection of this plot is 0.14, and the critical value for $\alpha = 0.05$ and $n = 40$ is 0.189. Therefore we would not reject the null hypothesis that the periodogram [$I(f_i)$] is that of a white noise process.

Periodicity

Rejection of the test for deviation from a white noise process for a stationary series implies that the data display some periodicity but does not reveal where that periodicity lies. Bolviken [6] suggests significance tests for the periodogram for equally spaced data. However, the accuracy of these tests for unequally spaced data is unknown. An alternative, computationally expensive but reasonable approach to determine if the power observed at particular frequencies is unusually high is to estimate the probability density of power at each frequency. Two similar empirical approaches based on the empirical distribution function of the sample data can be used. One is based on the principle of randomization [15, 21, 29] and the other on the procedure known as the bootstrap [16–18].

Both randomization and bootstrap techniques are data resampling schemes designed to estimate the standard error of a test statistic or its probability distribution. Both techniques, as applied to developing significance tests for the periodogram, involve randomly resampling from the observed distribution of the variable of interest to generate a time series of length equal to the observed series, and doing so repeatedly. Each of these resampled series is then subjected to the same sequence of analytical techniques (trend removal and Fourier transformation) and the distribution of power [$I(f_i)$] at each frequency is tabulated for a large number of such series.

The essential difference between the randomization and bootstrap approaches is that in the former data are resampled without replacement, while in the latter sampling occurs with replacement. The randomization approach generates the probability distribution of the test statistic across permutations of the time sequence of observations. The bootstrap approach generates the probability distribution of the test statistic across a larger array of data rearrangement since individual observations may be sampled repeatedly to generate each synthesized time series. When applied to the same empirical distribution function and test statistic, the bootstrap generally yields larger standard errors for the test statistic and thus a more conservative test. Both the bootstrap and randomization tests are based on the notion that the empirical distribution of the random variable is the best estimate of the probability distribution of that variable. However, the randomization approach fixes the sample distribution function as the population distribution function while the bootstrap treats the empirical distribution as a constrained estimate of the population distribution. Variations on the bootstrap, designed to remove the constraints imposed by the fact that any empirical distribution will be finite and discrete, have been suggested [17].

In their analyses of periodicity in extinction rates and the record of impact craters, Raup and Sepkoski [35], Sepkoski and Raup [41], and Alvarez and Muller [1] use randomization to assess the statistical significance of their analyses. The bootstrap may be a better approach because it allows events of different magnitude, order, and number than that obtained in the observed record. The randomization approach only permutes the order of the events.

To illustrate significance tests of the periodogram based on the bootstrap, I have conducted analyses of extinction rates in the fossil record, as described above, and of the section examined by Raup and Sepkoski [35]. The bootstrap requires that the random variables to be resampled be independent and identically distributed. Since the rates of extinction occurring in each stratigraphic stage may depend on rates in earlier stages, extinction rates do not meet this condition. However, we may consider the temporal ranges of families of marine organisms in the fossil record to be independent and identically distributed. Therefore, the bootstrap would be based on sampling from the observed temporal ranges of the 3708 marine families, independently, with each family having probability 1/3708 on each draw. Using this sampling scheme we generate a new sample of 3708 families and their temporal ranges. From this we know the stage of first occurrence and the stage of last occurrence for each family, and we can compute the numbers of families and the rates of origination and extinction in each interval. These data are then subjected to the same analyses as the observed time series, and the probability of obtaining the observed level of power, or greater power, at each frequency is estimated as the proportion of bootstrap samples equal to or more extreme than the observed power values.

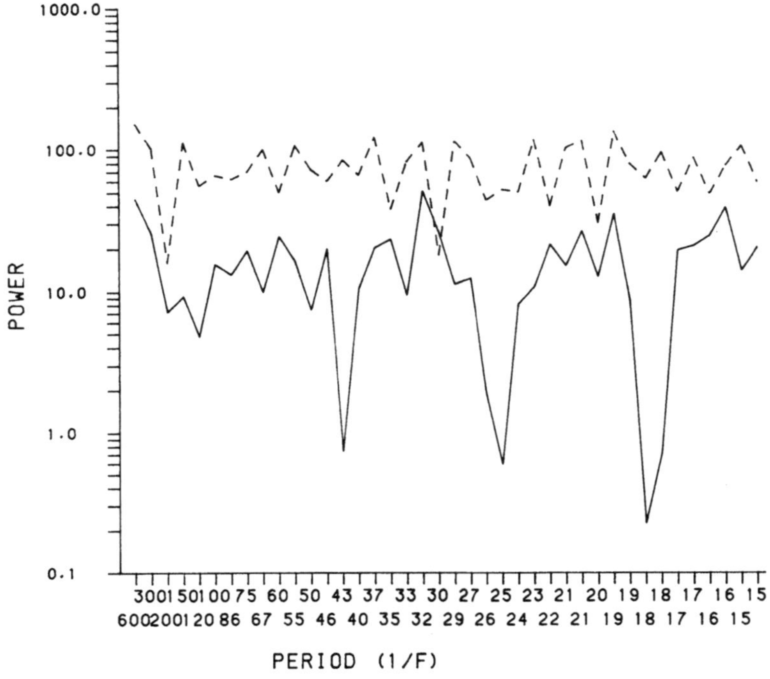

Fig. 8. Periodogram (*solid line*) and upper 95th percentile of the bootstrapped distribution of power for each frequency (*dashed line*) for 599.9 myr record with extant taxa removed

Figure 6 illustrates only the upper 95th percentile of the bootstrapeed distribution of power for the 600 million year extinction rate record of all taxa based on 1000 bootstrap samples. Figure 8 provides similar estimates for the entire record with extant taxa removed and Fig. 9 presents these same analyses confined to the section of the fossil record examined by Raup and Sepkoski [35]. For those frequencies at which the periodogram exceeds the critical percentile, one would conclude that significant periodicity is present. However, because this procedure involves conducting a statistical test for periodicity at each of forty frequencies with $\alpha = 0.05$, the experiment wide error rate (the probability of at least one Type I error) is 0.87. In other words, one would expect on average to exceed the upper 95th percentile of the bootstrapped distribution of power at two periodicities by chance alone.

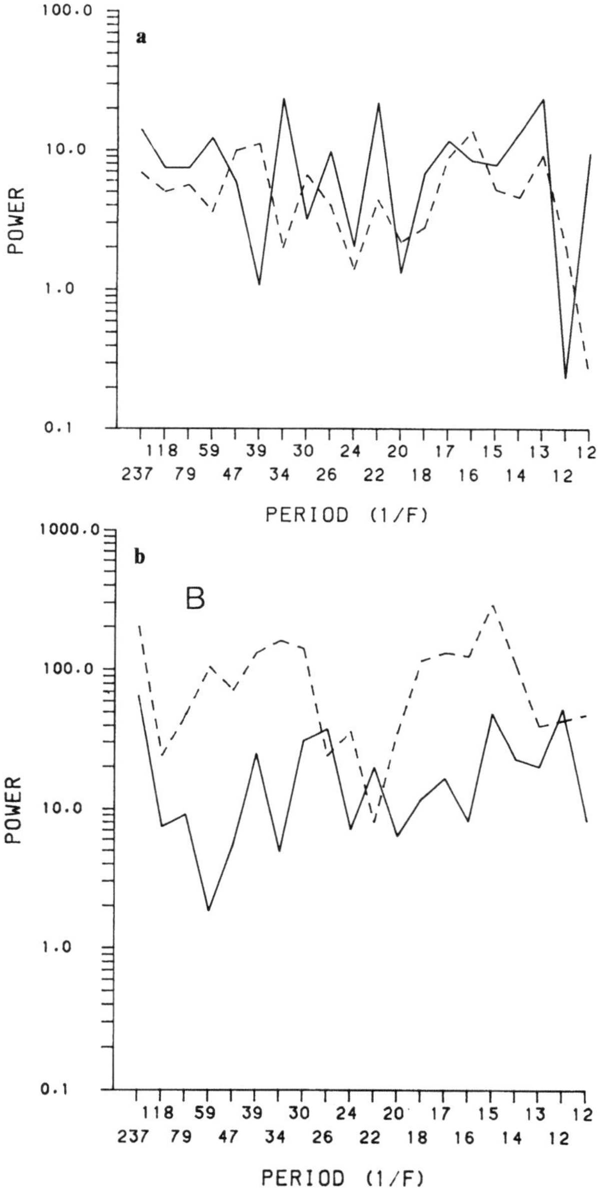

Fig. 9 a, b. Periodogram (*solid line*) and upper 95th percentile of the bootstrapped distribution of power for each frequency (*dashed line*) for section of record examined by Raup and Sepkoski [35]. **a** all taxa, **b** extant taxa removed

Furthermore, it is impossible without independent evidence to determine which periodicities are true periodicities and which represent Type I errors.

Examining the Relationship between Two Time Series

Even if a series of observations is not periodic it may still be related in a causal or correlational sense to processes that give rise to other time series of data. For example, Alvarez and Muller [1] suggest that the periodicity in extinction rates detected by Raup and Sepkoski [35] may be related in a causal way to global environmental changes associated with asteroid impacts. They examined the temporal record of large impact craters to determine if these impacts tended to occur with a periodicity similar to that detected by Raup and Sepkoski [35] or Fischer [19] for mass extinctions. Their conclusions were based only on a visual comparison of the shapes of the periodograms for these two series and not on an explicit, statistical assessment of the relationship between these two time series. This approach will certainly not be useful for non-periodic time series and can result in erroneous interpretations if two series have similar periodicities but are cycling out of phase.

The relationship between two statistical time series can be explicitly examined in both the time domain and the frequency domain. In the time domain, the cross-correlation function and transfer function models [8] can be used to establish whether a substantial component of the variation in one time series can be explained by another time series. Significance tests for cross-correlations are available. The percentage of the variation explained by a transfer function model $l(R^2)$ can be computed although its probability distribution in comparing time series is not established. However, the properties of these techniques when applied to an irregegularly spaced and sparse data record are unknown.

In the frequency domain, the cross-spectrum is used to examine the correlation between two time series. Apparently little effort has been made to examine the properties of the cross-spectrum for unequally spaced data. However, it is computationally simple to extend the Fourier transform for irregularly spaced data to cross-spectral analysis. If we denote the Fourier cosine and sine coefficients of two time series at frequency f as a_1, b_1 and a_2, b_2 and use $-j$ to represent the imaginary part, then the cross-spectrum $[C_{12}(f)]$ is simply

$$C_{12}(f) = \frac{1}{L} [(a_1 a_2 + b_1 b_2) - j(b_2 a_1 - b_1 a_2)], \tag{21}$$

where L is the temporal length of the record. The first half of the right side of Eq. (21) is the co-spectrum [$L_{12}(f)$] and measures the relationship between the in-phase components of two time series. The second half is the quadrature spectrum [$Q_{12}(f)$] which measures the relationship between the out-of-phase components of the two series. The co- and quadrature spectra can be combined in several ways to yield test criteria for a frequency domain test of the correlation between two times series.

Jenkins and Watts [27] suggest testing for correlation between two times series by examining the in- and out-of-phase components separately. They suggest examining a plot of the integrated sample co-spectrum, $J_{12}(k)$, to estimate the in-phase correlation between two series. This quantity is a normalized sum of the co-spectral estimates

$$J_{12}(k) = \frac{2}{L s_1 s_2} \sum_{i=0}^{K} L_{12}(f_i), \tag{22}$$

where s_1 and s_2 are the standard deviations of each series. If the series are uncorrelated $J_{12}(k)$ will be uniformly zero at all frequencies, but if the two series are correlated $J_{12}(k)$ will equal the value of this correlation at the Nyquist frequency. To examine the correlation between the out-of-phase components of two series, Jenkins and Watts [27] employ the phase spectrum

$$F_{12}(f) = \arctan\left(-\frac{Q_{12(f)}}{L_{12}(f)}\right). \tag{23}$$

If two series are uncorrelated the sample phase spectrum will be uniformly distributed in the range $-\pi/2$ to $\pi/2$. Therefore, a test based on the cumulative phase spectrum using the Kolmorgolov-Smirnov test outlined earlier [11, 27] can be used to detect correlation in the out-of-phase component of time series. If two series are correlated, one would expect both tests to indicate so.

Figures 10 and 11 illustrate the integrated co-spectrum and cumulative phase spectrum of the time series of origination and extinction rates presented in Fig. 5a. The integrated co-spectrum oscillates about zero and the maximum deviation of the cumulative phase spectrum from expectation is only 0.1. This indicates that these two series are uncorrelated.

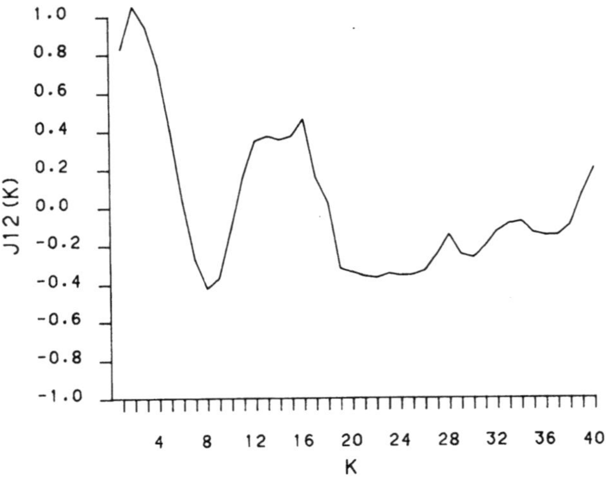

Fig. 10. Integrated co-spectrum of extinction and origination rates for 599.9 myr record with all taxa included

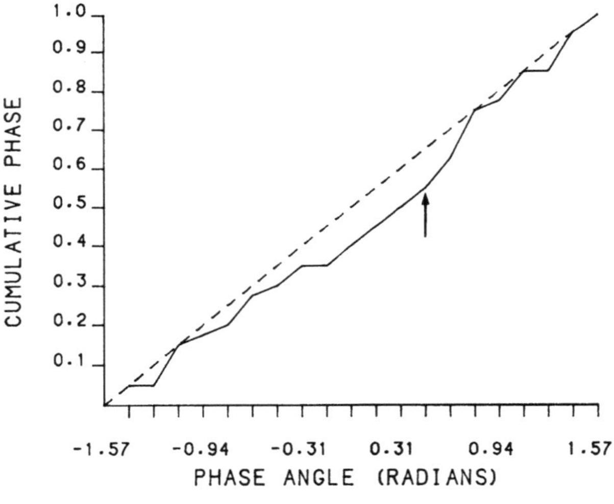

Fig. 11. Cumulative phase spectrum of extinction and origination rates for 599.9 myr record with all taxa included (*solid line*) and expected phase spectrum (*dashed line*). The maximum vertical deviation between these two distributions is indicated by the *arrow*

Prospectus

This paper has only scratched the surface of the ins and outs of analyzing the fossil record as a statistical time series. I have identified the uncertainty in estimates of the temporal occurrence of fossils as the primary limiting aspect of the data per se and the irregular spacing and sparseness of dates in the fossil record as the major obstacles to the time series analyst. I have recommended frequency domain approaches over time domain approaches because of the relatively simple extension of the Fourier transform to unequally spaced data and because inferential techniques are available or easily developed in the frequency domain.

The examples presented here tend to corroborate the findings of Raup and Sepkoski [35] and Fischer [19] and suggest that the extinction rate of families of marine organisms is periodic. However, because the experiment-wide error rate of power spectrum analysis is high when many frequencies are examined it is difficult to determine the true period. Furthermore, we can estimate this periodicity to no closer than ± 3 million years, and confidence in this estimate of periodicity is probably considerably less. A more precise estimate of this periodicity and our confidence in this estimate would be very useful in examining hypotheses concerning its causes.

Acknowledgements. Guidance from J. Cosby, G. Hornberger, D. Meeter, E. Rastetter, R. Ritter, and P. Young was instrumental in the preparation of this material. The manuscript benefitted greatly from careful reading by R. Dueser, E. Rastetter, and J. Trefil. I would particularly like to thank J. Trefil for contributing substantively to the manuscript and immeasurably to my understanding of the Fourier transform.

References

1. Alvarez W, Muller RA (1984) Evidence from crater ages for periodic impacts on the earth. Nature 308:718–720
2. Ashton JH, Rowell AJ (1975) Environmental stability and species proliferation in late Cambrian trilobite faunas: a test of the niche variation hypotheses. Paleobiology 1:161–174
3. Bambach RK (1977) Species richness in marine benthic habitats through the Phanerozoic. Paleobiology 3:152–167
4. Barham SY, Dunstan FDJ (1982) Missing values in time series. In: Times series analysis: theory and practice 2, ed Anderson OD, pp 25–42. Amsterdam: North-Holland
5. Bennett RJ (1979) Spatial time series: analysis, forecasting and control. London: Pion Press

6. Bolviken E (1983) New tests of significance in periodogram analysis. Scand J Stat 10:1–9
7. Box GEP, Anderson SL (1955) Permutation theory in the derivation of robust criteria and the study of departures from assumption. J Roy Stat Soc B 17:1–26
8. Box GEP, Jenkins GM (1976) Time series analysis: forecasting and control. San Francisco: Holden Day, Inc
9. Bradley JV (1968) Distribution-free statistical tests. Englewood Cliffs: Prentice-Hall
10. Cochrane D, Orcutt GH (1949) Application of least-squares regressions to relationships containing auto-correlated error terms. J Am Stat Ass 44:32–61
11. Conover WJ (1980) Practical nonparametric statistics, 2nd ed. New York: John Wiley and Sons
12. Cox DR, Stuart A (1955) Some quick sign tests for trend in location and dispersion. Biometrika 42:80–95
13. Daniels HE (1950) Rank correlation and population models. J Roy Stat Soc B 12:171–181
14. Dunsmuir W (1981) Estimation for stationary time series when data are irregularly spaced or missing. In: Applied time series analysis II, ed Findley DEF, pp 609–649. New York: Academic Press
15. Edgington ES (1980) Randomization tests. New York: Marcel Dekker, Inc
16. Efron B (1979) Bootstrap methods: another look at the jackknife. Ann Stat 7:1–26
17. Efron B (1979) Computers and the theory of statistics: thinking the unthinkable. Siam Rev 21:460–480
18. Efron B, Gong G (1983) A leisurely look at the bootstrap, the jackknife and cross-validation. Am Stat 37:36–48
19. Fischer AG (1981) Climatic oscillations in the biosphere. In: Biotic crises in ecological and evolutionary time, ed Nitecki MH, pp 103–131. New York: Academic Press
20. Fischer AG, Arthur MA (1977) Secular variations in the pelagic realm. In: Deep-water carbonate environments, eds Cook HE, Enos P, pp 19–50. Soc Econ Paleontol Min Spec Publ 25
21. Fisher RA (1936) The coefficient of racial likeness and the future of craniometry. J Roy Anthropol Inst 66:57–63
22. Flessa KW, Powers KV, Cisne JL (1975) Specialization and evolutionary longevity in the arthropoda. Paleobiology 1:71–81
23. Gaster M, Roberts JB (1977) The spectral analysis of randomly sampled records. Proc Roy Soc Lond A 354:27–58
24. Gottman JM (1981) Times series analysis: A comprehensive introduction for social scientists. Cambridge: Cambridge University Press
25. Harland WB, Cox AV, Llewellyn PG, Pickton CAG, Smith AG, Walters R (1982) A geologic time scale. Cambridge: Cambridge University Press
26. Hollander M, Wolfe DA (1973) Nonparametric statistical methods. New York: John Wiley and Sons
27. Jenkins GM, Watts OG (1968) Spectral analysis and its applications. San Francisco: Holden Day, Inc
28. Jones AH (1962) Spectral analysis with regularly missed observations. Ann Math Stat 33:455–461
29. Kempthorne O (1955) The randomization theory of experimental inference. J Am Stat Ass 50:946–967
30. Kitchell JA, Pena D (1984) Periodicity of extinctions in the geologic past: deterministic versus stochastic explanations. Science 226:689–691

31. Koopmans LH (1974) The spectral analysis of time series. New York: Academic Press
32. Mann HB (1945) Nonparametric tests against trend. Econometrica 13:245–259
33. Raup DM (1972) Taxonomic diversity during the Phanerozoic. Science 177:1065–1071
34. Raup DM (1976) Species diversity during the Phanerozoic: an interpretation. Paleobiology 3:289–297
35. Raup DM, Sepkoski JJ (1984) Periodicity of extinctions in the geologic past. Proc Natl Acad Sci USA 81:801–805
36. Raup DM, Stanley SM (1971) Principles of paleontology. San Francisco: WH Freeman and Co
37. Scheinok PA (1965) Spectral analysis with randomly missed observations: the binominal case. Ann Math Stat 36:971–977
38. Sepkoski JJ (1979) A kinetic model of Phanerozoic taxonomic diversity. I. Analysis of marine orders. Paleobiology 4:223–251
39. Sepkoski JJ (1982) A compendium of fossil marine families. Milwaukee Contrib Biol Geol 51:1–125
40. Sepkoski JJ, Bambach RK, Raup DM, Valentine JW (1981) Phanerozoic marine diversity and the fossil record. Nature 293:435–437
41. Sepkoski JJ, Raup DM (1986) Periodicity in marine mass extinctions. In: Dynamics of extinction, ed Elliott D. New York: John Wiley & Sons, pp 3–36
42. Simpson GG (1944) The major features of evolution. New York: Columbia University Press
43. Stanley SM (1973) An explanation for Cope's Rule. Evolution 27:1–36
44. Valentine JW (1973) Phanerozoic taxonomic diversity: a test of alternate models. Science 180:1078–1079

Patterns and Processes in the History of Life,
eds. D. M. Raup and D. Jablonski, pp. 149–165. Dahlem Konferenzen 1986
Springer-Verlag Berlin, Heidelberg

The Systems Approach: An Interface Between Development and Population Genetic Aspects of Evolution

G. P. Wagner

Max-Planck-Institut für Entwicklungsbiologie
7400 Tübingen, F.R. Germany, and
Zoological Institute, University of Vienna, Austria

Abstracts. The systems approach is an attempt to clarify the implications of morphological concepts such as homology and the natural system within the framework of the Modern Synthesis. The concept of homology implies the existence of invariant features of characters, most probably caused by developmental and functional constraints. The evolutionary history of homologues shows that the key events in transspecific evolution are changes in the system of developmental constraints: constraints caused by primitive developmental programs were overcome, leading to a higher flexibility of some characters, and new constraints acquired leading to the fixation of body design at a higher level of organization. Quantitative genetic theory shows that evolution of functionally coupled characters is highly dependent on an appropriate allocation of variance and thus depends on an appropriate pattern of developmental constraints. It is concluded that organismic evolution should be studied at two levels: a) the level of character evolution, and b) the level of the evolutionary modification of constraints.

Introduction

The systems approach to the study of organismic evolution is a conceptual framework designed by Riedl during the 1970s [31, 32]. Its aim is to clarify the relationship between the evolutionary implications of the morphological tradition and the population genetic approach of the Modern Synthesis. This research program soon turned out to cover a vast array of complicated questions, ranging from the problem of how to recognize lawful order in interspecific variation to the relationship between adaptation and constraints in organismic design and the role of development in evolution. In this contribution I will restrict myself to one question: What may the morphological tradition and the systems approach in particular contribute to the current discussion on the role of constraints and adaptation in organismic evolution?

Morphology, as the term is used here, is not merely the art of describing organic form, necessary as ancillary discipline for taxonomy, paleontology, and developmental biology. Morphology is the science of the patterns and principles of interspecific variation, the principles of organic design, including the methods used to reconstruct phylogeny. It deals with the static fingerprint, the history of life left in recent fauna. Thus the starting point of morphology is quite remote from the basic mechanisms of evolution which reside in molecular genetics, ecology, developmental biology, and population genetics. Nevertheless, the study of interspecific variation is the only direct empirical approach to broad patterns in the history of life.

If one asks what the contribution of morphology may be to the study of evolution, one has to look at the basic concepts of morphology and what they imply about the process of evolution. This question is discussed in the next section. It will lead us to the concept of constraint which is analyzed further in the following section. In particular the consequences for population genetics are discussed.

What Does Morphology Tell Us About Evolution?

The most widely acknowledged use of the morphological method is the reconstruction of phylogeny. The implications of this branch of morphological research are covered by Valentine (this volume). Another aspect emerges from the analysis of the methods successfully applied to reconstruct phylogeny. Riedl realized that the concepts and methods of morphology comprise a variety of implications about the process of morphological evolution that deserve closer examination [32].

The basic concept of morphology is homology, from which the concept of the "natural system of organisms" is derived. Homology is an equivalence relation between characters. It denotes a similarity most easily explained by inheritance from a common ancestor. All similarities not caused by inheritance are collectively called analogies, i.e., similarities due to independent evolution of similar features. These definitions of homology and analogy are causal explanations of similarities and as such they are not operational; they are not useful in recognizing homology and inferring its implications. But before we go on and see how homology is recognized, it may be useful to ask how the term "homologue" differs from less specific terms such as "character" or "trait."

A character or trait is any feature that characterizes an organism in some respect, such as body weight, length of tail, or color of the iris. A homologue is a more abstract concept derived from homology. A homologue is the class that comprises all structurally definable characters (e.g., bone elements, muscles, or inherited motor patterns), among which the homology relation holds true. Usually these characters bear the same name in all species where they occur, regardless of form and function. They are considered "to be the same" in two respects: first, homologous characters are considered as being derived from a character in a common (usually hypothetical) ancestor of those species that bear the homologue (historical continuity). Second, homologues are identifiable in different species, on the basis of features more or less invariant or variable in a regular manner. Historical continuity alone is not sufficient, as the following example will show.

There is little doubt that insects have evolved from annelid-like ancestors consisting of a number of identical segments plus nonsegmental anterior and posterior extremities. The ancestor of modern-day insects most probably had segments in historical continuity with segments 7, 8, and 9, which constitute the insect thorax. There is no indication that these segments become intercalated at the origin of insects. The insect thorax originates from a gradual differentiation of the respective segments. Nevertheless, the "thorax" is not homologous to the corresponding segments in centipedes or annelids. Only the individual segments of the thorax are homologous to any other arthropod segment in general. The thorax is the unit differentiated from the rest of the body in terms of appendages and internal anatomy, a condition not found in centipedes. The homologue "thorax" is an entity that originated from certain segments by synorganization and differentiation and is not identical with the elements it comprises.

How Homologues Are Recognized

This is the major theme of theoretical morphology [6, 29, 32, 33]. In the present context it is neither possible nor necessary to explain all the methodological problems involved, and the author wishes to apologize for the shortcut in advance.

Homology is inferred by means of several heuristic criteria (for instance, see [29]). The criteria are designed to estimate, in a qualitative way, the probability that an observed similarity is actually due to common descent. Only the two most important criteria will be mentioned here: a) the criterion of structure and position, and b) the criterion of coincidence.

According to the first criterion, homologues are expected to have similar position with respect to other structures, and their component parts are expected to have similar position with respect to one another. The greater the complexity and the degree of similarity, the greater the confidence in inferring homology. The next step should be to evaluate explanations of similarity other than common descent, e.g., similarity due to common function, the use of the same material, or the limitations of design [33].

In proceeding this way for each character in a group of organisms, a number of presumptive homology relations are established. Finally, these presumptive homologies are tested for consistency by the criterion of coincidence. Each presumptive homologue suggest a classification of the species in two groups: those with and those without the homologue. The criterion of coincidence now tests whether the system of all classifications suggested by all presumptive homologies gives a coherent pattern. This criterion is related to the principle of parsimony as used in cladistic analysis. At the supraspecific level, usually only hierarchical classifications are regarded as sensible because of the divergent mode of speciation. All classifications that disturb the hierarchical pattern will be considered as indications that some characters were erroneously considered homologous. This definition of homology is "self-referential" in a way common to all theoretical terms in the natural sciences, and logically valid ([45], pp 299–305).

In summary, homology is inferred form the degree and complexity of the similarities found among species and subsequently tested for consistency by comparing the distribution of the characters among species.

The Implications

The implications of the concepts of homology and homologues can be grouped according to the criteria utilized to identify them. The implications of the criterion of structure and position concern the role of constraints in evolution and are discussed in this section. The macroevolutionary consequences are related to the criterion of coincidence and will be discussed in the following section.

The goal of the criterion of structure and position is to find invariant features in the interspecific variation of characters in order to define the homologues in a way that makes them *identifiable* in different species. The invariant features used to define homologues should be historically contingent in the sense that reasons other than common function or material are responsible for the similarities. The search for homologies thus suggests a search for constraints which limit the power of natural selection to mold phenotypic traits and are the cause of the invariants. The concept of constraints is more deeply analyzed in the paper by Stearns (this volume).

The invariant features used in comparative anatomy can be of very different kinds: they could be simply a fixed character, such as the number of segments comprised in the thorax of Malacostraca. (This number is variable outside the class Malacostraca.) They could be limitations in the range of shapes that can be realized for fabricational reasons [36], e.g., the shells of early Cambrian gastropods are all uncoiled, displaying only two degrees of freedom in variation. Later the shells become planispirally coiled. The shapes of these gastropods have three degrees of freedom [40]. In the vertebrate central nervous system the long-range connections of a group of cells are very conservative, while position and differentiation of cells may change dramatically. An example is the "neocortical equivalent" in the telencephalon of nonmammalian vertebrates. It is an unlaminated nucleus, i.e., no "cortex," but has the same principal connections typical for the neocortex in mammals [13, 19].

To be sensible, the concept of homology requires a certain minimal degree of complexity, differentiation, and genetical autonomy of the characters. For instance, there are no problems in distinguishing the two bones of the zeugopod in all tetrapods in terms of structure and muscle attachment sites. But there is doubt concerning the identifiability of the tarsal and carpal bones across the tetrapod classes. The prechondrogenic patterns in Xenopus, Mus, and chick already exhibit the class-specific specializations [17].

The concept of homology thus is restricted to those invariants of form that have been acquired during evolution and which are therefore the best indication of historical relations among organisms. *Homology implies the existence of historically acquired and genetically determined developmental constraints*. Constraints of phenotypic evolution most probably are caused by the intermediate structure that mediates the transfer of genetic information to the phenotype, i.e., developmental mechanisms (see Stearns, this volume). Nevertheless, homology does *not* imply *identity* of developmental mechanisms [29, 35]. For instance, the coelomatic cavity in the tentaculate class Phoronidea can be built either by folding, where the tissue retrains an epithelial character, or by aggregation of mesenchymal cells [34]. Another example is metamorphosis in holometabolic insects. It is a radically different way to achieve the same basic end product as in hemimetabolic insects. Obviously, developmental constraints are not identical with immutability of the developmental processes. Hence there is good reason to distinguish between conservativism of developmental processes and the developmental constraints of a character. A character can be "developmentally constrained" if the developmental mechanisms generate a highly biased pattern of heritable phenotypic variation accessible to natural selection (see below). A biased phenotypic variation pattern, however, need not be associated with or caused by an immutable developmental process.

Implications Concerning Macroevolution

Macroevolution, as the term is used here, is "any evolutionary change in the biological properties of an existing higher taxon, or an evolutionary change bringing about the origin of a new higher taxon" [37, 48]. The relationship between the concepts of higher taxons and the concept of homology has been discussed above in connection with the criterion of coincidence. Higher taxons are defined by the distribution of homologues in such a way that the system of higher taxons gives a hierarchical classification.

Each group of organisms sharing a presumptive homologue (e.g., the chorda) is a presumptive higher taxon. An "established" higher taxon is characterized by a number of homologues that occur in the same set of species, i.e., homologues tend to be associated. For instance, the "vertebrata" not only share a chorda and a central nervous system derived from an epithelial tube, but also basically the same design in the vascular system, the same embryological derivation of mesodermal elements, and an internal skeleton. The rank of the taxon depends on the number and distribution of homologues that are not represented in all members of the taxon and which therefore can be used to define taxa of lower rank (e.g., not all vertebrates have a dentary-squamosal joint, but all recent mammals do).

Higher taxa are directly derived from the concept of homology. Organisms grouped in a higher taxon share at least one identifiable homologue. The identity of homologues resides in a common pattern of constraints and adaptive opportunities. Consequently, the key event in the origin of a higher taxon is a change or sequence of changes in the pattern of constraints. If one accepts this proposition, one reaches a definition of the biological nature of higher taxons: *a supraspecific taxon is a set of species shar-*

ing a common pattern of constraints and adaptive opportunities. This suggestion is supported by a statistical analysis of the co-variation of life-history traits in mammals which have not been used to define the taxa [39].

Paleontological evidence indicates that the origin of a homologue, such as the tetrapod extremity, does not come about by saltation but is a multistage process. Step by step, different features of the characteristic tetrapod pattern become established and remain (almost) fixed thereafter [32]. Among the oldest known vertebrates, the Agnatha, neither the existence of paired appendages nor their inner structure is determined. Paired appendages appear not at all, or as continuous fin folds (Anaspida), or as scale-covered flaps (some Osteostraci). The presence of paired fins is already obligatory in all archaic Gnathostomes (e.g., Placodermi), while the internal structure and the support on girdles remained variable. The next step towards the tetrapod condition was to narrow the base of the appendages such that only one bone remained in contact with the girdle.

This feature is already fixed in all Sacropterygian fishes. However, the typical tetrapod pattern with three segments along the proximodistal axis was not yet achieved: the stylopod with humerus or femur, the zeugopod with radius and ulnar or tibia and fibula, and the autopodium were not individualized. There is no recognizable homology between the bones of the dipnoian (lungfish) pectoral fin and the tetrapod extremity. No earlier than in the rhipidistian Crossopterygians were the bone elements of the stylo and zeugopod individualized sufficiently to be considsered homologues of the corresponding tetrapod bones. The bone elements of the autopodium remain variable until true amphibians have emerged.

Constraints, emerging step by step during phylogeny, withdraw certain degrees of freedom in phenotypic evolution from the access of microevolutionary mechanisms. Inter-individual variation becomes canalized, and evolution proceeds on the basis of interspecific variation. Control is passed up to higher levels of selection based on speciation and extinction (see Stearns, this volume). The critical *link between micro- and macroevolution* resides a) in the microevolutionary mechanisms that mold the system of constraints, and b) in the influence the constraints have on the dynamics of adaptive processes.

The coincidence criterion further implies that changes in the pattern of constraints and adaptive opportunities are rare and more or less irreversible. Otherwise, the pattern of presumptive homologues (recognized according to the criterion of structure and position) would not fit into a hierarchical scheme of classification. This conclusion is substantiated by the simulations of random phylogenies [28]. Phylogeny is reconstructable as long as the frequency of modifications is not too high compared with the speciation rate.

Evolutionary Biology of Constraints

Constraints are all mechanisms which limit the power of natural selection to mold the traits [38] at a given stage of the life cycle. Two possibilities may limit the power of natural selection. The first is that the traits or some allometric relation between them are under stabilizing selection, whatever the environmental conditions may be. Examples are the physical laws that require a certain relationship between body weight and the diameter of limb bones or between body weight and respiratory surface [14]. All these causes of stabilizing selection ("internal selection") are called *functional constraints*.

The second possibility is that features of the morphogenetic design limit the range of variation of the traits. These constraints are called *developmental constraints*. The range of variation at a stage of the life cycle may be limited either because the realization of variants is prevented [1] or because selection at earlier stages of development has eliminated variation [12, 32]. Selection at earlier stages is included here to cover those constraints where deviant morphology is developmentally realizable (e.g., homeotics) but is incompatible with the integrity of further development (lethality of "macromutations").

Even if functional and developmental constraints have totally different causes, their patterns are in partial correspondence with one another. This was first recognized by Olson and Miller [26] in the analysis of phenotypic correlation matrices. They found that functionally related characters are on the average correlated to a higher degree than functionally unrelated characters ("morphological integration"). This result has been confirmed for genetic correlation matrices [3, 11, 22]. The degree of integration is higher than expected by chance among functionally related characters and lower than expected by chance among functionally unrelated characters [44].

Further evidence that functionally coupled characters are also highly integrated developmentally comes from the analysis of experimentally induced atavisms. The system of muscles, connective tissue, and bones in the tetrapod limb is a functionally highly integrated system which is forced to evolve in concert. In addition, the pattern of muscle attachment sites and of muscle individuation can be influenced by modifying skeletal development experimentally [24]. In particular, experimentally induced elongation of the fibular in the chick hind limb, which is taken as an atavistic condition [16], leads to a modification of those muscles associated with the fibular such that muscle organization clearly resembles reptilian conditions [25].

However, one need not conclude that genetic correlations are necessary simply to allow the evolution of functionally coupled characters. The opposite is proven in hybridization experiments on the inheritance of courtship behavior in orthopterans. There is a strong functional coupling between the inherited signal sender and receiver mechanisms used by male and female crickets in courtship. Nevertheless, clear evidence exists for their independent genetic basis despite their co-evolutionary history [5].

During phylogeny, developmental constraints become modified in two directions: a) towards increasing versatility of form by overcoming archaic constraints, and b) towards increasing integration, as exemplified in the history of the vertebrate limb (see above).

The design of each organism is limited by archaic constraints, resulting simply from the physicochemical properties of the materials utilized and the geometry of tissue interactions. These archaic patterns sometimes are modified in a way allowing increased versatility in form as, for instance, in the evolution of gastropod shells. The evolution from uncoiled shells to planispirally coiled and finally to conispirally coiled shells is associated with an increasing number of possible solutions to any given mechanical problem [41]. Another example is the increased adaptive flexibility gained by temporal dissociation of developmental processes necessary to allow heterochrony [15, 27].

While it is easy to understand what the adaptive advantage of overcoming archaic constraints is, it is not so easy to see the advantage of acquiring constraints. To see this one must look more closely at the factors that determine the rate of multivariate phenotypic evolution.

The Influence of Developmental Constraints on the Rate of Multivariate Phenotypic Evolution

Intuitive reasoning about tempo and mode of evolution is usually based on the assumption that the rate of evolution is a monotonically increasing function of selection intensity and the amount of additive genetic variation. As a corollary, it is often concluded that stasis or bradytelic evolution has to be due either to the absence of directional selection (i.e., stabilizing selection) or to the lack of genetic variance on which selection can work. However, quantitative genetic theory shows that the rate of evolution is a systems property, depending not only on the variance of those characters currently under directional selection, but also on the total pattern of variation of the whole phenotype. This fact is especially relevant for the evolution of functionally coupled characters.

The evolutionary dynamics of functionally constrained characters can be modeled by so-called "corridor models" [43]. A corridor model is an adaptive landscape which looks like a ridge, i.e., a fitness function m(z) [($z=(z_1, \ldots, z_n)$), z_i being a quantitative polygenic character], where there is a "path" along which directional selection occurs and stabilizing selection is acting in all directions orthogonal to it. In adaptive landscapes of this type rather peculiar phenomena can occur – one of them may be called the "evolutionary window." If one analyzes, on the basis of quantitative genetic theory, how an increase in variance of all characters influences the rate of evolution (under constant heritability), one realizes that the rate of evolution can decrease if total variance exceeds a certain optimal level (Fig. 1). The term "evolutionary window" refers to the range of total phenotypic variance within which the evolution of functionally copuled characters can occur. If variance exceeds the limits of the evolutionary window, phenotypic evolution is inhibited ([43], and Wagner, submitted). In addition, the structure of the covariance matrix is of critical importance [9].

The reason for this phenomenon is that variation along the direction of the ridge has an effect on the rate of evolution opposite to that of variation orthogonal to it. It is *not* related to pleiotropy. Increased variation along the line of directional selection accelerates evolution, but variation ortho-

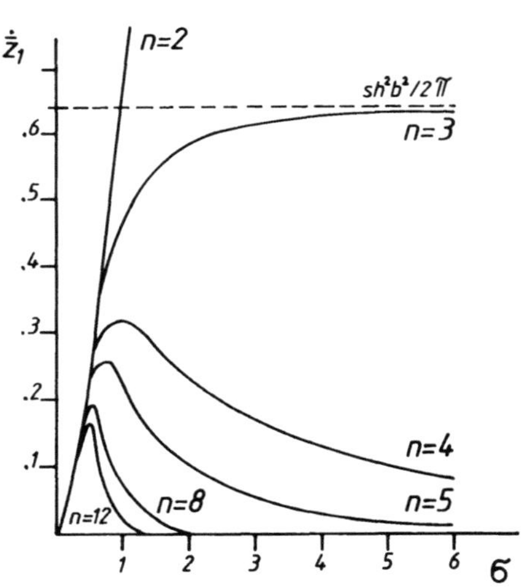

Fig. 1. Influence of variation σ of characters ($h^2=1$) on the rate of evolution $\dot{z}_1$ given the fitness function $m(z)=(sz_1$, if $|z_i| \leqq b/2$ for all $i \neq 1$, and $-q$, if there is at least one $i \neq 1$ with $|z_i| > b/2)$. n is the number of characters, $b=2$, and $s=1$. Note that the rate of evolution can decrease if variation of characters exceeds a certain optimal level

gonal to the direction of evolution ("lateral variation") can *inhibit* evolution. This phenomenon occurs in a large class of adaptive landscapes but not in all. The inhibitory effect never occurs in adaptive landscapes of functionally unconstrained characters which contain only one optimum but is frequent in adaptive landscapes of functionally coupled characters ([9, 43], and Wagner, submitted). These results suggest that the rate of evolution of functionally integrated phenotypes can be influenced very much by the pattern of *phenotypic* variation (in addition to the amount of additive genetic variation under directional selection). Restrictions in the amount of "lateral variation" of functionally coupled characters can be of adaptive significance. Constraints can accelerate the rate of phenotypic evolution if their pattern corresponds to the system of functional constraints.

It may be worth mentioning that the mechanisms described here are not equivalent to similar phenomena described, e.g., by Bossert [7] on the basis of mutation-selection models. Calculations of this type are irrelevant for quantitative characters in sexually reproducing populations.

The phenomenon of evolutionary windows suggests that stasis does not have to be due to the absence of directional selection, which is questioned, e.g., by Bakker [4] or the lack of genetic variation, which is questioned by the majority of geneticists (e.g., [23]). The presence of too much nonadaptive phenotypic variation is an additional possibility for explaining stasis.

This may be of interest because all attempts so far to pin down developmental constraints in terms of developmental mechanisms have been unsuccessful. For instance, the concepts of positional information and of heterochrony fail to provide an understanding of constraints on change [20, 47]. This is further underscored by the fact that stasis or invariance of some phenotypic features does not have to be correlated with immutability of the developmental processes [34]. It may well be that the explanation of constraint resides in the relation between three factors: a) the highly biased allocation of variance among the phenotypic traits (for persuasive evidence see [2]) caused by the epigenetic system, b) the fact that selection acts on whole organisms, and c) the coherence of sexually reproducing populations caused by the statistical laws of mating and recombination.

How Constraints Become Modified in Phylogeny

This is a largely unresolved question. It is a well established fact that the expression and distribution of heritable phenotypic variation is itself genetically regulated to some extent [10, 30]. Hence there is no problem in imagining that the pattern of heritable variation is subject to natural selection.

But it is not so easy to explain why selection should have led to a correspondence between developmental and functional constraints and whether this is related to the influence of developmental constraints on the rate of evolution.

Recently Cheverud [12] attempted to explain the correspondence between developmental and functional constraints by using results from Lande [21]. Cheverud argues that the pattern of pleiotropic mutations, described by the *U*-matrix in Lande's theory, will tend to mimic the pattern of stabilizing selection due to "internal selection." This idea is directly related to the suggestions of Riedl [32] who also predicted such a correspondence, although Riedl used other arguments (for a discussion see [12]). In this way the functional relationships between phenotypic traits will become "internalized" in the structure of the epigenetic system. The result would be that evolution becomes canalized by the developmental constraints acquired to cope with functional constraints in the past, whether these functions are still relevant or not [32].

Cheverud proposes that the evolution of the mutation matrix *U* is due to stabilizing selection on the characters. However, this proposition is not yet well understood for it is not clear what the selection intensity would be on a gene that modifies the *U*-matrix. By analogy with other modifier models, one would expect that the force of selection is extremely small, far below the threshold required to overcome random drift [8]. The effect of acquired constraints on the rate of adaptation would be secondary.

A possible solution could be provided by a principle which emerged from studies on dominance evolution, the simplest example of modifier evolution [8, 42, 46]. The slowness of modifier selection applies only to situations where the primary alleles are almost in equilibrium, but the rate of selection can be several times as large if the primary genes are not in equilibrium. This comes form the fact that genes which have a negligible effect on the phenotype and on the mean fitness in equilibrium may have an appreciable effect on the rate of adaptation of the entire genetic system, i.e., on the quantity

$$\int \bar{m}(t)\, dt\,.$$

This mode of selection has been called "feedback selection," as it is caused by the feedback between the modifiers and the primary genes [42].

One may therefore suggest that selection of modifiers of the *U*-matrix will be very much reinforced in periods of directional selection on the phenotype. However, a decision about the possible importance of nonequilibrium selection has to be postponed until a thorough nonlinear analysis of appropriate models is available.

A further unresolved problem is how genes acquired by feedback selection can be retained. Feedback selection is effective only during periods of directional selection. However, genes which may be more or less neutral at the onset of their history and which become selected by feedback selection need not remain neutral. Consider, for instance, two bones in a fin-flap of a Devonian fish. As long as these bones are not much differentiated in form and function their cells will not have to recognize specifically their positional information. However, without positional information no developmental autonomy is possible, as is given in most limb bones of tetrapods [47], and no genetic differentiation can occur. But a gene necessary to recognize positional information as such will have almost no influence on the phenotype. The phenotypic effect of this gene depends on its use by other genes that control the morphogenesis of the bone *after* positional information has been recognized during ontogeny. Hence, genes required for position-specific determination of blastema and genetic differentiation can only be selected in concert with those genes that utilize the different states of determination. As soon as positional information is utilized by other genes, a loss of the ability to recognize positional information will have a large phenotypic effect (e.g., homeotics). Genes which have a large effect in derived species need not have a visible effect on the phenotype of ancestral species at all.

Conclusions

Concerning the question of directionality and randomness in phenotypic evolution, the systems approach offers a research program rather than an answer. The concepts of morphology and of developmental constraints suggest that organismic evolution should be studied at two levels: a) the level of character modification, i.e., the level on which the adaptionist program is formulated, and b) the evolutionary history of constraints. Evolution is assumed to be driven by adaptive processes, but directionality is caused by constraints which can become changed under certain conditions in an unpredictable manner.

The systems approach suggests that the key events for macroevolutionary change reside in the interaction between these two levels of evolution. Constraints limit the possible outcomes of adaptive processes, and microevolutionary processes mold constraints under certain conditions. It is *not* concluded that macroevolution is causally decoupled from microevolution, but interaction between the two levels is limited. No additional evolutionary mechanisms are proposed. Nevertheless, macroevolution is not considered as a simple accumulation of gene substitutions. It is considered as an autonomic field of investigation because the dynamic of macroevolution is governed by "systems properties" not evident at the level of elementary mechanisms of evolution alone, such as mutation, selection, and recombination. The systems properties are expected to result from the fact that selection acts on whole organisms, from the coherence of sexually reproducing populations and from the constraints acting on the diverse levels of organic design.

In contrast to recent attempts to incorporate the concept of constraints into evolutionary biology (for instance, see [1]), the systems approach is not so much concerned with the nature of the constraints as such. It is much more an attempt to study their evolutionary history and their role in the origin and evolution of higher taxonomic categories. This approach allows one to define a number of specific research problems. The theoretical problems concern the question of how constraints can become modified during evolution by conventional population genetic mechanisms. One should also work out the statistical problems encountered in identifying constraints by a comparative method. The empirical problems mainly concern the comparative biology of constraints: what are the differences between major taxonomic groups of organisms in developmental terms? The most direct approach probably is the comparative anatomical and experimental analysis of induced atavisms [16, 24, 25].

All these problems necessitate a thorough revival of the comparative biological tradition, because the only direct empirical approach to be broad patterns of evolution is the comparative method.

Acknowledgements. The financial support of the Stiftung Volkswagen Werk is gratefully acknowledged. The author wishes to express his gratitude to S. Stearns, A. L. Panchen, D. Futuyma, and G. Müller for discussion and useful comments.

References

1. Alberch P (1980) Ontogenesis and morphological diversification. Am Zool 20:653–667
2. Alberch P (1983) Morphological variation in the neotropical salamander genus Bolitoglossa. Evolution 37:906–919

3. Atchley WR, Rutledge JJ, Cowley DE (1981) Genetic components of size and shape. II. Mulivariate covariance patterns in the rat and mouse skull. Evolution 35:1037–1055
4. Bakker RT (1983) The deer flees, the wolf pursues: Incongruencies in predator-prey coevolution. In: Coevolution, eds Futuyama DJ, Slatkin M, pp 350–382. Sunderland, MA: Sinauer
5. Barlow GW (1981) Genetics and development of behavior with special reference to patterned motor output. In: Behavioral development, eds Immelmann K, Barlow GW, Petronovich L, Main M, pp 191–251. Cambridge: Cambridge University Press
6. Bock WJ (1973) Philosophical foundation of classical evolutionary classification. System Zool 22:375–392
7. Bossert W (1967) Mathematical optimization: Are there abstract limits on natural selection? In: Mathematical challenges to the neo-Darwinian interpretation of evolution, eds Moorhead PS, Kaplan M, pp 35–46. Symposium Monograph 5. Philadelphia: Wistar
8. Bürger R (1983) On the evolution of dominance modifiers. I. A non-linear analysis. J Theor Biol 101:585–598
9. Bürger R (1986) Constraints for the evolution of functionally coupled characters. Evolution 40:182–193
10. Charlesworth B, Lande R, Slatkin M (1982) A neodarwinian commentary on macroevolution. Evolution 36:474–498
11. Cheverud JM (1982) Phenotypic, genetic, and environmental morphological integration in the cranium. Evolution 36:499–516
12. Cheverud JM (1984) Quantitative genetics and developmental constraints on evolution by selection. J Theor Biol 101:155–171
13. Ebbeson SOE, Schroeder DM (1971) Connections of the nurse shark's telencephalon. Science 173:254–256
14. Gould SJ (1966) Allometry and size in ontogeny and phylogeny. Biol Rev 41:587–640
15. Gould SJ (1977) Ontogeny and phylogeny. London and Cambridge, MA: Belknap Press
16. Hall BK (1984) Developmental mechanisms underlying the formation of atavisms. Biol Rev 59:89–124
17. Hinchliffe JR, Griffiths PJ (1983) The prechondrogenic patterns in tetrapod limb development and their phylogenetic significance. In: Development and evolution, eds Goodwin BC, Holder N, Wylie CC, pp 99–121. Cambridge: Cambridge University Press
18. Karlin S, McGregor J (1974) Towards a theory of the evolution of modifier genes. Theoret Pop Biol 5:59–103
19. Karten HJ (1969) The organization of the avian telencephalon and some speculations on the phylogeny of the amniote telencephalon. Ann NY Acad Sci 167:164–179
20. Kauffman SA (1983) Developmental constraints. Internal factors in evolution. In: Development and evolution, eds Goodwin BC, Holder N, Wylie CC, pp 195–225. Cambridge: Cambridge University Press
21. Lande R (1980) The genetic covariance between characters maintained by pleiotropic mutations. Genetics 94:203–215
22. Leamy L (1977) Genetic and environmental correlations of morphometric traits in random-bred house mice. Evolution 31:357–369

23. Maynard Smith J (1983) The genetics of stasis and punctuation. Ann Rev Genet 17:11–25
24. Müller G (1986) Effects of skeletal change on muscle pattern formation. Bibliotheca Anatomica 29:91–108
25. Müller G (1985) Experimentelle Untersuchungen zur Theorie des epigenetischen Systems. In: Evolution – Ordnung und Erkenntnis, eds Ott J, Wagner GP, Wuketits FM, pp 92–96. Hamburg, Berlin: P Parey Verlag
26. Olson EC, Miller RL (1958) Morphological integration. Chicago: University of Chicago Press
27. Raff RA, Kaufman TC (1983) Embryos, genes and evolution. The developmental-genetic basis of evolutionary change. New York: Macmillan
28. Raup DM (1977) Stochastic models in evolutionary paleontology. In: Patterns of evolution as illustrated by the fossil record, ed Hallam A, pp 59–78. Amsterdam, Oxford, New York: Elsevier
29. Remane A (1952) Die Grundlagen des natürlichen Systems, der vergleichenden Anatomie und der Phylogenetik. Reprint 1971. Königstein: Otto Koeltz
30. Rendel JM (1967) Canalization and gene control. London: Logos Press
31. Riedl R (1977) A systems-analytical approach to macro evolutionary phenomena. Q Rev Biol 52:351–370
32. Riedl R (1978) Order in living organisms. New York: J Wiley
33. Rieger R, Tyler S (1979) The homology theorem in ultrastructural research. Am Zool 19:655–664
34. Salvini-Plawen LV, Splechtna H (1979) Zur Homologie der Keimblätter. Z Systemat Evolutionsforsch 17:10–30
35. Sander K (1983) The evolution of pattering mechanisms: Gleanings from insect embryogenesis. In: Development and evolution, eds Goodwin BC, Holder N, Wylie CC, pp 137–159. Cambridge: Cambridge University Press
36. Seilacher A (1973) Fabricational noise in adaptive morphology. Syst Zool 22:451–465
37. Stanley SM (1983) Macroevolution and the fossil record. Evolution 36:460–473
38. Stearns SC (1982) The role of development in the evolution of life histories. In: Evolution and development, ed Bonner JT, pp 237–258. Dahlem Konferenzen. Berlin, Heidelberg, New York: Springer-Verlag
39. Stearns SC (1983) The influence of size and phylogeny on pattern of covariation among life-history traits in the mammals. OIKOS 41:173–187
40. Vermeij GJ (1971) Gastropod evolution and morphological diversity in relation to shell geometry. J Zool 163:15–23
41. Vermeij GJ (1973) Biological versatility and earth history. Proc Natl Acad Sci USA 70:1936–1938
42. Wagner GP (1981) Feedback selection and the evolution of modifiers. Acta Biotheoret 30:79–102
43. Wagner GP (1984) Coevolution of functionally constrained characters: Prerequisites of adaptive versatility. Bio Systems 17:51–55
44. Wagner GP (1984) On the eigenvalue distribution of genetic and phenotypic dispersion matrices: Evidence for a nonrandom organization of quantitative character variation. J Math Biol 21:77–95

45. Wagner GP (1984) The logical basis of evolutionary epistemology. In: Concepts and approaches in evolutionary epistemology, ed Wuketits FM, pp 285–307. Boston: D Reidel
46. Wagner GP, Bürger R (1985) On the evolution of dominance modifiers II. A non-equilibrium approach to the evolution of genetic systems. J Theor Biol 113:475–500
47. Wolpert L (1982) Pattern formation and change. In: Evolution and development, ed Bonner JT, pp 169–188. Dahlem Konferenzen. Berlin, Heidelberg, New York: Springer-Verlag
48. Wuketits FM (1982) Grundriß der Evolutionstheorie. Darmstadt: Wissenschaftliche Buchgesellschaft

Standing, left to right:
Bruce Runnegar, Steve Stearns, Robert Selander, Jim Valentine

Seated (center), left to right:
Jeff Levinton, Adam Urbanek, John Turner, Brian Charlesworth

Seated (front), left to right:
Walter Nagl, Karl Bandel, Gerd Müller

Patterns and Processes in the History of Life,
eds. D. M. Raup and D. Jablonski, pp. 167–182. Dahlem Konferenzen 1986
Springer-Verlag Berlin, Heidelberg

Organismic Evolution: The Interaction of Microevolutionary and Macroevolutionary Processes

Group Report

J. S. Levinton, Rapporteur
K. Bandel
B. Charlesworth
G. Müller
W. Nagl
B. Runnegar
R. K. Selander
S. C. Stearns
J. R. G. Turner
A. J. Urbanek
J. W. Valentine

The relationship between DNA sequence evolution and morphological evolution is not yet known in much useful detail. Empirical and theoretical population genetics are consistent with all we know about speciation and morphological variation in natural populations. We acknowledge, nevertheless, that we need a better understanding of gene expression and developmental mechanisms before we can understand in detail both the patterns and the mechanisms of morphological evolution. We therefore have chosen to identify those research programs and past accomplishments that should further our understanding of the evolutionary processes that produce differences of large taxonomic scale.

Origin of Baupläne

The definition of Bauplan is a contentious issue, partly because the term has a typological connotation. Practically speaking, Baupläne are identified as phylum- or class-level entities. Bauplan can be defined cladistically as an idealized set of characters that unites all members of a group. This may include derived characters defining the Bauplan in question, plus those that might be common to other groups. The more primitive characters are included with derived characters because we see the Bauplan as a useful concept, partly because it defines a structure that must be understood functionally. Function cannot be understood without understanding the role of all characters, ancestral as well as derived ones. Bauplan might also be defined as clusters in the phenotype space. In this case we would de-emphasize the unique cladistic significance of some of the characters but would emphasize the morphological distinctness among Baupläne.

The concept of Bauplan has meaning because of the observation that evolution has involved elaborations on a relatively small number of combinations of major body parts, rather than of all possible combinations of characters. Only a small number of early lineages has survived long enough to be recognized in the fossil record or to give rise to the extant diversity of phyla (Valentine, this volume).

The definition of Bauplan does not necessarily imply temporal stasis. Baupläne may be considered from two perspectives – synchronic and diachronic. From the synchronic (a single time plane) point of view, the differences among Baupläne may be defined by differences found at that time only. From the diachronic (temporal) point of view the differences among Baupläne may change steadily. Orders of mammals, for example, have become incrasingly morphologically distinct over time [25]. There is no reason that this might not have happened during the rise of the phyla, though we have no evidence for this and first see them in the record as distinct groups with little further addition of Bauplan-specific characters.

Not much is known about the times of origin of the major metazoan Baupläne. However, there is some evidence that animals with the grade of organization of modern annelids were in existence by about 650 million years ago [11] and it is assumed that almost all of the extant metazoan phyla had appeared by the Early Cambrian (550–570 myr). Furthermore, occurrences of soft-bodied fossils in late Precambrian strata and the Middle Cambrian Burgess Shale suggest a diversity of phylum-level taxa that no longer exists. Similarities in the amino acid sequences of the collagens of living cnidarians, platyhelminths, molluscs, and vertebrates suggest that metazoans are monophyletic [30]. Sampling further suggests that the ratio of species to higher taxa was far lower then than it ever has been since. The origin of nearly all the phyla early in the history of the metazoans is not just an artifact of systematics but involves a rapid early diversification of body plans under relatively low species diversity conditions, that has not been repeated. Indeed, the origination rate of higher taxa has declined over time [38].

The early rise of diversity of Baupläne, followed by the extinction of many, and a failure to re-elaborate any more over time, is amenable to explanation. The early origin of a diversity of Baupläne may be related to the low species diversity of the Precambrian and consequent latitude for diverse directions of natural selection. Some of us believe that the early state of the metazoans may have involved unusual modes of gene evolution while others of us see no reason to invoke anything more than ecological opportunity. The later reduction in the rate of production of new phyla may be due to a filling up of ecospace. Newly arising groups originating in more recent times would not have had the chance to spread geographically and invade, via adaptive radiation, a variety of habitats and therefore would probably have rapidly gone extinct if they arose at all. It is also possible that, later in the history of the metazoans, it was not genetically or developmentally possible to produce new Baupläne.

It is clear from other examples that ecological opportunity has played a major role in creating the conditions necessary for an adaptive radiation [33]. We need much more information documenting the early appearance of the major Baupläne to set limits on the rate of appearance and total number of early appearing, major structural types. It is also essential to apply as soon as possible the findings of DNA and protein sequence analyses to establish an accurate genealogy and dates of origin for the extant phyla. This will help us in understanding evolutionary transitions in morphology and function.

There is no reason to believe that Baupläne arose either a) only in single major steps, or b) only piecemeal as a series of changes. In some cases, a keystone change may have set the stage for a cascade of others. In the mollusks, for example, one hypothesis suggests that the dorsal cuticle of a flatworm-like ancestor might have been elaborated into a shell, which would have been the basis for the further molluscan radiation. In the origin of the mammals, however, it is quite clear that the Bauplan arose as a long progression of changes in many characters involving transitional forms that were functionally harmonious (summarized in [17]).

Some Genetic and Developmental Mechanisms of Change and Stasis

We have tried to define some properties of the internal organization of the organism that may constrain the direction of evolution. Physiological properties and structures, epigenetic programs, and various aspects of gene expression (e.g., pleiotropy) may all constrain the direction of evolution. Hall [13] has suggested that some tissue interactions involved in the control of developmental programs are sufficiently fundamental that their disruption will have drastically negative effects on fitness. Such constraints should be combined, not contrasted, with the effects of the natural environment that select on variation within populations. One might argue that the internal constraints correspond to Darwin's concept of "unity of type," which he used to define the phylogenetic background that a species had while natural selection was operating. He saw the "conditions of existence" as the other major factor that influence the form of an organism, as it determined the environmental component of the process of natural selection. The concept of unity of type is useful only if we avoid the fallacy of uniting a taxon with a given set of characters and then speculating as to why these characters have failed to change during the evolution of the group, without further reasons for associating the characters by some process. Appropriate factors would include genetic, developmental, or functional constraints [20].

Developmental Mechanisms of Evolutionary Change

The development of metazoan organisms results from the interaction of individual cells and mechanisms for successively activating different genes. Each of the cells potentially contains the same genetic information, but local effects activate these genes differentially. The genome primarily codes for proteins but the morphological result of development is only indirectly related to the genomic and behavioral repertoire of cells. By interfering with the same developmental mechanisms (e.g., proliferation rates of mesenchymal cells within the developing limb bud) in organisms of different lineages, different lineage-specific results are obtained [1]. These are comparable to the genetically caused differences in the same character among species of these lineages. Thus lineage-specific, epigenetic interactions constrain the pathways of phenotypic evolution [39].

Developmental constraints restrict the range of possible morphologies which could result from genetic variability, but developmental constraints do not necessarily explain parallel evolution. Similar phenotypes or even stages of development may be attained by different developmental pathways [31]. Phenomena such as gastrulation may reflect the necessity for folding and contact among tissues for spatial information in development and could have arisen many times in evolution due to natural selection for tissue complexity [3].

By experimental interference with specific developmental processes, the phenotypic effects of genes may be simulated. It can be demonstrated that slight changes in the developmental processes result in coordinated morphological change. Experimentally induced changes in the skeletal system of the limb result in changes in the muscular system which resemble the patterns of interspecific variation of these muscles [23].

We are interested in understanding why invariance is such a common theme of evolution. Certain DNA sequences are now being found in quite distantly related organisms. At the large-scale organismal level we note that essential features of the Baupläne have not been lost over long periods of geological history. At smaller scales, a remarkable amount of invariance has been found in clades of exceptional longevity. In some plethodontid salamander clades [41], extremely little morphological change has occurred over many speciation events and through millions of years. This small change contrasts with extensive divergence in allelic forms identified at enzyme loci. Identity of form clearly does not imply genetic identity. Indeed, hybrids between similar species often show strong deviation of form due to genetic incompatibility.

Experiments on characters that are invariant over groups of related species suggest that stasis may be enforced by selection for canalization. In Drosophila, scutellar bristles are invariant across species but variable within species. Selection experiments demonstrate that the genetic change required to change bristle number by a count of one can be modest or great, depending upon the nature of the change. Certain transitions in this character are strongly canalized against change [26]. Wright's [44] classic experiments on digit number determination in the guinea pig demonstrated a continuous genetic scale of effects, but a threshold located on this scale was found to determine digit number. Directional selection therefore is ineffective in changing digit number until the threshold is breached.

We find no reason to believe that such mechanisms of stasis are incompatible with natural selection. Indeed, they must be the result of long periods of stabilizing selection for a given morphology [7]. It is not necesssarily paradoxical that morphology remains constant despite some environmental change. While there is of course some evolutionary inertia built in by canalization, one can hardly say that the genetics involved in the determination of head number, which is probably of necessity highly canalized, is quite the same as that of digit number in guinea pigs. Most metrical and meristic traits are poor candidates for cases in which long periods of evolutionary stasis have been caused by developmental constraints, due to the demonstrable existence of genetic variability [6], but some such traits are strongly canalized [26].

Internal constraints influence the direction of evolution to some degree, and characters long established in the previous history of a clade may be too interactive in the developmental sense to be easily lost due to strong drops in fitness. Much stasis in characters seen at the species level, however, probably involve some degree of canalization, probably reinforced by natural selection. In such cases, strong directional selection would be required to change the phenotype significantly. Given some degree of habitat choice, an organism might to some degree choose its own conditions for existence. This factor, combined with functional interrelationships among body parts, provides an adequate Darwinian explanation for stasis of species-level morphological variation in a changing environment. Stasis is completely compatible with traditional neo-Darwinian evolutionary mechanisms.

Molecular Genetics and Organismal Evolution

A large number of molecular genetic mechanisms have been suggested in recent years as important sources of evolutionary change. We do not question the existence of the genetic processes proposed, but some of us wonder whether this general strategy of searching for novel mechanisms will be a useful substitute for genetic studies of variation in natural populations. No matter what mechanisms determine variation, the genetically based variation known in natural populations is more than sufficient to account for extensive morphological evolution. Others of us are not willing to preclude a search for additional mechanisms.

Lateral gene transfer from other species does occur but it must be fairly rare in nature. One might expect bacteria to experience transfer of genes carried on plasmids, but it is known that rates of transfer are no greater than the mutation rate. This is probably an overestimate of the natural condition, as wild strains are usually repressed for conjugative pilli formation. There is a remarkable degree of identity of gene order on the bacterial chromosome in E. coli and in Salmonella [28]. This suggests the great rarity of successful transfer in bacteria.

Laterally transferred DNA sequences can usually be considered as similar to mutations arising within a population. They are often incomplete genes or nonfunctioning repeated sequences and therefore usually disrupt gene action and, therefore, fitness. If they are improbable in the first place, and furthermore are of no more significance than ordinary mutations, they cannot be said to have much importance in morphological evolution.

There is an important exception to our argument that involves the unlikely but possible transfer of whole functioning genes. Over the span of hundreds of millions of years, the chance for such a transfer increases. The best potential example of the lateral transfer of an important gene between two distantly related lineages is the presence of genes for hemoglobins in some angiosperms (legumes, Parasponia, etc.), but not, so far as is known, in any other group of plants. Angiosperm globin genes are thought to be homologous to animal globin genes because they contain non-coding sequences (introns) that interrupt the coding regions (exons) in precisely equivalent positions to those of animal globin genes [5]. This could be explained by a) functional constraint of the location of the introns; b) inheritance of the gene from the common ancestor of animals and plants, with lack of expression in plant ancestors; and c) lateral gene transfer.

At present we do not have an adequate test to distinguish among these alternatives. The amino acid sequences of the plant globins are as different from the sequences of all animal globins as are the globins of different animal phyla. However, until a globin from each of the potential donor groops (e.g., Nematoda) has been sequenced, it will not be possible to exclude the possibility that lateral gene transfer has occurred in the relatively recent past, as opposed to some time before the origin of the angiosperms [30].

Transposable elements have been studied intensively in recent years and are now known to influence morphology, mutation rate, and hybridization among strains of the same species [32]. In general their effect is due to disruption of gene function; this casts some doubt on their likely role in morphological evolution. It is possible that their presence might alter the action of a gene to a limited degree and thereby influence its frequency and the course of evolution in a natural population. The introduction of transposable elements into new populations is a source of mutation, but evidence from the p element in Drosophila melanogaster suggests that such an introduction can be highly disruptive [18].

Structural and Regulatory Genes

A distinction can be made between structural genes, whose products are functioning proteins such as enzymes, and regulatory genes, whose gene products regulate gene expression by binding to DNA. Even structural genes can have regulatory function. In eukaryotes we still know too little about gene regulation to understand anything of much significance in morphological evolution. It is not even clear whether regulation of gene expression in eukaryotes can be ascribed primarily to regulatory genes as defined above. While no one doubts the power of action of regulation of gene expression at different points in development, there may not be a dichotomy of effects on development when a nucleotide substitution occurs in a true regulatory gene as opposed to a structural gene. In Drosophila, ecdysone is known to be an important intermediary in the regulation of gene expression at various stages of development [2]. AMP also is important in intercell communication and the indirect control of gene expression. Changes in a variety of structural genes could change the action of these developmental messengers.

Although it may well be that gene regulation and even regulatory genes per se may be the agents of morphological evolution, opinion was divided on whether it would be fruitful to focus on these factors when so much more about morphological variation is known from traditional studies of population genetics. The variation we observe and can study in the laboratory and in the field my be determined by either regulatory or structural genes, but this distinction may not be pertinent when one thinks about the rate of morphological evolution and the transmission genetics of traits.

Genomic rates of evolution have been measured extensively and present a picture consistent with the action of the constraint of natural selection, acting in concert with a background of stochastic fixation of alleles at many apparently neutral sites. This can be seen particularly in the relatively rapid evolution of the third nucleotide of each codon, which does not usually give any information in specifying an amino acid, and in the rapid and random evolution of pseudogenes, nonfunctional copies of genes. In some cases, such as fibrinopeptide amino acid sequences, evolution is sufficiently rapid to fit the expectations of the neutral theory [19]. In most proteins, however, sequence evolution is significantly slower, presumably being due to conservation of functionally crucial sites. This suggests that, at least in sequence evolution, there are no particular constraints restricting random evolution with the exception of alterations of gene product function.

Sequence evolution shows variations in rate among proteins and for the same protein in different lineages. The variance, however, is not sufficiently great to preclude sequence divergence as a crude molecular clock. The calibration of the various molecular clocks depends upon estimates of divergence times from the fossil record. In the future, studies of sequence divergence may provide invaluable evidence on times of divergence of major taxa, but they will definitely lend insight to the study of phylogenetic relationships.

Phenotypic stasis, as described above, occurs despite considerable genetic change. This is especially apparent in the highly conserved tertiary structure of many proteins, despite extensive sequence evolution. Morphological stasis can also be seen among species of the ciliate Tetrahymena, where extensive sequence divergence and even the presence of different proteins can be found, despite identical structure of the cytoskeleton [42]. In combination with the evidence cited above at higher taxonomic levels, this suggests that morphology is actively conserved in evolution, probably by natural selection.

Punctuations, Stasis, and Above-Species Level Properties

We suggest an ordering of problems of organismic evolution in terms of some hypotheses previously suggested in the literature. First, we must consider the pattern of phyletic evolution: is it punctuational (i.e., exhibits periods of morphological stasis, interspersed with brief periods of evolutionary change) or do we have a spectrum of patterns of rates? Second, is morphological change concentrated in speciation events, or can known patterns of phyletic change account for morphological divergence of the order of species differences? Finally, we must deal with the issued surrounding speciation and extinction.

If phyletic evolution is sufficient to account for extensive morphological change, then processes relating to speciation and species-level extinction may not create evolutionary diversity so much as they determine the relative abundance of different forms in nature and influence the pathways of evolution by determining which species (and their associated morphologies) survive to be the pool of ancestors for future evolution.

The punctuated equilibria theory has stimulated a concentrated examination of evolutionary change in fossil lineages and a reconsideration of the principles of population genetics that might apply to evaluating the theory. Several important biases have been discovered in the examination of fossil lineages. First, interruptions in sedimentation and erosion results in large segments of time being unrepresented, especially at time scales where microevolution is known to work [9]. Second, complete geographic sampling is usually not possible. These problems limit the measurement of rates of change and testing of theories concerning speciation in the fossil record.

The problem is exacerbated by the essential impossibility of identifying fossil species by anything other than morphology. With no criterion for inferring reproductive isolation, the testing of the proposition that morphological change is associated with speciation by the use of fossil data alone is impossible. When one cannot tell morphs of single species from different genera [34], and when one realizes that minor mutations often produce discrete changes that can easily be employed to recognize a new species [24], one becomes skeptical about the chances for tracking speciation in the fossil record. Although we do not believe that it is possible to study speciation in the fossil record, it is possible to study rates and patterns of morphological change in fossil lineages. The hypothesis that morphological change is concentrated in speciation events in extremely difficult to test in the fossil record and can only be approached by indirect means.

An important question is the genetic basis for morphological change seen in fossil lineages. While extensive morphological evolution has been inferred to occur in foraminifera, significant change has also been ascribed to ecophenotypic effects. A study done on rapid divergence in snails [43] has been criticized similarly. Variation and evolutionary change in fossil sticklebacks, however, seem to mimic variation known to have a genetic basis in living populations of congeners [4].

Studies of fossil lineages in recent years have produced a diversity of phyletic patterns, some in concordance with the punctuational pattern. These show a jump in morphology after a period of stasis. Unfortunately, they are difficult to interpret, because a single jump with a significant difference in mean could imply that the rate of evolution changed from very low to very high during that interval. In some cases, patterns of stasis are misleading [26] as they may merely indicate the presence of a predominant morph stabilized by canalization. Such cases must be distinguished from those where continuous or meristic variation is possible.

There are also many cases showing extensive phyletic change involving species and generic level transitions [12, 21, 27, 29]. There is no evidence that a lack of statistically significant temporal variation is common in the record, even if sustained directionality may be uncommon. An analysis [6] of six available fossil data sets demonstrated significant variation in evolutionary rates in all but one. The pattern seen in a foraminiferan lineage is illuminating. The direction of change continually switches, even if there is an overall trend in one direction. This even occurs during a period of rapid evolution that gives rise to a species-level transition. Many cases can be found in the fossil record, however, where no net directional change occurs, or where no net change at the phenetic species level can be detected for long periods (e.g., [8]).

It may be typological thinking that separates many paleontological species, with no perceived transitions. In a graptolite lineage, Skevington [34] reckoned members of two different form genera to be morphs of the same phyletic species! As mentioned above, intergeneric transitions are commonly observed to be the result of phyletic evolution. This casts doubt on the proposition that phyletic evolution is insufficient as a process to generate the morphological diversity we see in living organisms, as suggested by Stanley [36].

The general power of phyletic evolution can be seen in a lineage of linograptid graptolites in the Upper Silurian of central Europe [37]. The evolution of the Linograptinae illustrates a phyletic change from one-stiped monograptids into multiramous, compound colonies. These large structural, taxonomic, and adaptive results were attained within a single lineage represented by a sequence of four chronospecies. Two of them appeared after a bottlenecking of the lineage, implying an evolutionary event in a geographically restricted region or small population. The last species succeeded the ancestral one as it was undergoing a rapid increase in abundance. This implies transformation in large populations. This sequence undermines the belief that phyletic evolution has little or no significance for macroevolution.

The assertion of a pattern of punctuation was done partially to counter a supposed expectation of neo-Darwinian population genetics theory that evolutionary change should be continuous and of constant rate. But the classic population genetics literature suggests neither point. Recent reexplorations of the subject [7] suggest that, if anything, sustained constant rates of evolution are the least plausible prediction from theory. Theory, furthermore, does not restrict the likelihood of extensive evolution to small peripheral populations. If anything, extensive evolution under natural selection is to be expected in large panmictic populations. Large populations, if geographically widespread, may not be truly panmictic and may be subject to spatially varying, sometimes opposing, selection pressures, with occasional bursts of gene flow. This might slow down the rate of evolutionary change. Artificial selection experiments have demonstrated that long continued responses to selection are much more easily attainable in large than in small populations [15].

As a result of these problems in theory and practice, it is not possible, using fossil material, to test the hypothesis that morphological change is concentrated in speciation events. This hypothesis is, however, quite testable in living organisms, as Turner (this volume) clearly shows. In this context, the weight of the evidence is against the hypothesis.

Vrba [40] has discussed the possible fortuitous spread of a character or clade via random extinction-speciation processes. To provide some focus we distinguish among species selection, species drift, and hitchhiking. First, we use *species selection* to identify cases where some species-level property nonrandomly influences speciation rate or extinction rate. In contrast, *species drift* is the case where no species-level properties determine extinction or speciation, but random or non-species level properties result in a change of the proportion of species of different clades. *Hitchhiking* is the fortuitous spread of a character in a clade as a result of its association with other characters undergoing species selection, or by its association with a clade increasing by means of species drift. For example, the single stamen in orchids may not contribute to high speciation rates, yet high speciation rates in the group have spread this property over many species. The spread of the single stamen would involve hitchhiking. Similarly, gastropod speciation might be more rapid in the tropics, while spines might also be selected as a trait to resist predation that is coincidentally intense in the same regions. The spread of spines over many species would constitute hitchhiking of spination. Note that hitchhiking of a character can occur either in species drift or in species selection, Hitchhiking is an instance of downward causation, where a process in one level of a hierarchy affects levels beneath.

While higher level processes such as species selection and species drift are possible, there seems to be no large number of documented cases. A. Hallam's work (personal communication) on Jurassic bivalves shows a predominance of lineages with the punctuational pattern, but little evidence for cladogenesis. This pattern also predominates in Miocene U.S. Atlantic coastal plain bivalves (e.g., [22]). Nevertheless, Turner's study of mullerian mimetic butterflies suggests that mimetic patterns in Heliconius may predominate among some groups of butterfly species due to relatively rapid speciation in this genus. This would be a case of hitchhiking as Turner's evidence (this volume) demonstrates that mimetic patterns do not affect the probability of speciation. Fürsich and Jablonski [10] inferred the probable appearance of an early clade of boring snails. Its extinction was in all probability not due to a lack of prey, and the loss (by extinction) of the characters relating to shell boring constitutes a case of hitchhiking. We emphasize that these processes are not likely to be involved in evolutionary processes associated with the adaptive origin of complex traits such as eyes and limbs.

Under the operation of species selection, one would expect that a species-level property such as speciation rate might be increased over time. Several have demonstrated in fossil mollusk morphospecies a relationship between dispersal, geographic range, and geologic range [14, 16, 35]. These differences might influence the fate of different groups, producing true species selection. It should be borne in mind, however, that in order to demonstrate an effect of species selection, it is not adequate simply to demonstrate a statistically significant change in frequency of the same character state among the species existing at the two successive times. It must be shown that the change in frequency cannot be accounted for by purely stochastic fluctuations in frequency due to the accidents of speciation and extinction, i.e., by species drift. In conclusion, there is some cause for belief in the efficacy of both species selection and species drift in shifting the relative abundance of species with given properties. We look forward to more intensive research in this area.

Conclusions and Directions for Future Research

Much of our time was spent in questioning the power of the fossil record to test a series of models requiring knowledge of rates of evolution and speciation. It would be salutary for the field if more limits were set on the degree to which paleontological data can be used to discriminate among alternatives, and to set limits on measures of rate and time of appearance. Paleontologists have argued rightly that environmental conditions in previous parts of geological history may have been quite different from those today. This has led to a host of reasonable conclusions such as the episodic nature of evolutionary radiations and the relative suddenness and coordination of extinctions among unrelated taxa. But there may also be a tendency to invoke exotic genetic mechanisms in evolution where none may be necessary.

We suggest a series of research problems, designed to amplify the data base of paleontology and to focus on testable approaches.

1. Taxonomic and molecular data bases. We clearly need many more studies of taxonomy, particularly of early Cambrian and late Precambrian forms, and studies of molecular genealogy of living phyla. The former will give us a clearer picture of the diversity and morphological relationships of early metazoans while the latter will establish phylogenetic relationships and perhaps timing. These data will set constraints on the timing and directions of evolution.

2. We should identify those living systems where speciation and morphological change can be related, in order to study the processes of species selection and species drift.
3. We need more developmental experiments to determine the role of developmental mechanisms in evolutionary change.
4. Much more attention should be paid to small-scale changes in the fossil record and what they mean. If we cannot identify biological species, then this should be admitted and morphological entities should be studied. Understanding the limits of inference will help us define what we can test. For neontologists, it would help greatly if more attention were focused on the genetics and morphological changes accompanying speciation. Research programs combining studies of genetics of natural populations with studies of clades of closely related species would be more helpful in understanding the interface between within-species and between-species processes.
5. We should be studying morphological evolution in natural populations and especially interactions among characters. Such studies have been pitifully few and, as a result, we know little or nothing about interactions of characters during evolutionary changes. The interactions of canalization with stabilizing selection should be studied.
6. Some larger-scale approaches would also be useful. In cases where the fossil record is relatively complete, it would be useful to quantify the evolution of body plans and to document the appearance of characters that we regard as essential to Baupläne. In this regard, it would be useful to have more understanding about similarities and differences in development and genetics between different evolutionary groups.

Acknowledgements. I thank the members of our group who criticized and, after reaching a consensus, helped me rewrite crucial passages. J. W. Valentine read and criticized a penultimate version, while B. Charlesworth, K. Bandel, B. Runnegar, and S. C. Stearns also provided helpful criticism. I am very grateful to P. Winchester who patiently helped me with various drafts. We all thank the organizers, D. M. Raup and D. Jablonski, and especially S. Bernhard, for their skillful organization. We are equally grateful to our own moderator, S. C. Stearns.

References

1. Alberch P, Gale E (1982) Size dependence during the development of the amphibian foot. Colchicine-induced digital loss and reduction. J Embryol Exp Morph 76:177–197
2. Ashburner M (1980) Chromosomal effects of ecdysone. Nature 285:435–436

3. Ballard WW (1976) Problems of gastrulation: Real and verbal. BioScience 26:36–39
4. Bell MA, Haglund TR (1982) Fine-scale temporal variation of the Miocene stickleback *Gasterosteus doryssus*. Paleobiology 8:282–292
5. Brown GG, Lee JS, Brisson N, Verma DPS (1984) The evolution of a plant globin gene family. J Molec Evol 21:19–32
6. Charlesworth B (1984) Some quantitative methods for studying evolutionary patterns in single characters. Paleobiology 10:308–318
7. Charlesworth B, Lande R, Slatkin A (1982) A neo-Darwinian commentary on macroevolution. Evolution 36:474–498
8. Cope JCW, Skelton PW (1985) Evolutionary case histories from the fossil record. Spec Paper Palaeontol 33. Palaeontological Association of London
9. Dingus L, Sadler PM (1982) The effects of stratigraphic completeness on estimations of evolutionary rates. Syst Zool 31:400–412
10. Fürsich FT, Jablonski D (1984) Late Triassic naticid drillholes: carnivorous gastropods gain a major adaptation but fail to radiate. Science 224:78–80
11. Glaessner MF (1984) The dawn of animal life. Cambridge: Cambridge University Press
12. Grabert B (1959) Phylogenetische Untersuchungen an *Gaudyrina* und *Spiroplectinata* (Foram.) besonders aus dem nordwestdeutschen Apt und Alb. Abh Senckenb Naturf Ges 498:1–71
13. Hall BK (1984) Developmental processes underlying heterochrony as an evolutionary mechanism. Can J Zool 62:1–7
14. Hanson TA (1980) Influence of larval dispersal and geographic distribution on species longevity. Paleobiology 6:193–207
15. Hill WG (1982) Predictions to response to artificial selection for new mutations. Genet Res 40:255–278
16. Jablonski D (1986) Larval ecology and macroevolution in marine invertebrates. Bull Mar Sci, still in press
17. Kemp TS (1983) Mammal-like reptiles and the origin of mammals. London: Academic Press
18. Kidwell MG, Kidwell JF, Sved JA (1977) Hybrid dysgenesis in *Drosophila melanogaster:* a syndrome of aberrant traits including mutation, sterility and male recombination. Genetics 86:813–833
19. Kimura M (1983) The neutral theory of molecular evolution. Cambridge: Cambridge University Press
20. Levinton JS (1986) Developmental constraints and evolutionary constraints: a discussion and critique. In: Genetics, development, and evolution, eds Stebbins GL, Ayala FJ, Gustafson JP. pp 253–288 New York: Plenum Press
21. Malmgren BA, Berggren WA, Lohman GP (1983) Evidence for punctuated gradualism in the Late Neogene *Globorotalia tumida* lineage of planktonic foraminifera. Paleobiology 9:377–389
22. Miyazaki JM, Mickevich MF (1982) Evolution of *Chesapecten* (Mollusca: Bivalvia, Miocene-Pliocene) and the biogenetic law. Evol Biol 15:369–409
23. Müller G (1985) Effects of skeletal change on muscle pattern formation. In: Development of the skeletal muscles, ed Crist B. Basel: S Karger
24. Palmer AR (1985) Quantum changes in gastropod shell morphology need not reflect speciation. Evolution 39:699–705
25. Radinsky LB (1982) Evolution of skull shape in carnivores. 3. The origin and early radiation of the modern carnivore families. Paleobiology 8:177–195

26. Rendel JM (1959) Canalization of the scute phenotype. Evolution 13:425–439
27. Reyment RA (1982) Analysis of trans-specific evolution in Cretaceous ostracodes. Paleobiology 8:293–306
28. Riley M (1984) Arrangement and rearrangement of bacterial genomes. In: Microorganisms as model systems for studying evolution, ed Mortlock RP, pp 285–315. New York: Plenum Publ Co
29. Rose KD, Bown TM (1984) Gradual phyletic evolution at the generic level in early Eocene omomyid primates. Nature 309:250–252
30. Runnegar B (1986) Molecular palaeontology. Palaeontology 29:1–24
31. Sander K (1983) The evolution of patterning mechanisms: gleanings from insect embryogenesis and spermatogenesis. In: Development and evolution, eds Goodwin BC, Holder L, Wylie CC. London: Cambridge University Press
32. Shapiro JA (1983) Mobile genetic elements. New York: Academic Press
33. Simpson GG (1953) The major features of evolution. New York: Columbia University Press
34. Skevington D (1967) Probable instance of genetic polymorphism in the graptolites. Nature 213:810–812
35. Spiller J (1977) Evolution of turritellid gastropods from the Miocene and Pliocene of the Atlantic coastal plain. Ph D Dissertation, State University of New York at Stony Brook
36. Stanley SM (1979) Macroevolution: pattern and process. San Francisco: WH Freeman & Co
37. Urbanek A (1963) On generation and regeneration of cladia in some Upper Silurian monograptids. Acta Palaeontol Polonica 8:135–258
38. Valentine JW (1969) Patterns of taxonomic and ecological structure of the shelf benthos during Phanerozoic time. Paleontology 12:684–709
39. Vavilov N (1922) The law of homologous series in variation. J Genet 12:47–69
40. Vrba ES (1983) Macroevolutionary trends: new perspectives on the roles of adaptation and incidental effect. Science 221:387–389
41. Wake DB, Roth G, Wake MH (1983) On the problem of stasis in organismal evolution. J Theoret Biol 101:211–224
42. Williams NE (1984) An apparent disjunction between the evolution of form and substance in the genus Tetrahymena. Evolution 38:25–33
43. Williamson PG (1981) Palaeontological documentation of speciation in Cenozoic molluscs from Turkana basis. Nature 293:437–443
44. Wright S (1934) An analysis of variability in number of digits in an inbred strain of guinea pigs. Genetics 19:506–551; 293:437–443

Patterns and Processes in the History of Life,
eds. D. M. Raup and D. Jablonski, pp. 183–207. Dahlem Konferenzen 1986
Springer-Verlag Berlin, Heidelberg

The Genetics of Adaptive Radiation: A Neo-Darwinian Theory of Punctuational Evolution

J. R. G. Turner

Dept. of Genetics
University of Leeds
Leeds LS2 9JT, England

> ... these few insects ... show that the same species varies in a different way in different localities, therefore that natural selection does not explain everything; the local conditions determine what variations shall arise – natural selection *creates nothing*.
>
> H. W. Bates (the discoverer of mimicry),
> letter to D. J. Hooker, 1862.

> ... not the wing of a butterfly can change in form, or vary in colour, except in harmony with, and as a part of, the grand march of nature.
>
> Wallace 1865 [33].

> The function of natural selection is selection and not creation.
>
> Punnett 1911, p 132 [22].

Abstract. It has been claimed that the punctuational pattern of evolution observed in the fossil record demands a new theory of evolution, largely replacing neo-Darwinism. It is shown here that the processes of microevolution already known in actual populations will account for punctuational evolution; this assertion is supported by an in-depth study of the genetics of adaptive radiation in butterflies of the genus *Heliconius*.

Punctuational Evolution: An Analysis

Gould has suggested that the neo-Darwinian or Modern Synthesis view of evolution, in certain forms, is "effectively dead" [8]. That is to say that the theory was not so much wrong as inadequate; it describes only a limited part of the evolutionary process, the adaptive adjustment of existing populations; it accounts adequately neither for the origin of species nor for evolution in the long term [8, 15]. In particular, it is held that the pattern of evolution apparent in the fossil record, an alternation of stasis with extremely rapid change accompanied, so it seems, by branching of the evolutionary tree, is at best not predicted by neo-Darwinism and at worst incompatible with it. A "new and general theory of evolution" is therefore needed to account for the facts. Briefly, this new theory, known rather confusingly as "punctuated equilibrium" – the same term as is applied to the pattern of stasis and change actually observed in the fossil record – holds that evolution is to be accounted for not by natural selection and the other recognized processes of microevolution, but rather by a series of processes collectively called "macroevolution." These are chiefly an evolutionary jerk which relatively suddenly generates new species with new and constant morphologies – say, in less than fifty millennia – and a process of species selection which determines the long-term trends of evolution by selecting among the species so formed; the species themselves undergo relatively little evolution. This theory is represented by the well-known Gould-Eldredge "candelabra" version of the evolutionary tree shown in Fig. 5a, below. Hence natural selection and the other microevolutionary processes have only a minor part to play in determining the major features of evolution. One especially unorthodox outcome, but one with very wide implications, is that evolution cannot be seen as predominantly a matter of individual adaptation.

It is easy enough to espouse this theory, or to disagree with it; my present aim is the harder one – to synthesize the new theory with neo-Darwinism. To this end, I find it convenient to present punctuated equilibrium as a three-pronged fork:

Prong 1: *punctuational pattern of evolution* – the fossil record reveals a pattern of stasis alternating with periods of extremely rapid change;

Prong 2: *evolution by jerks* – this pattern arises because species appear rapidly, morphologically distinct from their closest relatives, and remain more or less unchanged until they become extinct; such large morphological changes are (almost) always cladogenic – that is, they occur *only* when the evolutionary tree branches – and therefore very little significant evolutionary change occurs by the neo-Darwinian processes of gradual or rapid change *within* species; various subsidiary theories (Wrightian peak-shifts, Mayrian founder-effects, Goldschmidtian macromutations, ecological release) have been proposed to account for this 1:1 correlation of morphological change and speciation;

Prong 3: *the second process* – long-term trends and patterns of evolution are therefore created not by evolution *within* species, which remain relatively static in their morphologies, but by a form of natural selection acting *between* species, involving three processes: a) a differential rate of death (extinction), b) a differential rate of birth (speciation), and c) a bias in the direction of the jerk.

Prong 1, I suggest, is an observation which, even if not universal, is common enough to be taken as a phenomenon in need of explanation. I shall however argue that *prong 2* is a simple misinterpretation: the phenomenon of punctuational evolution as revealed by the fossil record can be fully explained by neo-Darwinian microevolutionary theory. No new theory over and above what is already known from the disciplines of evolutionary genetics and the Modern Synthesis is needed to explain these observations on fossils. The model of evolution by jerks is at the very least superfluous and is in fact less compatible with our overall knowledge than is the conventional theory. (It is important to note that rapid-change-and-stasis is not the essential feature of *prong 2*; the crucial point is the very strong association of rapid and large changes with speciation and/or cladogenesis [10, 25]). Nonetheless, I shall then further argue that *prong 3* is a correct and valuable extension of evolutionary theory, for while in no way incompatible with neo-Darwinism, it discusses matters not dreamed of in the older philosophy. Thus with the tricky central prong removed, both the observations (*prong 1*) and the theory (*prong 3*) of macroevolution can be painlessly joined onto conventional neo-Darwinian microevolutionary theory to produce a new synthetic theory of evolution.

It would be possible to argue this case purely theoretically, but being an empirical scientist addressing empirical scientists, I shall use a real example: a group of organisms which I believe show evolution occurring according to this synthetic model – a real case study of adaptive radiation.

Heliconius:
The Genetic Basis of Adaptive Radiation

The problem with joining macroevolution and microevolution has always been that of time: it is impossible to do population genetics with fossils; it is difficult to observe living populations for long enough to see the "macro" changes. We have over the last decade or so discovered a system which to some extent overcomes this problem, an example of adaptive radiation and convergence, in the strict sense that this is understood by comparative morphologists, that has taken place in a period of time so long as to match the shortest of the periods considered by paleontologists, but whose most recent products are still closely enough related to be studied by conventional genetic crosses. Although we have no fossils we do have a way round this problem: a surprising amount about the past can be inferred by using cladistic techniques. The butterflies involved have the further advantage that they ar mimetic. As mimicry is one of the best understood of all adaptations, we can draw on the considerable understanding of its ecology and dynamics which has been built up over the last fifty years.

The South American genus *Heliconius* can be classified into morphological groups, which are probably close to being actual phylads, and also be the color patterns exhibited by the adults. When this is done (Table 1) a very interesting pattern emerges: there is a rather low correlation between the two methods of classification, resulting in the filling of many cells off the diagonal of the table. Many of the phylads have radiated into a rather large number of the patterns and have therefore converged on each other in the process. The main, although possibly not the only, function of this convergence has been muellerian mimicry – the mutual resemblance of warningly colored species. The butterflies are known to be distasteful to bird predators, and their resemblances are good enough to fool not only the birds but experienced entomologists – I have yet to find a major collection which did not have specimens placed in the wrong species, sometimes by the very taxonomists who had prepared the original species description! (The references relating the empirical evidence on these points can be found in various reviews [3, 26, 27]; I shall omit detailed citations in this paper.)

Table 1. Adaptive radiation, or convergent and divergent evolution, in *Heliconius* and the related genera: classification by morphology (genera and subgenera) on one axis and color pattern on the other shows that morphological groups have radiated into different color patterns. Examples of these are shown beneath. pt = in part. Updated by Brown from Turner [3]. Reproduced, with permission, from the Annual Review of Entomology, vol 26. Copyright 1985 by Annual Reviews Inc

Genus/group	Dennis-Ray (red, yellow)	Red on FW, (yellow on HW)	Red on HW, yellow on FW	"Tiger" ithomiine	Other mimetic*	Blue and yellow, FW	White or yellow on HW & usually FW	Black with yellow bars
Eueides	*eanes* (pt) *vibilia* (pt) *tales*	*eanes* (pt)	*procula* (pt) *tales* (pt)	*isabella* *lampeto*	*procula*[a] *vibilia*[b] *pavana*[b]	*eanes* (pt)	—	—
Neruda	*aoede*	—	—	—	*godmani*[c]	*metharme*	—	—
Laparus	*doris* (pt)	—	*doris* (pt)	—	—	*doris* (pt)	*doris* (pt)	*doris* (pt)
Heliconius I	*xanthocles* (pt)	—	*xanthocles* (pt) *hierax*	—	*hecuba*[d]	*xanthocles* (pt)	—	*hecuba* (pt)
Heliconius II	*burneyi* *egeria* *astraea*	—	—	—	—	*burneyi* (pt) *wallacei*	—	—
Heliconius III	*melpomene* (pt) *timareta* (pt) *elevatus*	*melpomene* (pt) *besckei*	*melpomene* (pt) *timareta* (pt)	*nattereri* ♀ *numata* *hecale* *ethilla*	*atthis*[d] *cydno* (pt)[c] *cydno* (pt)[d]	*hecale* (pt) *timareta* (pt) *cydno* (pt) *luciana* (pt)	*cydno* (pt)	*nattereri* ♂ *pachinus* *luciana* (pt)
Heliconius IV	*erato* (pt) *demeter*	*erato* (pt) *hermathena* *telesiphe*	*erato* (pt) *clysonymus* *hortense* *ricini*	*hecalesia* (pt)	*hecalesia*[c] *charitonia* (pt)[d]	*leucadia* *sara* (pt) *antiochus* (pt) *congener*	*erato* (pt) *sara* (pt) *antiochus* (pt) *sapho* *eleuchia*	*hewitsoni* *sara* (pt) *charitonia* (pt
Other Genera		*Podotricha* *telesiphe*						*Philaethria* (3 species)

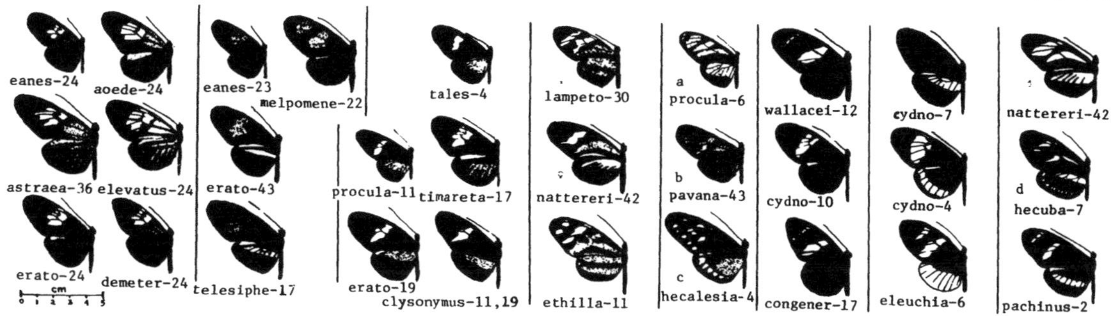

* a = various primitive Ithomiinae, b = *Actinote* (Acraeinac), c = *Tithorea* (Ithomiinae), d = *Elzunia* (Ithomiinae).

The trouble with most such adaptive radiations is that while they look pretty, our lack of knowledge of their genetic basis puts an effective stop to further enquiry. We can overcome this problem in *Heliconius*, for the phenomenon occurs also *within* some of the species: Fig. 1 shows the most spectacular example, the parallel radiation of the races of *Heliconius melpomene* and *Heliconius erato*. Both species have radiated into a considerable number of different patterns while continuing to remain (or to become)

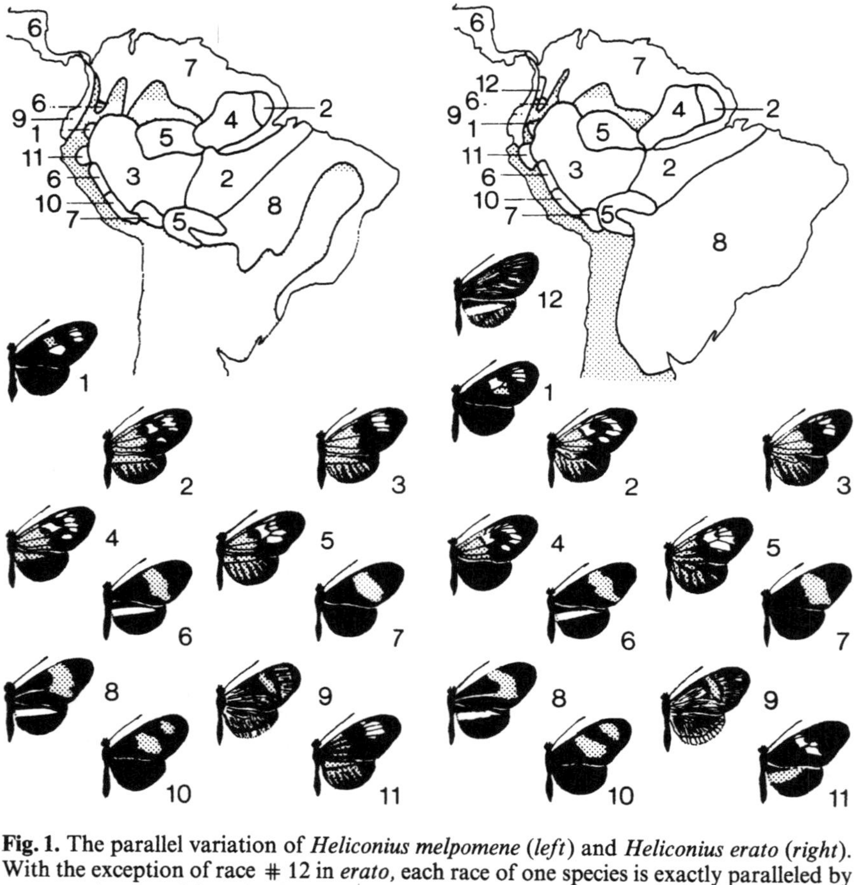

Fig. 1. The parallel variation of *Heliconius melpomene* (*left*) and *Heliconius erato* (*right*). With the exception of race # 12 in *erato*, each race of one species is exactly paralleled by a sympatric race of the other [23, 27]. Reproduced, with permission, from the Annual Review of Ecology and Systematics, vol 12. Copyright 1985 by Annual Reviews Inc. and J. R. G. Turner

parallel muellerian mimics in all the areas in which they occur together. Each race of one is matched by an almost identical race of the other (with just one exception). Were their races to speciate, they would obviously generate a pattern of parallel species such as can be seen in any pair of rows in Table 1; it therefore seems legitimate to conclude that we are seeing an early stage of the process which generates the adaptive radiation and convergence of the whole group. The genetic basis of the process can be studied then by crossing various races of *Heliconius erato* and *Heliconius melpomene* (within the species only; crosses between them do not succeed).

Rather surprisingly, in view of what most evolutionists have come to expect about racial differences and about muellerian mimicry, the large alterations in pattern have been produced not polygenically, but by the alteration of a relatively small number of genes of rather large effect: whole red patches and yellow bars are taken out or put in at one go, or in two or three rather large chunks, and their shapes are likewise altered by single mutations of comparatively large effect, breaking a solid yellow patch into a group of dots, for example [23, 31]. Evolution appears to have been much less gradual than one might expect.

The actual course of evolution, in the absence of fossils, can only be inferred. Luckily, a combination of genetics and cladistics puts the inference on an exceptionally firm footing. It is an established theorem of population genetics, known from its derivation by Haldane in 1924 as Haldane's Sieve, that when new mutations become established in populations under natural selection there is a large or overwhelming tendency for dominant rather than recessive mutations to establish themselves. The reason for this is the simple one that recessive mutations cannot be expressed, and therefore selected, until they have risen to a rather high frequency by random drift, whereas dominant mutations are steadily selected when they are still at very low frequencies. In this way, most of the genes producing industrial melanism in moths are dominant to the original form of the species. It is not that recessive melanic mutations do not occur – they do; they merely seldom establish themselves in populations. It is thus a fair inference, when we have a number of genetic variants in a species, that the dominant alleles are usually the derived ones, and the recessive alleles usually the ancestral condition. This allows us [23] to put a direction on the evolution of the genes and to perform cladistic reconstructions without the usually questionable technique of designating an out-group [27, 30].

Such a piece of genetic cladistics is shown in Fig. 2. The result surprised us, as we had initially been skeptical about the power or accuracy of this kind of reconstruction. The patterns of the overall ancestors of both *melpo-*

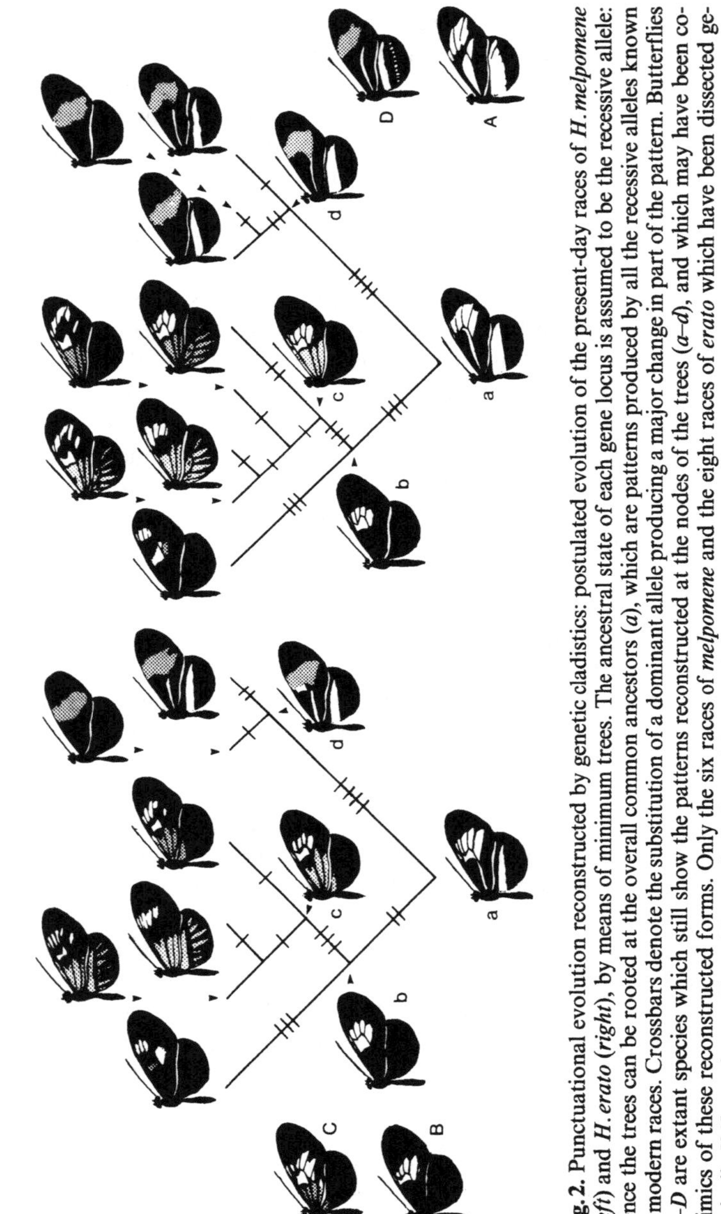

Fig. 2. Punctuational evolution reconstructed by genetic cladistics: postulated evolution of the present-day races of *H. melpomene* (*left*) and *H. erato* (*right*), by means of minimum trees. The ancestral state of each gene locus is assumed to be the recessive allele: hence the trees can be rooted at the overall common ancestors (*a*), which are patterns produced by all the recessive alleles known in modern races. Crossbars denote the substitution of a dominant allele producing a major change in part of the pattern. Butterflies *A–D* are extant species which still show the patterns reconstructed at the nodes of the trees (*a–d*), and which may have been co-mimics of these reconstructed forms. Only the six races of *melpomene* and the eight races of *erato* which have been dissected genetically [23] are shown. From Turner [31]. Copyright 1985 by J. R. G. Turner

mene and *erato* are both unexpected and convincing (Fig. 2a): they are entirely black and yellow, which is not what anyone would, I think, anticipate from looking at the modern races, which all (with the exception of a single race of *erato*) have quite extensive red or orange marks, sometimes without any yellow at all. They are, furthermore, excellent mimics one of another, suggesting that the parallel mimicry between the two species is older than the recent burst of race formation. They are also corroborated by the outgroup: both *melpomene* and *erato* have close relatives which are nonmimetic and which may therefore be expected to evolve rather slowly in comparison with mimetic species. They might have patterns rather close to the ancestral. Both these species (one of them is shown in Fig. 2A) have patterns of yellow bars very like the reconstructed ancestors of *melpomene* and *erato*.

The cladograms suggest also that this close mimicry, starting with the overall ancestors, has been continued and maintained throughout the several rounds of race formation which have ended with the present-day races: ancestors reconstructed at the nodes of the trees bear a close resemblance to one another; they would probably be even closer if our knowledge of the genetics of *erato* were more complete. With the wisdom of hindsight, this would indeed seem to be the only explanation for the almost total parallel variation between the modern races. It is passing belief that two species, each represented by some two dozen races but having totally different patterns, should somehow have converged to become muellerian mimics not just in one area, but throughout all of their diversified races. Evolution in parallel is the only plausible route to this end. (To the question which will naturally arise in the reader's mind, are the two species using the "same" genes, we have only a partial answer: there is good evidence from functional and linkage relations [30, 31] that *some* of the genes producing parallel patterns in the two species are homologues; there is clear evidence that others are not [23].)

Punctuated Phyletic Evolution?

Now imagine what the fossil record of these butterfly races would look like if butterfly patterns fossilized. Would it be gradual or punctuated? We can answer that from standard population genetics. The cladograms, it will be remembered, are constructed not in the conventionel way from phenotypic characters, but from the genetic composition of the races; in each branch, therefore, the crossbars indicate the substitution of a gene altering a major part of the color pattern (it is possible in most cases to see which parts are altered by comparing the butterflies at the opposite ends of the branch). The bars are placed conventionally at the centers of the branches but could of course represent gene substitutions at any position in the branch. The interesting question is how much of the total branch each gene substitution occupies. We can show from Haldane's equations that to change the composition of a population from having 10^{-x} individuals with a new pattern to having $1-10^{-x}$ individuals with it (say from one in 1000 to 999 in 1000, where $x=3$) takes a period of roughly $t=10^{x/2}/s$ generations, where s is the selection coefficient in favor of the new phenotype [30]. Thus for $x=3$, which is a generous margin for detecting a minority form in a fossil (or even living) population, the change will take 311 generations for a selection coefficient of 0.1; it will take around 3160 generations for a selection coefficient only one tenth as large. In *Heliconius*, with around ten generations a year, these are very short times indeed (three decades or three centuries, probably faster, as the equation assumes that s is constant, whereas in a muellerian mimic it will in fact increase as the new pattern becomes more common). Perhaps a millennium is the longest period that can reasonably be assumed.

Just how long a period the whole tree spans we do not know. The most likely historical explanation of the races is that they formed more or less allopatrically when the rain forests where the butterflies live became fragmented during the cool dry periods which accompanied the Quaternary glaciations at higher latitudes and altitudes (Fig. 3) (reviews in [20]). This would put a period of over 30,000 years on the complete tree. In that case, the evolutionary changes occupied only a small part of the total time: hundreds of years out of tens of thousands.

The conclusion is simple: it is likely that the butterfly patterns have remained static for long periods, interspersed with relatively rapid bursts of change when one major part of the pattern has been altered by the substitution of one, or a few, major genes. In other words, in playback or fossils, these butterflies would show a very fine example of punctuational evolution.

To what extent do these inferences about *Heliconius* back the theory of punctuated equilibrium? To answer this it is necessary again to consider the three prongs of the fork. The evolutionary reconstruction is clearly consistent with *prong 1,* that evolution may be punctuated rather than gradual. It might be consistent with *prong 2*, that punctuational events are normally evolutionary jerks in which the large changes are intimately associated with branching of the tree. But I suggest that there is no strong reason for supposing this to be so and that there are some good reasons against it. First,

Fig. 3. Probable geographical origin of the extant races of *H. melpomene* and *H. erato:* approximate location of South American rain forests at the peak of the last glaciation, ca 18,000 years BP, deduced form a combination of biogeographical and paleoecological data. After K. S. Brown, redrawn [20]

if we regard speciation as the event which has to be associated with the punctuational change in order to produce a jerk, then *Heliconius* clearly are not following this pattern, as the various races have not achieved the status of full species: they hybridize readily in the laboratory and produce extensive hybrid zones where their borders touch in the wild [1, 27]. If simple racial isolation is held to be the criterion, then *Heliconius* do not conform to this either, because populations which are geographically isolated from others, and have apparently been so for some considerable time, do not necessarily undergo punctuational change: the populations of *melpomene* and *erato* on Trinidad are distinguishable from those of the adjacent Venezuelan mainland in minor features such as size and the precise hue of their red markings (and, as is the way with butterfly taxonomists, have been given separate subspecific names), but to the untutored eye they are simply representatives of the Venezuelan race. No major gene alterations have occurred in the pattern during what is presumably a fairly long period of separation. A slightly firmer case could be made for supposing that when major phenotypic changes occur they tend to lead to speciation in these butterflies; it may be that changing the color pattern leads to selection for reproductive isolation because hybrid, recombinant patterns are at a disadvantage in the mimicry system (rare warning patterns are normally relatively poorly protected), and the altered patterns themselves can be used as courtship recognition signals [27]. However, this has clearly not happened yet in the zones of natural hybridization which fringe all the races at their contact with others, and this has lasted for probably eight millennia (80,000 generations) since the end of the last glaciation. The best we can do is to suppose that a punctuational change of pattern will *eventually* lead a race to become a full species, but this puts a rather stretched meaning onto the proposal that punctuational change occurs "at the same time" as cladogenesis.

We can try to save *prong 2* by supposing that although not closely associated in terms of human life spans, cladogenesis and punctuational change occur together within a short geological period, say, within fifty millennia, and that this is followed by long periods in which neither process occurs. But while it is possible, but unproved, that the punctuational change will always lead to speciation within this period, there is something decidedly odd about the hypothesis that punctuational change cannot occur without leading to cladogenesis i.e., theory of the absence of punctuational phyletic change [8, 24]). Suppose that the Trinidad populations of our butterflies, isolated but essentially unaltered, undergo a punctuational change in the year 5000 AD. This would be an evolutionary jerk if the remaining populations of the species still existed on the mainland. It would not be an evolutionary jerk if the remaining populations had become extinct. In short, the theory of the absence of punctuated phyletic evolution imposes a very strange restriction: once a population has become the sole representative of its species, it cannot undergo a large evolutionary change! Yet what is there about the existence of other populations, somewhere else in the range, which can advance or hinder the evolution of a population? I suggest that punctuational change is "associated" with cladogenesis only in the tautological sense that until there are two or more branches of a tree, we cannot see them becoming different from one another. But there seems to be nothing to stop a branch that is now the sole representative of its species from undergoing evolutionary change of any kind, except that by definition it cannot become reproductively isolated.

The final way of trying to preserve *prong 2* is to suppose that what we are seeing in *Heliconius melpomene* and *erato* is not in fact the prelude to the formation of a parallel set of different species, and that when species in this group arise, they do so, complete with new patterns, by some different mechanism: the evolutionary jerk of *prong 2*. There is no way of disproving this but no evidence for it. We do, on the contrary, have some evidence that the genetic changes seen between the races of *melpomene* and *erato* are of the same kind as the genetic differences seen between them and their close relatives. Figure 4 shows a close relative of each species, and alongside them a *melpomene* and an *erato* which are strikingly similar in several features,

Fig. 4 a, b. Differences between the patterns of related species of *Heliconius* appear to be of the same kind as those which differentiate races of *melpomene* and *erato*. **a** The pattern of *Heliconius ethilla* (*left*) approximated by a *melpomene* pattern produced by combining various alleles known from various races of that species. **b** The pattern of *Heliconius telesiphe* (*left*) approximated by a specimen of *erato* produced by interracial hybridization. From Turner [30]. Copyright 1985 by J. R. G. Turner

even though the relatives have patterns which, in different ways, are markedly changed from those of normal *melpomene* and *erato*. *Heliconius ethilla* belongs to a completely different mimicry group, and yet it is possible to construct a *melpomene* very like it simply by choosing the right genes from the array presented to us within the known races. This has been done only on paper, as we have never managed to carry out the appropriate breeding program, but the *erato* specimen shown which resembles *Heliconius telesiphe* in the very notable taxonomic character of the ragged edge to the yellow hindwing bar, completely unknown in wild *erato* but characteristic of many other species, has actually appeared in one of our laboratory crosses.

There is thus reasonable ground for supposing that the genes accounting for differences between species are of a similar kind to those which account for racial differentiation, and that no further process of change in the patterns takes place during speciation.

What Causes Stasis and Change?

The causes of the punctuational changes in *Heliconius* are more a matter for theoretical discussion than for empirical demonstration. The chief function of the butterflies' patterns appears to be warning color and mimicry, although it may not be the only function, and the most likely explanation for the changes is therefore that they have been altered in mimicry of other butterflies flying in the same habitat. It is a property of warning color that the most common pattern tends to be the best protected, and it therefore follows that there is an advantage to any warningly colored species in copying a more common (or of course more distasteful) species if it can do it. This leads to the extensive muellerian mimicry found among distasteful tropical butterflies as well as among such groups as temperate bumblebees [19]. However, it is clear that a warningly colored species cannot always achieve this: if it could, then there would be only one warning pattern at any one place. In fact there are something like a half dozen warning patterns among the forest butterflies of tropical South America, and also in the tropical forest of west Africa [17, 18].

We believe [23] that this results from two restraints. First, there is a limit to what predators will regard as the "same" pattern; once two patterns are sufficiently different, there is no chance that a minor variant of one will be mistaken for the other. This puts an unbridgeable gap between the two patterns, a concept familiar as the maladaptive valley between two adaptive peaks. Second, the gap might sometimes be bridged by a single large mutation which took the pattern of the less protected species clear across the gap into rough mimicry of the better protected. In this way, patterns will undergo punctuational evolution; this is our explanation for the major gene changes seen in *melpomene* and *erato*. But there is a limit to what mutation can achieve; it may be impossible to alter extensively the pattern of a butterfly by a single mutation in such a way as to achieve mimicry of a very different pattern. Thus the mimetic patterns in any one area remain distinct and diverse, but from time to time individual species will "switch" patterns when they fly with one that is better protected and which is within striking distance of a single mutation. We have given an account of the rare species *Heliconius hermathena* doing just this at one place in the Amazon basin; where it comes into contact with populations of *erato* and *melpomene* which are not too unlike it, it takes to mimicking them by means of a single gene which wipes out most of its yellow marks ([23]; see also [31]). The evolution of muellerian mimicry is thus limited by the restraints of the external environment and the constraints of the genetic and developmental material.

We have here a possibly very general explanation for the alternation of stasis and punctuational change during evolution. If the races of *Heliconius melpomene* and *erato* really did chiefly arise in forest refuges, cut out from the formerly, as now, continuous Amazonian rain forests by unfavorable conditions in the Quaternary, we have to ask what caused their patterns to switch. I suggest [28] that the chief cause was the long-term changes in faunal and perhaps even floral composition that must inevitably occur on such an ecological island as the result of the progressive extinction of other plant and animal species. As species die out, the whole ecological structure is changed: some mimetic butterflies may disappear; others may have their abundance, in the long term, radically changed by the extinction of food plants, competitors, competitors of parasites, etc. What was formerly a rare and not well protected pattern may become the best protected in that particular refuge. Any species that can reach it by means of a single mutation (and the waiting time needed for the mutation is an important factor in making *long-term* changes of abundance, such as occur on an isolated island, the crucial factor) will do so. As studies of present-day refuges have shown [2], each island tends to develop a different faunal composition as a result of the disorderliness (not randomness) of the extinctions – different species become extinct on different islands. As a result, the locally best protected and hence most mimicked pattern will differ from refuge to refuge, and races of the type seen in *melpomene* and *erato* will appear.

I believe this to be a general explanation for punctuational change, because it must also occur in other aspects of the species' ecology besides mimicry. In an old and established fauna, most of the ecological niches become filled, just as the main mimetic patterns become established, and rather little evolution is possible: species constrain each other's evolution by packing as tightly as possible into the available ecological space. This generates stasis. Punctuational change will occur, first, if a species invents a new way of earning its living and hence of opening up a previously unoccupied part of the ecological space; and, second and much more commonly, when an existing species becomes extinct, leaving its part of the ecological space vacant. There will then be comparatively rapid evolution on the part of any species which are able to alter themselves so as to exploit that unoccupied part of the ecosystem. The analogy with the mimetic butterflies should be obvious; the only major difference is that the mimetic patterns tend to evolve toward those parts of the mimicry space which are well occupied, not those that are empty.

Many evolutionary geneticists will wish to question the likelihood that such punctuational changes will be produced by single mutational changes as they are in *Heliconius* [4, 5]. This is a question I leave open; as far as the overall features of evolution are concerned, it makes little difference whether the changes are produced by single large mutations or by several smaller ones, for the speed of evolution depends much more upon the strength of selection (and the generation time) than it does on the genetic architecture of the trait selected. We have simulated two-locus systems with various kinds of rather strong epistasis and have found that even with very strong inter-locus epistasis and loose linkage, the time required to alter the frequency of the *phenotype* is seldom altered by more than one order of magnitude from that predicted if the trait is governed by a single gene (Turner and Mukherjee, unpublished). In fact, it is easy enough to show that except for very large populations of big organisms with long generation times, changes of more than 50,000 years demand selection coefficients so small that the directional change would be swamped by random drift. Except for the tracking of very slow secular changes in the environment, population genetics predicts that evolution by natural selection will be punctuational. Punctuated equilibrium does indeed arise as a valid prediction from population genetic theory.

I therefore suggest that punctuated evolution is created by changing patterns of faunal balance: tight packing of the niche space produces evolutionary stasis; extinction of species tends to create ecological opportunities, which are exploited often by rapid, punctuational change.

Major Evolutionary Trends

There is therefore an acceptable neo-Darwinian explanation for punctuational change in *Heliconius*. The butterflies show the phenomena described by *prong 1* but do not demand evolution by jerks, the theory of *prong 2*. Does neo-Darwinism therefore account for the whole of their evolution? If the explanation is correct, then nothing should happen in the longer term that is contradictory to neo-Darwinism, but equally that is not to say that nothing further can happen, or that phenomena do not occur at a different level [9–11]. I believe that we can see the phenomena subsumed under *prong 3* occurring in this group. It is a notable feature of the butterflies, when the related genera are included, that there are far more species with the black, red, yellow, orange, and blue patterns characteristic of *Heliconius* than there are with the extensive orange (with only a little black) patterns of the related groups *Agraulis, Dione,* and *Dryas*. This surely does not indicate an overall evolutionary trend within the species of the group toward adopting the *Heliconius* pattern, but much more the fact that there are more species of *Heliconius* than there are species in the related genera. The overall trend toward the *Heliconius* pattern (if we assume, as is probably correct, that the orange one is ancestral) is very probably more a matter of this kind of *species selection* than it is of individual selection toward the black patterns.

Neo-Darwinism and the Fossil Record – A Synthesis

The result of this form of evolution, an interaction between the constraints of the environment and the restraints of the genome, is therefore a pattern of punctuational change. *Heliconius* butterflies present a real live example of the outcome of punctuational evolution. But this pattern differs, of course, in one very important respect from the pattern of punctuational change described by Gould and Eldredge [12] in the fossil record: the punctuational changes in *Heliconius* are *phyletic* changes; except in the tautological sense that we cannot detect such a change unless we have two branches of the cladogram to compare (in other words, in that we would have no means of reconstructing the evolutionary past of a *Heliconius* species which had only one present-day race and which therefore presented us with nothing to compare or breed it with), the punctuational changes are not associated in any strict way with the branching of the tree. As far as we know, such a change can occur at any time along one of the branches.

There seem to be two conclusions to be drawn from this: either what we have in *Heliconius* is an example of a phenomenon different from the usual punctuational theory of evolution, or it is indeed an example of it, and there is a flaw in that theory. It is the second point that I shall try to argue. Conventional punctuated equilibrium theory has been widely represented to the scientific and lay public in the form of the present Fig. 5a (and its various two-dimensional versions), as asserting that morphological change and reproductive isolation are correlated 1:1 during evolution: no large morphological change occurs without reproductive isolation, and no species of any evolutionary significance becomes isolated from its sister species without undergoing a large morphological change. With this assumption, an evolutionary tree is drawn (Fig. 5a) which differs radically from the conventional tree, and this tree is then used to show that evolution is of a radically different nature from that supposed by the Modern Synthesis. Sibling species, whose existence is of course acknowledged, vanish in the presentation not, as the skeptic might imagine, by some sleight of hand, but because they literally have no interesting part to play in evolution – if the vertical "candles" of Fig. 5a were each to consist of between two and twenty morphologically sibling species, this would make not one iota of difference, either to the appearance of the fossil record or, more importantly, to the dynamics of evolution [25]. Within classical punctuational theory, sibling species are a genetical detail of no more interest than silent variations in the base sequence of the DNA. But this interesting (*se non è vero, è ben trovato*) conclusion is based purely on the *assumption* that morphological change cannot occur without the reproductive or geographical isolation that initiates a branching of the tree. This assumption is pivotal to the acceptance of punctuated equilibrium as a new and "different" theory and is the essential feature of the middle prong of the fork. But it is an assumption, not an empirical observation. As Gingerich [7] has recently observed, "paleontological studies ... provide no evidence of change coinciding with branching speciation... For all the attention punctuated equilibrium has received as a model for speciation, there is still no clear evidence supporting it" (also [14]). In fact, we have enough examples of sibling species, which are reproductively isolated without having undergone morphological change, to vitiate it in one direction (this of course is common ground [25]), and a good enough argument that a single phyletic line should be able to change without undergoing branching at the same time (what forces a line to branch when it changes?) to make it suspect in the other direction. This pivotal feature of the full punc-

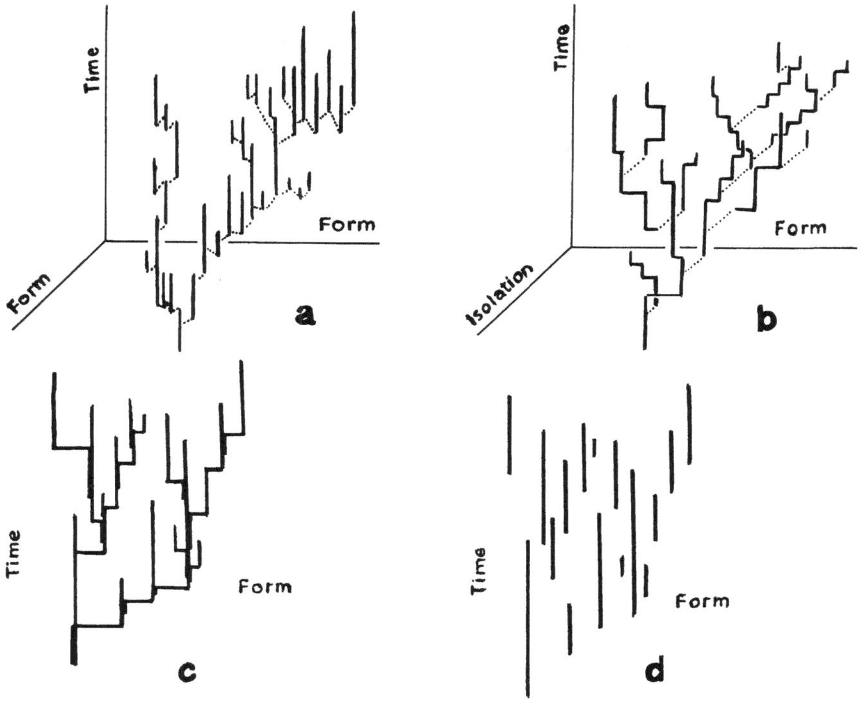

Fig. 5 a–d. Various ways of depicting and explaining punctuational evolution. In the Eldredge-Gould [6, 12] "candelabra" tree **a** large morphological changes and speciation are assumed to be highly correlated. In **b**, the "not-the-punctuational-model" tree [29], it is assumed that speciation-cladogenesis and rapid morphological change are not strongly correlated. Evolution of type **b**, if we are unable to detect the branching events (for example, if the species are sibling species or if the morphological changes which occur do not fossilize), will appear as in **c**, that is, as if the tree were viewed directly from in front. If large morphological changes occupy only relatively short periods of time, as is suggested in the study of *Heliconius* (Fig. 2), then the fossil record will present the picture shown in **d** – an illusion of evolution occurring without intermediate stages, in which branching and morphological change appear to take place simultaneously. This leads to the misinterpretation of the evolutionary process seen in **a.** Figs. **b, c, d** copyright 1985, by J. R. G. Turner

tuated equilibrium package-offer remains a decidedly questionable assumption.

What happens if we try to interpret the evolutionary tree without this assumption? There are then indeed two processes occurring side by side, but they are not microevolution and macroevolution, as assumed in punctuated equilibrium theory, but simply the old and familiar processes of phyletic change and speciation (Fig. 5 b). Phyletic change, we can assume from the empirical evidence I have presented and from what is known in the fossil record, is a mixture – just what mixture we do not know – of gradual and punctuational change; in order to avoid ambiguity of perspective in Fig. 5 b, only the punctuational changes are shown. The crucial distinction between this, my version of the evolutionary tree, and the Gould and Eldredge version is therefore that mine separates morphological change and reproductive/geographical isolation on two axes. Phyletic evolution and the branching of the tree are continual but largely unrelated processes; while the organisms evolve, the tree branches.

What we have here then is a fairly standard neo-Darwinian view of the evolutionary tree, with the less conventional addition of more or less frequent punctuational changes, of the type observed in *Heliconius* perhaps. How will this neo-Darwinian process reveal itself in the fossil record? Neither reproductive nor geographical isolation is readily visible: it is as if the tree in Fig. 5 b were drawn as a three-dimensional computer graph and then rotated so that the dotted lines which represent the branching of the tree were end-on toward us. What we would see is shown in Fig. 5 c; the punctuational events are clear enough, but the morphologically sibling species cannot be detected: many of the vertical lines of the tree simply hide behind each other, revealing themselves only when one of them undergoes an evolutionary change that the other does not suffer, and so pops out from behind. The sibling species, which are indeed a piece of genetic trivia if we accept the assumptions of *prong 2,* now have a significant part to play in generating the evolutionary pattern. If the punctuational events are in fact so rapid as to be narrower than a bedding plane, as there is every reason to believe they are, then what we would see is the fossil record shown in Fig. 5 d, which is the one that punctuationists have told us is frequently observed: morphological species, rather distinct from those already in existence, appearing suddenly, remaining more or less static, and then disappearing. This picture is likely to be even more common than I have indicated, for when it comes to the fossil record, a great many species must be sibling species: with only a limited part of the morphology preserved in the fossils, a great many organisms that would have been readily distinguishable to a prehistoric biologist by their different color, soft parts, or behavior will present the paleontologist with identical skeletons. (How many species of birds or butterflies would we recognize if all we could see were the skeletons?) This of course accounts for the apparent absence (or at least scarcity) of punctuated phyletic change in the fossil record: naively, we would expect this to be revealed by the disappearance of the ancestral morphological type at the very moment the new morphological type appears [25]. If the ancestral type is hidden among its morphological siblings, we will be totally unable to detect its disappearance. Just how common such siblings are in the groups favored by paleontologists is a matter of dispute. Stanley [25] believes them to be largely absent in such groups as echinoids and mammals. Hecht and Hoffman (in preparation) are able to quote counterexamples such as the sibling species of the House Mouse.

The neo-Darwinian version of the evolutionary tree therefore explains the fossil record as well as does the Gould and Eldredge candelabra version. I have two reasons for preferring the neo-Darwinian version to the punctuational one. First, it abandons the unfounded assumption that – as morphological change cannot occur without cladogenesis, and hence sibling species are a trivial fluctuation – morphological change and cladogenesis are in effect correlated 1:1. I believe that this assumption has substantially confused the discussion on both sides. Second, it follows that no special mechanisms above the known microevolutionary processes are required to explain evolution at this level. Writers on punctuationism have of course differed (both a between- and a within-writer variance!) according to how exciting or orthodox they wanted to be, on the extent to which non-neo-Darwinian mechanisms might be needed to explain *prong 2*. What has not generally been stated is that *prong 2* at the very least requires a special "catch and release" mechanism that permits large and rapid evolutionary changes only when the tree branches. For the rest of evolutionary history, species are somehow restrained from evolving except slightly and gradually. No amount of talk about "no special speciation mechanism being required" for *prong 2* [10, 11, 25] will make this problem, that somehow branching releases the evolutionary restraints, go away (this perhaps is how the "misunderstanding" [11] over Goldschmidtian macromutations arose). The implication has been largely overlooked perhaps because of the normally unstated assumption – as Gould has often said, it is the unstated beliefs that are the most interesting and the most questionable – that large populations cannot undergo large or rapid evolutionary changes. Hence in the original Mayrian version of the punctuational theory [6], only small populations could undergo significant evolution, and these were of course also populations prone to both geographical and genetic isolation, and hence at a high risk of speciating. The problem for neo-Darwinists is that there is no such prediction about stasis in large populations, and therefore a need for a new and unexplored mechanism operating the catch and its release. And to be fair to the punctuational school, when they were in an "exciting" mood they have not been wanting is suggestions for such a mechanism [8]. Abandoning *prong 2* overcomes this difficulty. Whether once we admit punctuational *phyletic* change in this way, species still remain static for longer than we expect in view of the probable changes in their physical and biotic environments (how often are niches actually opened by changes in the ecological balance?) [25] is a question I must leave for future discussion (e.g., [32]).

In summary, I suggest that *prong 2* – and the candelabra tree – results from a naive and oversimple misinterpretation of the fossil record.

Species "Selection" and Macroevolution

Population geneticists, understandably nettled at the public announcement that they are a threatened species, have been prone to reject the whole punctuated equilibrium package. As somebody said, everything in history happens twice. Without judging which is the tragedy and which the farce, I suggest that the current debate between punctuationists and neo-Darwinists is a replay of the schism between the Mendelians and the Darwinists at the turn of the century [21, 29]. In that debate the Mendelians overestimated the magnitude and accuracy of evolutionarily useful mutation, and in their enthusiasm they encouraged the Darwinists in their belief that they could reject the whole mutationist package, thus preventing them from seeing, at least at first, that the Mendelians were filling a yawning gap in their own theory. In the present debate it is the third prong of the punctuated equilibrium fork that presents a valuable extension of evolutionary theory, by suggesting that there are patterns and processes that occur at a higher level and in the longer term than those that are studied by population geneticists.

First, the punctuationists have performed a valuable service in emphasizing that cladogenesis or speciation has to be seen, along with mutation and random genetic drift, as one of the nonadaptive processes of evolution. This was a familiar enough feeling to many population geneticists, but others, Dobzhansky in particular, have tended to see species as individually adapted entities, protected by their "isolating mechanisms" against the invasion of foreign genes as an individual is protected from foreign DNA by its immune system. Species, I agree, are not individual entities in that sense, and isolation itself not an adaptive process. Second, their suggestion should be well-taken that when whole species are extinguished this may generate patterns and trends in morphological evolution that have nothing to do with natural selection at the individual level. Processes may occur in evolution which are no more predicted by population genetic theory than is the behavior of organisms from theories of chemistry, or the properties of metabolic pathways from particle physics. If it happens that species of horse which are individually large become extinct at a slower rate than species which are individually small, then we will observe a trend toward increasing size in fossil horses. This will be true no matter how the differences between the species are generated, whether by gradual evolution, or by neo-Darwinian punctuational change, or by the evolutionary jerks of standard punctuated equilibrium theory [25]. What is more, this trend will be truly nonadaptive in every sense and should be regarded not as is conventionally stated as "species selection," but as a form of random evolutionary drift, or (as came out during discussion at the workshop) as a kind of evolutionary hitchhiking.

"Selection" implies a functional relationship between the selecting agent and the character selected [13, 16, 29]. If species are indeed selected, they will be selected for their properties as species, such as their breeding system or population structure, not for the possession of a mimetic pattern or the number of toes they stand on. (Body size is an interesting intermediate case: it is an individual character, but on account of its inverse relationship with population number, it might appear to be "species selected.")

The Not-The-Punctuated-Equilibrium Theory

Evolutionary trends are therefore likely to be generated by a particularly knotty mixture of adaptive and nonadaptive processes: Gould [9, 10] has identified all three. Different rates of extinction and of speciation in different branches of the tree will create nonadaptive trends which we can describe, in general, as due to random species drift. Individual natural selection will generate slow gradual changes. Trends will also, Gould [9, 10] points out, be generated by any bias in the direction of the evolutionary jerks or punctuational events. The cause of such a bias in direction is, if punctuational events are generally of the type I have demonstrated in *Heliconius*, likely to be none other than the process of natural selection as population geneticists have always understood it. Gould's "bias theory," I believe, flies the theory of punctuational evolution firmly home to the ark of neo-Darwinism. Whatever pattern is imposed on diversity by the "second process" of *prong 3,* the diversity itself, generated by the familiar process of individual selection and whatever other processes it interacts with (chiefly mutation and random drift), must be predominantly the outcome of individual adaptation.

Of the three prongs of punctuated equilibrium theory, then, the first is acceptable as far as observation of the fossil record shows it to be true. The second is a naive misinterpretation of this fossil evidence, which can be interpreted more satisfactorily in neo-Darwinian terms. The third is an important extension of evolutionary theory into a hitherto little explored territory [11]. The punctuationists are not the first nor the last explorers in human endeavor to have discovered something important without quite realizing what it was they had found.

Acknowledgements. I am most grateful to all workshop participants who discussed this paper and caused significant improvements in its presentation, particularly to J. S. Levinton, D. J. Futuyma, and D. Jablonski. It is also a pleasure to acknowledge my debt to many years of pleasurable collaboration with the late P. M. Sheppard, and with K. S. Brown and W. W. Benson, none of whom is however directly responsible for the more controversial parts of this paper.

References

1. Benson WW (1982) Alternative models for infrageneric diversification in the humid tropics: tests with passion vine butterflies. In: Biological diversification in the tropics, ed Prance GT, pp 608–640. New York: Columbia University Press
2. Brown JH (1971) Mammals on mountaintops: Non-equilibrium insular biogeography. Am Nat 105:467–478
3. Brown KS (1981) The biology of *Heliconius* and related genera. Ann Rev Entomol 26:427–456
4. Charlesworth B (1982) Neodarwinism – the plain truth. In: Darwin up to date, ed Cherfas J, pp 23–26. London: IPC Magazines Ltd
5. Charlesworth B, Lande R, Slatkin M (1982) A neo-Darwinian commentary on macroevolution. Evolution 36:474–498
6. Eldredge N, Gould SJ (1972) Punctuated equilibria: an alternative to phyletic gradualism. In: Models in paleobiology, ed Schopf JM, pp 82–115. San Francisco: Freeman Cooper
7. Gingerich PD (1985) Species in the fossil record: concepts, trends, and transitions. Paleobiology 11:27–41
8. Gould SJ (1980) Is a new and general theory of evolution emerging? Paleobiology 6:119–130
9. Gould SJ (1982) Punctuated equilibrium – a different way of seeing. In: Darwin up to date, ed Cherfas J, pp 26–30. London: IPC Magazines Ltd
10. Gould SJ (1982) The meaning of punctuated equilibrium and its role in validating a hierarchical approach to macroevolution. In: Perspectives on evolution, ed Milkman R, pp 83–104. Sunderland, MA: Sinauer Associates
11. Gould SJ (1985) The paradox of the first tier: An agenda for paleobiology. Paleobiology 11:2–12
12. Gould SJ, Eldredge N (1977) Punctuated equilibria: The tempo and mode of evolution reconsidered. Paleobiology 3:115–151

13. Hodge MJS (1986) Natural selection as a causal probabilistic and empirical theory. In: The probabilistic revolution. Ideas in the sciences, ed Gigerenzer G, Krüger L, Morgan M, vol 2. Cambridge, MA: MIT Press/Bradford Books, in press
14. Levinton JS, Simon CM (1980) A critique of the punctuated equilibria model and implications for the detection of speciation in the fossil record. Syst Zool 29:130–142
15. Lewin R (1980) Evolutionary theory under fire. Science 210:883–887
16. Maynard Smith J (1983) Current controversies in evolutionary biology. In: Dimensions of Darwinism, ed Grene M, pp 273–286. New York: Cambridge University Press; Paris: Editions de la Maison des Sciences de L'homme
17. Owen DF (1974) Exploring mimetic diversity in west African forest butterflies. Oikos 25:227–237
18. Papageorgis C (1975) Mimicry in neotropical butterflies. Am Sci 63:522–532
19. Plowright RC, Owen RE (1980) The evolutionary significance of bumblebee color patterns: a mimetic interpretation. Evolution 34:622–637
20. Prance GT (ed) (1982) Biological diversification in the tropics. New York: Columbia University Press
21. Provine WB (1971) The origins of theoretical population genetics. Chicago: Chicago University Press
22. Punnett RC (1911) Mendelism, 3rd ed. London: Macmillan
23. Sheppard PM, Turner JRG, Brown KS, Benson WW, Singer MC (1985) Genetics and the evolution of muellerian mimicry in *Heliconius* butterflies. Phil Trans Roy Soc Lond B 308:433–607
24. Stanley SM (1979) Macroevolution. San Francisco: Freeman
25. Stanley SM (1982) Macroevolution and the fossil record. Evolution 36:460–473
26. Turner JRG (1977) Butterfly mimicry: the genetical evolution of an adaptation. Evol Biol 10:163–206
27. Turner JRG (1981) Adaptation and evolution in *Heliconius*: a defense of neoDarwinism. Ann Rev Ecol Syst 12:99–121
28. Turner JRG (1982) How do refuges produce biological diversity? Allopatry and parapatry, extinction and gene flow in mimetic butterflies. In: Biological diversification in the tropics, ed Prance GT, pp 309–335. New York: Columbia University Press
29. Turner JRG (1983) "The hypothesis that explains mimetic resemblance explains evolution": the gradualist-saltationist schism. In: Dimensions of Darwinism, ed Grene M, pp 129–169. New York: Cambridge University Press; Paris: Editions de la Maison des Sciences de L'homme
30. Turner JRG (1984) Darwin's coffin and Doctor Pangloss – do adaptationist models explain mimicry? In: Evolutionary ecology, ed Shorrocks B, pp 313–361. Oxford: Blackwell Scientific Publications
31. Turner JRG (1984) Mimicry: the palatability spectrum and its consequences. In: The biology of butterflies, eds Vane-Wright RI, Ackery PR, pp 141–161. New York: Academic Press
32. Turner JRG (1986) The evolution of mimicry: a solution to the problem of punctuated equilibrium. Am Nat, in press
33. Wallace AR (1865) On the phenomena of variation and geographical distribution as illustrated by the *Papilionidae* of the Malayan region. Trans Linn Soc Lond 25:1–71

Patterns and Processes in the History of Life,
eds. D. M. Raup and D. Jablonski, pp. 209–222. Dahlem Konferenzen 1986
Springer-Verlag Berlin, Heidelberg

Fossil Record of the Origin of Baupläne and Its Implications

J. W. Valentine

Dept. of Geological Sciences
University of California
Santa Barbara, CA 93106, USA

Abstract. Animal Baupläne and Unterbaupläne appear abruptly in the fossil record, mostly developing early in Phanerozoic time; metazoan radiations near the Precambrian/Cambrian boundary may have produced scores of phylum-level taxa, and during Cambrian and Ordovician time, hundreds of class-level taxa. Ancestors and intermediates are unknown or are conjectural as fossils. About one in ten invertebrate species is a known fossil, and sampling of the Cambrian and Ordovician faunas appears at least average for the Phanerozoic, so that many species representing ancestral and intermediate lineages should have been discovered if they could be fossilized as easily as the average species which is found. Transformations between Baupläne most likely occurred in small localized populations which evolved rapidly, perhaps the restructuring of a partitioned genome via regulatory changes, commonly creating heterochronies, can account for the large-scale but rapid morphological changes required. Virtual restriction of Bauplan origin to the early Phanerozoic may be owing to the open adaptive space and to simpler metazoan genomes at that time.

Introduction

At the upper levels of the taxonomic hierarchy, phyla- or class-level clades are characterized by their possession of particular assemblages of homologous architectural and structural features; in this paper, it is to such assemblages that the term Bauplan is applied. The Bauplan contains both ancestral and derived characters, the former commonly indicating alliances within the next higher taxon, the latter indicating membership in the clade. The derived characters of these Baupläne also suggest by their pattern of modification and their associations with novel characters, the pattern of relatedness among clade members at lower taxonomic levels; for these lower levels the modifications and novel characters are themselves derived features. For both ecological and evolutionary purposes it is best to regard the concept of Baupläne as embracing entire ontogenies rather than only the adult stages, and indeed, schemes of relatedness among phyla and classes traditionally employ developmental features.

Among the problems associated with the origin of Baupläne are questions as to how rapidly their characteristic features may be assembled and the extent to which these features are ancestral or derived. The answers to these questions are particularly important in placing constraints upon the evolutionary processes which are inferred to underlie cladogenesis at this high level. On these points the fossil record is of some help.

Pattern of Appearance of Baupläne in the Fossil Record

The Precambrian/Cambrian boundary, taken here as the base of the Tommotion stage and its equivalents, marks the first appearance of abundant remains of durably skeletonized organisms; skeletons representing several living phyla and probably some extinct ones are found in the earliest Cambrian stages. Only a single living phylum with a mineralized skeleton is not known from Lower Cambrian rocks (the Bryozoa, first found in the early Ordovician).

It is sometimes maintained that the Precambrian/Cambrian boundary represents only a time when mineralized skeletonization became possible and widespread, therefore giving rise to a Phanerozoic type of fossil record, but that it does not correspond with a particularly important phase of evolutionary history [12]. The distinctive approaches to mineralization which are pursued by different phyla are wholly consistent with the hypothesis that each higher clade evolved a skeleton independently, and thus the early Cambrian is certainly a time of widespread acquisition of this feature. However, there is strong evidence that the latest Precambrian and early Cambrian witnessed the most important evolutionary event during the entire history of the Metazoa [16, 20].

Durably skeletonized forms make up only about a third of the marine invertebrate species today. What was happening to the soft-bodied forms during the early Cambrian? A suggestion comes from the trace fossils, which includes record traces of the activities of vermiform and other organisms not likely to be numbered among the skeletonized fossils now known. The traces increase significantly in diversity and abundance across the Precambrian/Cambrian boundary [5]. Even more impressive evidence is furnished by the contrast in soft-bodied forms known from the Precambrian with those found in the Cambrian. Late Precambrian (Ediacaran or Vendian) faunas are composed chiefly of Cnidaria or of fossils which are plausibly interpreted as being of Cnidarian grade of organization. In addition there are a number of enigmatic forms which are probably triploblastic but which do not fit comfortably within modern phyla; some of these are annulated and elongate or shield-shaped and have been assigned to the annelida or arthropoda by some workers [8]. Our earliest glimpse of the soft-bodied fauna of the Phanerozoic is afforded by the spectacular Burgess shale fauna of Middle Cambrian age, which is teeming with phyla. In addition to undoubted representatives of six durably skeletonized and six soft-bodied living phyla, there are about a dozen distinctive forms which cannot be readily assigned to living phyla [4]. These unusual forms represent Baupläne of which we would not otherwise know.

It thus appears that the Precambrian/Cambrian boundary marks a major watershed in the history of life, a time when a radiation of metazoan Baupläne occurred in unprecedented and unsurpassed numbers. Some of these Baupläne involved mineralized skeletons, as organ systems coadapted with the other major Bauplan systems; these created the conventional base of the Phanerozoic. Others of the new Baupläne were soft-bodied; they fossilized only under exceptional conditions and we have no real idea as to how many of them there might have been. It is certain that some phyla which were in existence at that time have not yet been found. The triploblastic stock from which mollusks and eucoelomate phyla arose must have been present before the Cambrian began; this stock is usually envisaged as flatworm-like. Yet no credible records of proto-flatworms or other possible ancestral triploblasts are known from the Precambrian or the Cambrian (indeed there are no verified fossil flatworms at all). If we add up the extinct and the extant animal phyla, both soft-bodied and durably skeletonized, which are known or must have been present in Cambrian time, the number comes to about thirty. If we suppose that some of soft-bodied phyla without Cambrian records (such as the ctenophora and the pseudocoelomates) were also present in Cambrian time but have escaped detection since they fossilize so poorly, then the number of Cambrian phyla rises to about forty-five. We have surely not managed to discover representatives of every last Bauplan from that period; the Burgess shale fauna alone has greatly increased our appreciation of the anatomical diversity achieved early in the history of the Metazoa; there is no telling what other faunas were in existence, though the many enigmatic small shelly fossils from the early Cambrian suggest that there was a richness of form then as well [13]. It is not unreasonable to postulate that sixty or seventy phylum-level Baupläne appeared during the late Precambrian/early Cambrian radiation; one hundred is by no means out of the question; hundreds, however, would seem to be too many.

Many of the Baupläne which appeared in the early Cambrian are represented then by two or more major subtaxa – classes or orders – with distinctive Unterbaupläne, even though peak diversity in classes and orders occurs later. For example, there were numbers of arthropod types which had novel combinations of features which ordinarily characterize class-level taxa in that phylum, numbers of echinoderm and mollusk classes, of brachiopod orders which might usefully be considered as classes, and so on; most phyla with good records had several such subtaxa [9]. To estimate the number of such subtaxa (and therefore Unterbaupläne) in the early Cambrian, it would probably be conservative to multiply the number of phyla by three or four. Thus the number of Lower Cambrian Unterbaupläne may have been a couple of hundred to a few hundred.

A striking aspect of the Lower Cambrian faunas is that although diversity of higher taxa was great, species diversity was low. This appears not to be an artifact of the record but rather to be representative of a biosphere with few provinces and relatively low species packing within ecosystems. Reconstructions of diversity patterns provide estimates of standing species diversities of only a few thousand during the earlier Cambrian stages [24]. Based on such figures, it has been pointed out that on the average every fortieth species or so represented a new class or phylum [23].

It is instructive to compare the early Cambrian radiation with that of the Triassic, because at the beginning of each of these periods there were relatively low species diversities. If estimates of the severity of the Permian/Triassic extinctins are approximately correct (between 91% and 96%), then there actually may have been fewer species in the early Triassic than in the early Cambrian. Yet the Triassic (and subsequent Mesozoic) radiations of metazoans produced no phyla and no classes, characteristic products of the Cambrian radiation, and far fewer orders as well; these radiations were simply not productive of Baupläne. We will return to these points later.

Completeness and Resolution in the Fossil Record

The fossil record is certainly both incomplete and incompletely known, and the available resolution of events must be rather coarse. Some feeling for the completeness and resolution of the Cambrian fossil record may be derived from data on the completeness of the stratigraphic record. Empirical studies have demonstrated that the average rate of deposition of stratigraphic sections varies inversely with the time span of the sections [17]. This is because the stratigraphic record contains a variety of gaps of different time values, and the greater the time span represented by sediments, the more chance that more and larger gaps are present; obviously, the more and longer the gaps, the lower the average rate of deposition of a section. It also stands to reason that longer events are more likely to be represented in the record than shorter events. It has been empirically determined (from measurements on thousands of sections) that for the sorts of shelf-depth rocks which yield early Cambrian fossils, a section representing 25 million years is on average about 53% complete at the 5 million year level; that is, events lasting 5 million years will be resolved about 53% of the time. Events lasting only 500,000 years will be resolved only 16% of the time, and those lasting 50,000 years, only 6% of the time [17]. The average metazoan species lasts perhaps 5 million years, so we can say that in a section representing 25 million years, there will be at least some sediment preserved which was deposited during the lifetime of about half the species which lived at that site. On the other hand, if speciations occur in relatively short time intervals, say below 50,000 years, we will rarely find any sedimentary record which intersects such a brief event.

An important feature of sedimentary completeness is that it is time-independent; environment for environment, a Cambrian section of a given time span will be just as complete on average as, say, a Cenozoic section [17].

It is clear that most sediments are unfossiliferous, so that merely having some sediment from the appropriate time is no guarantee of recovering fossils, let alone an adequate sample of the fauna. As yet there are no empirical methodologies for the analysis of fossil completeness, but some idea of the average level that may be expected can be garnered from estimates of Phanerozoic marine species diversity. The number of durably skeletonized invertebrate species which lived during the Phanerozoic may be about 2,000,000 [24]; there are about 200,000 such species described [14]. Thus we probably know about one in ten of the species which can be readily fossilized.

Is it likely that the completeness of the fossil record is time-dependent, with older fossils having a greater chance of being lost to science than younger ones? For one thing, we have in general more younger rocks exposed, since the older ones are frequently blanketed by the younger, and therefore more samples of younger faunas are available. On the other hand, for times when the biosphere was relatively homogenous (as, for example, when there were few provinces and faunal associations were widespread), it requires many fewer samples to represent the fauna than it does for times of biotic heterogeneity. Now, it happens that the Cambrian was a time of relatively low provinciality [24], with perhaps twenty fewer provinces than the late Cenozoic, for example, and therefore we may know the Cambrian fauna better than the faunas of many later times when provinciality was higher. At any rate, it is reasonable to believe that the durably skeletonized elements of the Cambrian faunas are at least as well-known as the Phanerozoic average, that is, one in ten species. The low species diversity of the early Cambrian faunas and the high ratios of Baupläne to species then imply that the production of Baupläne was not tied in with the production of species. For example, if we imagine that Baupläne were built up from species to species, and it took, say, 50 species to create the morphological distance required of a novel phylum, then we should see on average about five intermediate species per phylum, since we have one in ten species. But in fact we see no intermediate species at all; those species that we have are radiating within the Baupläne, not crossing from one to another. The same observation may be applied to classes; there are no confirmed (and hardly any possible) intermediates known. For some reason, the lineages leading to phyla and classes (and orders, for that matter) did not have a normal chance of fossilization.

It is useful to briefly evaluate the major factors which militate against lineages being represented in the fossil record. These include a) lack of preservation of sediments representing the lineage's habitat, which might be erosional, for example; b) lack of a durable skeleton; c) minute body size; d) small population size; e) localized geographic range; and f) rapid evolutionary change. Some of these factors can be eliminated as general causes of the nonappearance of intermediates.

Most of the late Precambrian and Cambrian rocks which have yielded the earliest records of Baupläne are interpreted as representing level-bottom, shallow-water environments [15]. The adaptations of most of the phyla and classes appearing during this interval are plausibly interpreted as suited for life in shallow water; the skeletonized phyla in particular seem primitively adapted for epifaunal life in shallow benthic communities. The evidence from trace fossils also suggests an early onshore radiation [18]. There is some evidence that important elements of each of the major faunas of the Phanerozoic first occupied inshore habitats and then later spread offshore to dominate major portions of shelves and platforms [10]. Thus, in the stratigraphic sections available we are sampling the habitats and communities wherein the origins and radiations of most of the durably skeletonized clades and, for that matter, of most invertebrate Baupläne are likely to have occurred.

While a case can be made that many of the phyla appearing in the early Cambrian had soft-bodied ancestors that may have been minute, there are many clades for which this cannot be true but which still lack any record of ancestral or intermediate forms. It has been argued [3, 22] that the brachiopod Bauplan cannot function without a durable skeleton, so that soft-bodied ancestors of this clade were *not* brachiopods, and that the earliest brachiopods had durable skeletons. The six classes of durably skeletonized Blastozoans which radiated during the Early Ordovician are another case in point. Indeed, we see no intermediates during the first 50 million years of the Ordovician although at least nineteen durably skeletonized classes appeared then. At least twenty-one durably skeletonized classes appeared during the Lower Cambrian, and a number of these (under current phylogenetic schemes) had skeletonized ancestors, but there are few if any intermediates known. Evidently even non-minute and durably skeletonized lineages which were ancestral to novel Baupläne or Unterbaupläne did not have an average chance of being fossilized. Thus we are left with a few properties which may account for the nonappearance of ancestors: small and/or localized and/or rapidly evolving populations would tend to have a less than average chance of fossilization, and if the ancestors combined some or all of these properties, they would produce the sort of fossil record which we have. Furthermore, ancestral alliances must have been of low diversity. One might expect lineages that were actively exploring novel Baupläne to undergo extensive radiation, but if they did so we should have more of them as fossils. The intermediate stocks must not have produced long-lived branches either.

Evolutionary Mechanisms

In the light of the pattern of appearance of Baupläne, and of our expectations concerning the completeness of the record, it is possible to test various hypotheses of Bauplan origin to see if they are consistent with the facts and quality of the fossil record. A widely held view has been that higher taxa, and therefore Baupläne, are created by the same processes of microevolution which are responsible for cladogenesis on the species level, but acting over long periods of time to produce great morphological divergence. Some idea of the plausibility of such an idea can be gained by considering rapid modes of morphological change in populations evolving via microevolutionary processes.

Certainly the problems involved in explaining the origin of a new Bauplan have similarities with those involving the origin of new species. The new Bauplan must originate in response to an ecological opportunity, and the major morphological changes involved imply that the opportunity lies on a different adaptive peak (or perhaps range would be better) than that occupied by the ancestor. To imagine an adaptive peak or range is to imply the presence of an adaptive valley or pass separating it from other prominces on the adaptive landscape. Such adaptive valleys have also been pictured as boundaries between adaptive zones; in either case they are thought of as barriers to evolutionary access. Selection will act to move populations higher onto adaptive peaks or to maintain them at local adaptive summits, but it cannot create pathways leading down the peaks and across the valleys. Therefore it is necessary to appeal to some nonselective source of evolutionary change. Usually genetic drift is the source invoked in these circumstances.

Drift is most effective in small populations, wherein high rates of evolutionary change are possible [11]. As most morphological characters are highly polygenic, mutation rates per character are much higher than are rates per locus, perhaps reaching 10^{-2} per character per generation [11]. Such high rates provide abundant potential variability and permit the creation of novel morphotypes, through drift, in hundreds to thousands of generations [11] – say, thousands to tens of thousands of years for an average invertebrate. Such a morphotype would be about as distinctive as the average sister morphospecies as normally recognized by taxonomists. If a lineage were to evolve continuously and unidirectionally at this rate, one may estimate that phylum- and class-level morphological divergence could be achieved in between one and two million years or so. However, continuation of unidirectional change via drift (without selective guidance) is not at all likely, and indeed, long continuation of rapid change in small populations involves difficulties. Deleterious mutations are common and in a small population they could not be scrutinized by selection but would accumulate in the gene pool until they constituted an unbearable drift load, when extinction would occur. From experimental evidence, this is likely to happen in hundreds to thousands of generations [11]. Thus, if a small population does generate the morphological distance normally associated with morphospecies, it must then grow significantly in size to permit selection to winnow the deleterious alleles from the gene pool.

Large populations can evolve rapidly, not via drift but under the influence of selection, and it is easier to account for a prolonged and rapid unidirectional change via selection as long as it involves a climb towards an adaptive peak. Large populations, however, are often widespread and certainly are more likely to enter the fossil record than are small localized ones. To generate a novel Bauplan, then, evolutionary processes must move lineages down from an adaptive peak into and across an adaptive valley while generating enough morphological distance that when their populations do enter the fossil record they are considered to have attained a new Bauplan.

Partly because of the difficulty of accounting for the lack of a fossil record of the origin of Baupläne, a macroevolutionary explanation for Bauplan origins has been proposed [19]. Evidence for this process is based on the observation that some species originate quite rapidly (we cannot resolve the events in the record) but yet last with little morphological change for millions of years, only then to disappear abruptly without direct evidence of their phyletic transformation into any surviving forms [6]. This is the well-known punctuational pattern. Since speciation occurs within an ambient environmental mosaic, the new species are only by chance preadapted to oncoming conditions. Differential extinction of the less preadapted occurs, and only the survivors can produce still more species, some of which will turn out to be adapted to the next set of oncoming conditions. This differential survival of fortuitously adapted lineages is termed species selection [19], and it has been suggested that it can lead rapidly to novel morphologies. There is little support from the fossil record for this conjecture, however. As mentioned earlier, the time of origin of new phyla is a time of low species diversity, and the rates of speciation were low in absolute terms as far as we can tell; species selection requires high rates. Furthermore, the numerous intermediate steps postulated, implied to be represented by successful (and presumably large) populations in evolutionary stasis, are not to be found [23].

There are some sources of genetic variation other than recombination and structural gene mutation, and these may result in large morphological shifts but which do not create permanent loss of fitness or adaptation (some bottlenecking is permissible). One possible source of rapid morphological change is mutation in genes having regulatory effects. Such mutations may create heterochronic changes in development which would have little effect on fertilization and create little genetic load. It is possible that some such mutations could result from transpositions or from lateral transfers of control sequences or of genes. Lateral transfers between species are now well documented within members of the same class [2], and even transfers between different kingdoms are strongly suspected [1]. In any case, the likelihood that there are mechanisms which can rapidly produce major alterations in genome structure and expression and yet not break the continuity of descent has greatly increased as the complexities of gene regulation, processing, mobility, transfer, and mutability have come to be appreciated. It seems likely that the operation of processes which reorganize the pattern of gene expression, perhaps in small populations, could rapidly propel lineages across both morphological and adaptive space. The working out of any such processes awaits the unravelling of eukaryotic patterns of gene regulation in development.

Baupläne on Other Levels

Do the patterns associated with the origins of marine invertebrate Baupläne resemble patterns associated with other major origins and radiations? It is interesting to examine this question as it applies to two cases: the origin of kingdoms, involving the highest taxonomic level of organisms; and the origin of mammalian orders, the best-known of the major terrestrial radiations.

Prokaryotes existed alone for over 2 billion years, finally giving rise to eukaryotes near 1.5 billion years ago. So far as we can tell, eukaryotes began to produce truly multicellular clades almost immediately, geologically speaking, and continued to do so for some time. At least seventeen multicellular clades (chiefly algae) arose independently from unicellular eukaryotes. The potential of eukaryotes to produce multicellular life forms, the seeming lack of this ability in prokaryotes, demand explanation. One possibility is that the prokaryote genome lacked the regulatory power and sophistication required to specify the differentiating and morphogenetic processes required for multicellularity, while the origin of the eukaryotic genome provided such regulatory power. Perhaps the original adaptive significance of the eukaryotic genome was to regulate symbiontic and eventually organellar functions harmoniously with host cell functions. If so, this regulatory ability can be regarded as a key innovation which led first to the radiation of multicellular algae and fungi, and eventually to animals and (finally) vascular plants. The radiations within the kingdoms – for example, the metazoan radiation – testify to the potential for modification of developmental patterns to produce distinctive body plans, which resided in the eukarotic genome.

At a much lower taxonomic category, mammalian orders exemplify the morphological potential of their grade and Unterbaupläne – from bats to whales – and repeat the pattern or rapid radiation and abrupt appearance of major groups at this lower level. Although one expects the gaps in the terrestrial fossil record to be more frequent and larger at a given frequency than in the marine record, it is still interesting that the phylogenetic relations among mammalian orders are largely unknown. Again, it appears that the extent and rapidity of the morphological modifications require relatively abrupt genome restructuring involving changes seemingly more profound than changes of structural gene functions alone could provide in a short time, and also that during whatever process occurred, the intermediate forms had less chance of being fossilized than the average species lineage. Curiously, the origin of the class Mammalia is represented in more detail by fossils than are the origins of the mammalian orders. Perhaps this is because the class originated as a novel physiological grade but not as a major morphological innovation.

Constraints on the Origin of Baupläne

Finally we may ask, if powerful processes of genomic and morphologic change exist, why have most Baupläne originated so early? Why have new ones not appeared during the Neogene, for example? Two sorts of explanations have been proffered. One is that any novelty which arises by essentially nonselective processes is likely to be poorly adapted, and therefore a significant amount of relatively open adaptive space is required to permit a novel Bauplan to become established [7, 20, 21]. During the late Precambrian, metazoan adaptive space was relatively unoccupied – there were many vacant peaks in the adaptive landscape – and novel adaptive types found little interference. Furthermore, as the first metazoans radiated into this adaptive space they had to "invent" new Baupläne in order to be able to exploit the adaptive opportunities which became available. Afterwards, when an array of Baupläne was established, it was commonly possible merely to modify one of the existing plans in order to occupy open adaptive space. Two lines of fossil evidence suport this view. First, it is when we judge adaptive space to have been most open, as during the early Phanerozoic, or following the invasion of a new ecological realm (such as the terrestrial environment), or following a major extinction that eliminated the occupants of an adaptive zone (such as the Paleozoic corals), that the more distinctive Baupläne or Unterbaupläne appear. Second, even when species diversity is low, as during the early Triassic, we do not find that many major new Baupläne (phyla or even classes) necessarily appear. It is clear that the early Triassic species, though few in number, were scattered through adaptive space, so that there was not the necessity of "inventing" entirely new Baupläne during the rediversification. Instead, adaptive space was refilled by elaboration and modification of the varied groups which were still represented, and novelties were restricted to cases involving the reinvasion of adaptive zones which happened to become entirely emptied.

A second sort of explanation for the restriction of major evolutionary "inventions" to early times is that genomes may have evolved so as to become less forgiving of major restructuring [25]. It is certain that many genes, including many multigene families, originated after the early metazoan radiations. We do not yet know just how much simpler early metazoan genomes might have been; this is an extremely important subject which may well be within the reach of experiment soon. If those genomes were significantly more compartmentalized, with fewer epistatic complications, than those of today, then major alterations in a given part of the developmental system would have been less likely to invoke deleterious side effects in other parts. Such compartmentalization would tend to permit extensive changes to occur with less cost to fitness and adaptation and, therefore, more rapidly.

If these explanations are on the right track, then there is a possibility of understanding the factors which have determined the number of Baupläne produced. Basically, Bauplan numbers may be determined by the rate at which new Baupläne are developed relative to the rate at which "old" Baupläne are modified and diversified, as adaptive space fills [21]. If diversification of established Baupläne proceeded rapidly, adaptive space would soon be filled enough that the potential for the appearance of novel Baupläne would soon be lost, whereas if established Baupläne diversified only slowly, ne Baupläne might appear over extended periods of time. The factors which govern these rates are probably associated with the strength and number of barriers or boundaries between adaptive zones – in other words, with the structure of the environment – and one imagines that the evolutionary response to these boundaries is highly stochastic. If the metazoan radiation were to be repeated from some datum in the Precambrian, it would seem unlikely that the same number of Baupläne would result.

If major Baupläne can be assembled rapidly, perhaps within hundreds of thousands of years or less, then they have a certain objectivity which they lack if they are members of a gradually evolving continuum. Furthermore, many of the features which characterize Baupläne can then be considered as derived features, assembled as coadapted complexes and not necessarily inherited via a long series of ancestors among which they were gradually introduced. Finally, while the fossil record does seem to suggest that evolutionary mechanisms have the creative power to produce novel Baupläne in little more time than most speciation events require, this does not mean that all or even most Baupläne have arisen at some such maximum theoretical pace. On the contrary, it seems likely that Baupläne, like species, have come into existence via a variety of modes.

Acknowledgements. Work on which this paper is partially based has been supported by the National Science Foundation (EAR81-21212) and NASA (NAG2-73). Thanks are due D. H. Erwin and T. D. Walker for many a stimulating discussion; this paper incorporates unpublished research in which D. H. Erwin collaborated.

References

1. Bannister JV, Parker MW (1985) The presence of a copper/zinc superoxide dismutase in the bacterium *Photobactorium leiognathi*: A likely case of gene transfer from eukaryotes to prokaryotes. Proc Natl Acad Sci USA 82:149–152
2. Benveniste RE (1985) The contribution of retroviruses to the study of mammalian evolution. In: Molecular evolutionary genetics, ed MacIntyre RJ. New York: Plenum
3. Cloud PE (1949) Some problems and patterns of evolution exemplified by fossil invertebrates. Evolution 2:322–350
4. Conway Morris S (1979) The Burgess Shale (Middle Cambrian) fauna. Ann Rev Ecol Syst 10:327–349
5. Crimes TP (1974) Colonisation of the early ocean floor. Nature 248:328–330
6. Eldredge N, Gould SJ (1972) Punctuated equilibria: an alternative to phyletic gradualism. In: Models in Paleobiology, ed Schopf TJM, pp 82–115. San Francisco: Freeman, Cooper
7. Erwin DH, Valentine JW (1984) "Hopeful monsters," transposons, and Metazoan radiation. Proc Natl Acad Sci USA 81:5482–5483
8. Glaessner MF (1984) The dawn of animal life, a biohistorical study. Cambridge: Cambridge University Press
9. House MR (ed) (1979) The origin of major invertebrate groups. London: Academic Press

10. Jablonski D, Sepkoski JJ Jr, Bottjer DJ, Sheehan PM (1983) Onshore-offshore patterns in the evolution of Phanerozoic shelf communities. Science 222:1123–1125
11. Lande R (1980) Genetic variation and phenotypic evolution during allopatric speciation. Am Nat 116:463–479
12. Lowenstam HA, Margulis L (1980) Evolutionary prerequisites for early Phanerozoic calcareous skeletons. Biosystems 12:27–41
13. Matthews SC, Missarzhevsky VV (1975) Small shelly fossils of Late Precambrian and Early Cambrian age: A review of recent work. J Geol Soc Lond 131:289–304
14. Raup DM (1976) Species diversity in the Phanerozoic: A tabulation. Paleobiology 2:279–288
15. Rozanov AY et al. (1969) Tommotski Yarus i problema nizheni Grantitsy Keembriya [Tommotion Stage and the Lower Cambrian boundary problem]. Trudy Geol Inst Akad Nauk SSSR 206:1–380 (in Russian)
16. Runnegar B (1982) The Cambrian explosion: animals or fossils? J Geol Soc Austral 29:395–411
17. Sadler DM (1981) Sediment accumulation rates and the completeness of stratigraphic sections. J Geol 89:569–584
18. Seilacher A (1964) Biogenic sedimentary structures. In: Approaches to Paleocology, eds Imbrie J, Newell ND, pp 296–316. New York: Wiley
19. Stanley SM (1979) Macroevolution, pattern and process. San Francisco: Freeman and Co
20. Valentine JW (1973) Evolutionary paleoecology of the marine biosphere. Englewood Cliffs, NJ: Prentice-Hall
21. Valentine JW (1980) Determinants of diversity in higher taxonomic categories. Paleobiology 6:444–450
22. Valentine JW (1981) The lophophorate condition. In: Lophophorates, eds Dutro JT Jr, Boardman RS, pp 190–204. University of Tennessee Department of Geological Sciences Studies in Geology 5
23. Valentine JW, Erwin DH (1983) Patterns of diversification of higher taxa: A test of macroevolutionary paradigms. In: Modalites, rhythmes et mecanismes de l'evolution biologique: Gradualism phyletique on equilibres ponctues?, ed Chaline J. Coll Int CNRS 330:219–223
24. Valentine JW, Foin TC, Peart D (1978) A provincial model of Phanerozoic marine diversity. Paleobiology 4:55–66
25. Wright S (1982) Character change, speciation, and the higher taxa. Evolution 36:427–443

Patterns and Processes in the History of Life,
eds. D. M. Raup and D. Jablonski, pp. 223–232. Dahlem Konferenzen 1986
Springer-Verlag Berlin, Heidelberg

Molecular Phylogeny

W. Nagl

Dept. of Biology, University of Kaiserslautern
6750 Kaiserslautern, F. R. Germany

Abstract. The evolution of the genome and its coding and non-coding components is reviewed, and their possible relationship to phenotypic changes in the phylogenetic process is discussed. Emphasis is given to the origin of a new scientific paradigm, shifting the importance from the Darwinistic view (random mutations, selection, adaptation) to a physical view (thermodynamics, determinism).

Introduction

The study of evolution, which ought to unify life sciences, has alsways been a center of conflict [44]. The reason for that conflict may be found in the fact that today most scientists are specialists with a very limited knowledge of biology on the whole, who criticize evolutionary theories from the generalist point of view. Therefore, I think that there is a strong imperative to make a synthesis, although I necessarily also shall take sides. The main problem is that evolution can take many forms, and it is just the question as to which one is seen to be general validity and importance. The central aim of this essay is to show that molecular phylogeny requires the acceptance of a new paradigm: it is not so much random mutation and adaptive selection which lead evolution at the basis, but some internal, physical determinism (constraints which still allow the development of unexpected novelty and of human freedom). Life – and the manifold forms of genomes and organisms – evolved of necessity on Earth due to causal natural laws as a specific feature of matter.

Physical Aspects of Evolution: the New Paradigm

External versus Internal Factors

Darwin relied heavily on Newtonian views of physics and of science in formulating his ideas about biological evolution. Biophysicists as early as von Bertalanffy (1933; in [42]) and Schrödinger [38] questioned this view, and today it is clear that there is a simultaneous increase of information and entropy with time. Because at the time at which Darwin proposed his theory of evolution there was no concept of intrinsic factors which could evolve, he postulated a process of extrinsic effects – natural selection, with the consequence of functionalism rather than structuralism. I recent years, however, evidence has accumulated that nonequilibrium evolution can be explained by intrinsically generated change and that extrinsic factors may just affect the resultant evolutionary pattern ("round off the edges" [18]), but the latter are neither necessary nor sufficient for evolution to occur, or to explain it. Evolution of matter (inanimate and living) is a mode of matter *constraining itself by itself,* not an outcome selected by something else [19]. It involves material self-assembly or self-organization in which the elements are constrained and regulated internally (this is an important point in understanding somatic differentiation of organisms).

Determinism

A breakthrough in understanding important aspects of evolution (and ontogenesis) arose with the mathematic formulation of self-assembly processes and self-organization (including autocatalysis) on the basis of "hypercycles" [6] and on the basis of nonlinear thermodynamics of open systems far from equilibrium [31]. As those self-assembly processes occur according to physicochemical laws, a principal difference from Darwinian ideas is that nonrandomness and determinism (at the physical, chemical, and organismic levels) represents the mainroot of evolution, while chance is just setting a fluctuation that allows the system to change into the next higher level (of energy, complexity, etc.). *This has nothing to do with "teleology,"* because it is not an aim but history, which determines the process.

The necessity to include such aspects in an evolutionary theory comes from the following: so far, biology has not found objective criteria for the evolutionary progress, but such a criterion may be given by physics on the basis of statistical interpretation of the external dissipation function. The latter is inversely related to the probability state of a thermodynamic system and can be taken as approximately equal to respiration intensity. On this basis Zotin, Konoplev, and others (in [17]) introduced the criterion of orderliness of biological systems. This criterion can well be used for the classification of organisms according to their progressive level of organization and complexity [18]. A similar approach was undertaken by Haken [9] with the concept of synergetics. A synergetic system composed of a high number of units can exhibit properties not exhibited by the single units alone. However, the nature of these properties is determined by the properties of the individual units, but in the complex, self-regulating system, cybernetic feedback operations and couplings operate which are not existent in the single units. Such principles of self-organization occur at the physical, chemical, and biological levels (e.g., laser, oscillating reactions, biological evolution) and are based on a few common principles of open systems with nonlinear dynamics, crossing critical values of parameters and cooperation of microprocesses, which can then be amplified to the macroscopic level (e.g., [14, 32]). Random events are, for those systems, just the trigger of fluctuations which shift the system (in a deterministic way of amplification) to a higher level of structure and order (energy). In this sense, quantum mechanics can describe how novel properties emerge in the course of time. The internal regulation that brings about material self-organization places a certain restriction upon the kinetic freedom the constituent elements can maintain. The decrease in the amount of internal freedom induces, however, a change in physical boundary condition [19].

Constraints: DNA Conformation

The constraints that govern evolution and lead to determinism are represented by the generally valid physicochemical laws and by the hierarchic systemic conditions of complex living systems. The system's conditions constrain the freedom of direction in which a system can evolve without losing its properties, the function of all parts (or subsystems such as organs, tissues, cells) together (see [36, 42]).

At the molecular level, DNA conformation and its basis, the lows of thermodynamics, electrodynamics, and stereochemistry, ultimately dictate constraints on the structure and function of genes and hence of cells and organisms. It has been well established that macromolecules, including DNA, carry the information required for their elements. The structure can be predicted by applying topological and thermodynamic rules for finding the energetically most favorable structure for a given sequence, whereby the solvents plays an important role. In addition, interplanar interactions of nucleic bases, the so-called (electronic) stacking interactions, make an important contribution to the stability of the structure and functional properties of DNA. Stacking interactions are characterized by negative values of change of entropy (ΔS), enthalpy (ΔH), and partial volume (ΔV). As in other systems, therefore, thermodynamic (and quantum mechanical, see below) constraints may have given, and may further give, a "direction" to evolution. If the evolutionary changes in the nucleotide sequence organization and DNA conformation is nonrandom, then this fact should be reflected in higher-order structures of organization. Actually, karyotype evolution exhibits not randomness but highly specific patterns according to the so-called "chromosome field" [18]. These findings led again to the suggestion that evolution cannot proceed by random mutation and selection but is governed by physical necessities as was the origin of life. The evolution of multicellular, highyl differentiated organisms required control systems which were effective at higher levels than the operon model mechanisms in order to achieve division of labor under very similar environmental conditions. It has been pointed out [23] that gene products cannot control differentiation and morphogenesis as basic mechanisms as their transcription requires a regulator itself, which is again a gene product and therefore requires regulators, etc., *ad infinitum*. Also, hormones can only "regulate" (better "induce" or "amplify") some events if the cell provides a receptor, i.e., the cell is already differentiated. These and many other aspects of developmental biology and regeneration indicate the necessity of regulatory information which is different from the genetic information. This opinion does not exclude the fact that hormones induce competence and hence the formation of further receptors, and that, in a determined or predifferentiated cell, autoregulative mechanisms are effective, as in bacteria.

Diversification of living matter during phylogenesis and ontogenesis is therefore thought to be the consequence of the evolution of a hierarchy of regulatory mechanisms rather than that of protein-coding genes. Such mechanisms involve promoter regions, TATA boxes, replication origins, transposable elements, splicing sequences, etc. The biological significance of the huge mass of non-protein coding DNA may, however, be seen in its (and the chromatin's) conformation changes according to thermodynamic and electrostatic rules. Hence, the information within the "conformational DNA" (see also [49]) can be calles the "thermodynamic code" [24, 25]. The non-protein coding or "conformational" DNA can no longer be interpreted as "selfish" or "parasitic" or "junk" DNA, the more so because the "conformational" DNA may have still more physical functions which are related to its excited states.

Excited States of DNA

It is generally acknowledged that geometrical and conformational properties of biopolymers have an important effect on their biochemical behavior. It is less easily recognized that these properties also depend on their macromolecular electronic characteristics [33]. Great biological significance can be seen in the fact that DNA acts as a photon trap and photon emitter [29]. In the short time between the absorption of a photon (quantum) by one of the bases of DNA and the final storage or emission, many electronic rearrangements can take place, various excited states of different multiplicities may be populated, energy may be transferred between different regions of the DNA, and some hydrogen bonds and event covalent bonds may be formed while others may be broken. The bases have much stronger interactions in the excited states than in their ground states, and these interactions can again profoundly affect the excited states.

It has now been well established that polynucleotides emit photons (due to the decay of excimers) at room temperature. Excimers (excited dimers) are complexes of two molecules, with an attractive binding potency, because at least one molecule is excited by photons. The most interesting feature is that excimer emission can compete with singlet emission and in some cases predominates. This emission can lead to structural information on the polynucleotide in solution, but it is also important since it is known that excimer formation represents an excitation energy trap.

These aspects have been extensively studied and reviewed by Popp and his co-workers [30]. The ultraweak photon emission evidently exhibits coherence. This laser-like radiation from DNA has been considered as an elementary process of matter in nonequilibrium states anad hence as an important basis of living systems. In this respect, the "conformational" DNA may also have an important function as a photon trap and hence may control many metabolic events via its physical property by emission of coherent electromagnetic waves.

The Driving Forces

If it is not adaptation that drives evolution, what is it? Gladyshev [7] suggested that it is tandemism that can be envisages as the "motive force." Tandemism means that all processes in nature are thermodynamically and kinetically interrelated with all types of quasi-equilibria being interdependent.

According to the biophoton hypothesis, increase in exciplex formation (and hence structural complexity) is based on the driving force of the Bose-

Einstein condensation. This means that the probability that a second boson (photon) will be scattered into the same quantum state as a first one is somewhat higher than its being scattered into a different quantum state [21].

Actually, our measurements have proven that the excitation temperature increases proportionally with the energy of photons [30]. This increase does not automatically lead to higher complexity of systems, but it represents the basis therefor by increasing its probability. Photon-induced high electron temperatures have been also envisaged as the driving force for the origin of life and increasing complexity of the cell [39]. It is time to overcome the arrogance of ignorance of these fundamental aspects of evolution! All these reasons make it necessary to see evolution from a physical point of view. In addition, the advantage of a physical view lies in the fact that physicists try to see things simply, to understand a great many complicated phenomena in a unified way, in terms of a few simple principles [45].

Finally, a personal word on how I see science today. Many modern biologists are dissatisfied with the reductionism existent in many disciplines. For those who are not *a priori* enamored of the paradigms of mechanistic causality and reductionism and who agree that theories of evolution (and differentiation, morphogensis, etc.) must operate within the framework of quantum mechanics, some "holistic" attempts are open for discussion. Structure alone – although perhaps the most important aspect – is insufficient to find the relevant macroconcepts when the system is far from equilibrium and in a highly excited state. From this viewpoint, the reductionist and holistic approaches supplement rather than contradict each other. What we really need is a new synthesis.

The New Synthesis

Charles Darwin was a unique genius with a very broad interest and knowledge. This is visible in his many books on plants and animals which, unfortunately, did not receive the same attention as his *Origin of Species*. If one reads the latter book carefully, it is evident that Darwin recognized some ploblems in his ideas, because he evidently knew nothing about genes nor, of course, about DNA. It seems to me that the overgeneralization of the mutation-adaptation-selection concept is the work of numerous evolutionary biologists who later referred to Darwin. The more the Darwinian view of the world was challenged, the more the neo-Darwinists struck back – and still do (e.g., [8]). Other authors, however, do not really find a contradiction [40, 41, 43]. Although I emphasized more the physical theories, I agree with those who share the opinion that we do not need a controversy on "-isms" but a new synthesis that includes both the truth of Darwinism and recent results. We will have to include new aspects and mechanisms if the evolutionary theory is to further evolve.

Stebbins [40] stated that "while change is always necessary, we need not cast aside the theory that was developed on the basis on half a century of research, and is essentially a synthesis of Darwin's insight, coupled with the experimental research of leading geneticists as well as of the more progressive systematists, paleontologists and molecular biologists. What we need, rather, is a shift in emphasis, and a new basis for collaboration." Before any such collaboration can be fruitful, both microevolutionists and macroevolutionists must be fully aware of the enormous difference in the time scales to which each of them is accustomed and of the completely different meanings of such words as "sudden" and "gradual." DNA and karyotype changes in cell cultures occur within months, but skeletal changes occur over thousands and millions of years! Moreover, "gradual" and "punctual" (sudden, saltatorial) speciation are not mutually exclusive. As in other aspects of phylogeny, different mechanisms can lead to the same goal [22]. In addition, there is still a lack in agreement on what a species is. Normally, discussions on the nature of speciation assume that species have an objective existence. But if species cannot be objectively defined and are seen merely as artificial constructs of subjective figments of the imagination of taxonomists, then speciation can hardly be said to be real process at all [28].

In any case, what has been found about DNA (genome) evolution, intrinsic directionality, and karyotype evolution forces us towards a consideration of hierarchy in the control of evolution by physicochemical constraints up to organismic systemic conditions. If so, then the question, "is a new evolutionary synthesis necessary?" asked by Stebbins and Ayala [41] must be answered yes. Such a "neo-synthesis" may contain additional theories of process and of new combinations of long recognized processes [43]. Some such new theories have been discussed in this essay. Another can be seen in that of the "molecular drive" [5]. In the case of many families of genes and non-coding sequences, fixation of mutations within a population may proceed as a consequence of molecular mechanisms of turnover within the genome. These mechanisms can be random and directional in activity. There are circumstances in which the unusual concerted pattern of fixation permits the establishment of biological novelty and species discontinuities in a manner not predicted by the classical genetics of natural selection and genetic drift.

Conclusions

A good scientific hypothesis should make predictions which can be falsified or verified by subsequent observations and experiments. This is difficult with any hypothesis on evolution, but it should be possible at the molecular level by the aid of molecular biology and gene technology. In addition, biophysics has developed new routines which allow the understanding of the increase in complexity of organisms, the underlying constraints which give a direction to this increase, and other aspects.

In his article "Some Physics Aspects for 21st Century Biologists," Rowlands [37] stated that something substantial is missing from the way in which we view cellular function (and evolution). The stability of the complex structure of cells and organisms is maintained not only against the statistical odds of entropy but also against the constant exchange of molecules. Every argument of the science of thermodynamics tells us that a living system is an extremely unlikely one. The hypothesis that the movement of biological molecules by diffusion and their interaction by chemical laws will provide the sufficient as well as necessary cause of all cellular mechanisms and of evolution is obsolete. We will have to reformulate many problems in terms of the collective and nonlinear phenomena now well-known to physics. Evolutionary biology can then be examined with respect to the kinds of laws that are possible in a domain where thermal fluctuations (mutations) have macroscopic effects and shift the system to a higher order.

Last but not least, we have to see that every theory on evolution is incomplete: "it is the theory which decides what we can observe" [10]; "the way we see things is affected by what we know or what we believe" [13]. Therefore, we should not be so arrogant as to emphasize "as soon as the total human genome is sequenced, we know that's a human being" (to avoid polemics, I will not give the reference). In contrast, we should be modest and honest enough to confess with Socrates, "I know that I don't know anything" – and that is particularly true for evolution!

Acknowledgements. I thank several colleagues, especially A. L. Panchen, F. A. Popp, and S. C. Stearns, for helpful discussion.

References

1. Ayala FJ (ed) (1976) Molecular evolution. Sunderland, MA: Sinauer
2. Britten R (1982) Genomic alterations in evolution. In: Evolution and development, ed Bonner JT, pp 41–64. Dahlem Konferenzen. Berlin, Heidelberg, New York: Springer-Verlag

3. Caporale LH (1984) Is there a higher level genetic code that directs evolution? Molec Cell Biochem 64:5–13
4. Davidson EH (1982) Evolutionary change in genomic regulatory organization: speculations on the origins of novel biological structure. In: Evolution and development, ed Bonner JT, pp 65–84. Dahlem Konferenzen. Berlin, Heidelberg, New York: Springer-Verlag
5. Dover G (1982) Molecular drive: a cohesive mode of species evolution. Nature 299:111–117
6. Eigen N, Schuster P (1978) The Hypercycle. Berlin: Springer
7. Gladyshev GP (1982) Classical thermodynamics, tandemism and biological evolution. J Theoret Biol 94:225–239
8. Grant V (1983) The synthetic theory strikes back. Biol Zbl 102:149–158
9. Haken H (1980) Dynamics of synergetic systems. Berlin: Springer
10. Heisenberg W (1971) Physics and beyond. London: George Allen and Unwin
11. Hinegardner R (1976) Evolution of genome size. In: Molecular evolution, ed Ayala JF, pp 179–199. Sunderland, MA: Sinauer
12. Ho M-W, Saunders PT (1984) Beyond Neo-Darwinism. London: Academy Press
13. Hughes AJ, Lambert DM (1984) Functionalism, structuralism, and "ways of seeing." J Theoret Biol 111:787–800
14. Jetschke G (1983) Prinzipien der spontanen Strukturbildung in Physik, Chemie und Biologie. Biol Rdsch 21:73–92
15. Kimura M (1968) Evolutionary rate at the molecular level. Nature 217:624–626
16. King MC, Wilson AC (1975) Evolution at two levels. Molecular similarities and biological differences between humans and chimpanzees. Science 188:107–116
17. Lamprecht I, Zotin AI (eds) (1978) Thermodynamics of biological processes. Berlin: Walter de Gruyter
18. Lima-de-Faria A (1983) Molecular evolution and organization of the chromosome. Amsterdam: Elsevier
19. Matsuno K (1984) Protobiology: a theoretical synthesis. In: Molelcular evolution and protobiology, eds Matsuno K, Dose K, Harada K, Rohlfing DL, pp 433–464. New York: Plenum
20. Mayr E (1984) Die Entwicklung der biologischen Gedankenwelt. Berlin: Springer
21. Mishra RK, Bhaumik K, Mathur SC, Mitra S (1979) Excitons and Bose-Einstein condensation in living systems. Intl J Quant Chem 16:691–706
22. Nagl W (1978) Endopolyploidy and polyteny in differentiation and evolution. Amsterdam: Elsevier/North-Holland
23. Nagl W (1979) Seach for the molecular basis of diversification in phylogenesis and ontogenesis. In: Genome and chromatin: organization, evolution, function, eds Nagl W, Hemleben V, Ehrendorfer F, pp 3–24. Vienna: Springer
24. Nagl W (1982) Condensed chromatin: species-specificity, tissue-specificity, and cell cycle-specificity, as monitored by scanning cytometry. In: Cell growth, ed Nicolini C, pp 171–218. New York: Plenum
25. Nagl W (1983) Evolution: theoretical and physical considerations. Biol Zbl 102:257–269
26. Nagl W, Jeanjour M, Kling H, Kühner S, Michels I, Müller T, Stein B (1983) Genome and chromatin organization in higher plants. Biol Zbl 102:129–148
27. Ohno S (1982) Evolution is condemned to reply upon variations of the same theme: the one ancestral sequence for genes and spacers. Persp Biol Med 25:559–572

28. Paterson HEH (1981) The continuing search for the unknown and unknowable: a critique of contemporary ideas on speciation. S Afr J Sci 77:113–119
29. Popp F-A, Becker G, König HL, Peschka W (eds) (1979) Electromagnetic bio-information. Munich: Urban and Schwarzenberg
30. Popp F-A, Nagl W, Li H-H, Scholz W, Weingärtner O, Wolf R (1984) Biophoton emission. New evidence fo coherence and DNA source. Cell Biophys 6:33–52
31. Prigogine I (1955) Thermodynamics of irreversible processes. New York: J Wiley
32. Prigogine I, Stengers I (1984) Dialog mit der Natur. Munich: Piper
33. Pullman B (1981) Aspects of the macromolecular structure of the nucleic acids. Ann NY Acad Sci 367:182–191
34. Raff RA, Kaufman TC (1983) Embryos, genes, and evolution. New York: Macmillan
35. Ridley M (1985) The problems of evolution. London: Oxford University Press
36. Riedl R (1977) A systems-analytical approach to macro-evolutionary phenomena. Q Rev Biol 52:351–370
37. Rowlands S (1983) Some physics aspects for 21st century biologists. J Biol Phys 11:117–122
38. Schrödinger E (1945) What is life? London: Cambridge University Press
39. Smith TF, Morowitz HJ (1982) Between history and physics. J Molec Evol 18:265–282
40. Stebbins GL (1982) Perspectives in evolutionary theory. Evolution 36:1109–1118
41. Stebbins GL, Ayala FJ (1981) Is a new evolutionary synthesis necessary? Science 213:967–971
42. von Bertalanffy L (1970) Gesetz und Zufall: Systemtheorie und Selektion. Vienna: Molden
43. Vrba ES (1982) Darwinisms in 1982: triumph and the challenges. S Afr J Sci 78:275–278
44. Watts RL (1971) Genes, chromosomes and molecular evolution. In: Biochemical evolution and the origin of life, ed Schoffeniels E, pp 14–42. Amsterdam: North-Holland
45. Weinberg S (1980) Conceptual foundations of the unified theory of weak and electromagnetic interactions. Rev Mod Phys 52:515–523
46. Williamson PG (1981) Palaeontological documentation of speciation in Cenozoic molluscs from Turkana basin. Nature 293:437–443
47. Wilson AC (1976) Gene regulation in evolution. In: Molecular evolution, ed Ayala FJ, pp 225–234. Sunderland, MA: Sinauer
48. Wilson AC, Carlson SS, White TJ (1977) Biochemical evolution. Ann Rev Biochem 46:573–639
49. Zuckerkandl E (1981) A general function of noncoding polynucleotide sequences. Molec Biol Rept 7:149–158

Standing, left to right:
Wolfgang Stinnesbeck, Wayne Sousa, Michael Soulé, Jack Sepkoski, Karl Flessa

Seated (center), left to right:
Dave Raup, Tony Hallam, Gerrat Vermeij, Heinz Erben

Seated (front), left to right:
Martin Hüssner, Ken Hsü, David Jablonski

Patterns and Processes in the History of Life,
eds. D. M. Raup and D. Jablonski, pp. 235–257. Dahlem Konferenzen 1986
Springer-Verlag Berlin, Heidelberg

Causes and Consequences of Extinction

Group Report

K. W. Flessa, Rapporteur
H. K. Erben
A. Hallam
K. J. Hsü
H. M. Hüssner
D. Jablonski
D. M. Raup
J. J. Sepkoski, Jr.
M. E. Soulé
W. Sousa
W. Stinnesbeck
G. J. Vermeij

Mass Extinctions, Extinction Events, and Background Extinctions

At the present time the criteria used to distinguish categories of extinction (e.g., mass vs. background) are arbitrary (Sepkoski, this volume). For a variety of reasons, not the least of which are the difficulties associated with the precise resolution of the duration of events in the fossil record, our understanding of the range of variation in extinction intensities is inadequate. Until such time that the durations of episodes of extinction are accurately known, the expression of extinction intensity as a rate (number or proportion/duration), while both possible and desirable, can be misleading. Consider a hypothetical case of two extinction events, each resulting in the extinction of ten species. If one event occurred within a stratigraphic interval of one million years while the other event fell within an interval estimated to be two million years in duration, a twofold variation in calculated rates would result. In any event, extinction intensities calculated as the magnitude of the event divided by the interval's duration will always be underestimates. Instantaneous events are constrained to appear as protracted events if their effect is averaged over a long sample interval.

The possibility remains that the record of Phanerozoic extinctions may be one of a spectrum of extinction intensities ranging from mass extinctions down to background extinction levels; this spectrum may or may not be continuous. A consensus was reached that the term mass extinction should be set aside for extinctions characterized by substantial magnitude and global extent, broad taxonomic effect, and relatively short temporal duration. We recognize that this definition is not an operational one inasmuch as it fails to specify precisely the magnitude, extent, breadth, and duration needed to qualify. Nevertheless, our failure to provide an operational definition is not likely to retard research into the subject. Research strategies will be governed by the nature of the extinctions themselves, not by our definition.

There is widespread agreement that the five major extinctions of the Phanerozoic – those occurring in the end-Ordovician, Late Devonian, Late Permian, Late Triassic, and end-Cretaceous – were phenomena that stood above and apart from the rest of Phanerozoic extinctions (see Fig. 1).

Episodes of extinction intensity that are intermediate between these five and background levels may be best thought of as "extinction events." Such extinction events may often be taxon- or region-specific and are exemplified by extinctions such as those of Cambrian trilobites at biomere boundaries, Mesozoic ammonites at various horizons, Late Pleistocene mammals of

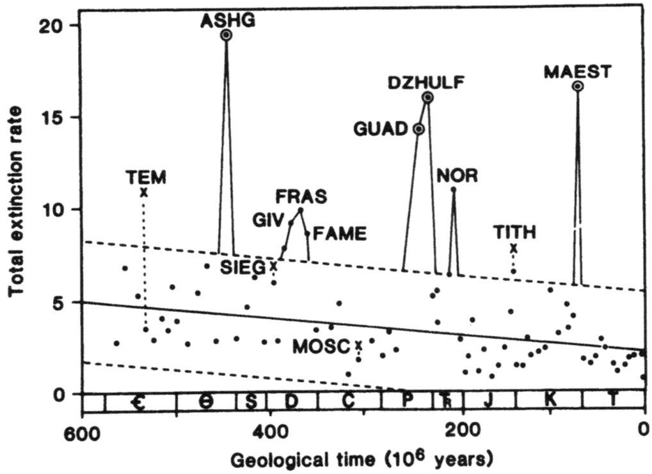

Fig. 1. Extinction rate (families per million years) of marine invertebrates and marine vertebrates through time. Mass extinctions occur in the Late Ordovician (*ASHG*), Late Devonian (*GIV, FRAS, FAME*), Late Permian (*GUAD, DZHULF*), Late Triassic (*NOR*), and Late Cretaceous (*MAEST*). X indicates extinction rate of stage if rarely preserved animals are included. So-called background rates occur within the dashed line on either side of the solid regression line [31]

North America, and Pliocene molluscs of the North Atlantic. Although such extinctions are clearly important events in the history of life, their apparently restricted taxonomic scope, geographic extent, and magnitude set them apart. Some of these taxon-specific extinction events may also be characterized by high origination rates, thus resulting in high evolutionary turnover with little effect on standing diversity. These events may be phenomena that will yield additional insight into those environmental factors that regulate evolutionary rates.

The time span over which extinctions are observed is important at many scales. Even so-called background extinctions, while seemingly continuous when compared to the five mass extinctions of the Phanerozoic, may be episodic if viewed at increasingly finer levels of temporal resolution. Considered on the year-by-year basis available in historic time, the extinctions of the past 300 years may also be episodic.

Less agreement characterized our search for qualitative differences between mass and background extinctions. Although it seems clear that mass extinctions often differ in their effects when compared to background extinctions, such differences could be attributed either to the operation of a threshold effect of to the imposition of causes of a fundamentally different nature. This is an issue to which we return later in this report.

We identified two areas in which future research would be likely to yield insight into the distinctions (if any) between and among mass extinctions, extinction events, and background extinctions.

The Nature of Variation in Extinction Intensity

Is the variation continuous or discontinuous? Critically important for such analyses are data derived from within as detailed a chronostratigraphic context as possible. Particularly useful analytical techniques for the study of the variation in extinction events may be those used in the study of flood and earthquake frequency. In addition, the application of polycohort survivorship techniques (Fig. 2) (for example, see [14, 28]) to taxon-specific analyses of extinctions through time may yield insight into the variation in extinction magnitude.

The Geographic Extent of Extinction

Several extinction events can be considered as "candidates" for mass extinction, yet their geographic extent is poorly documented (see Sepkoski, this volume). Attention should be directed toward such extinction events as the Late Cambrian biomere events, the Fammenian Tournasian, the Pliensba-

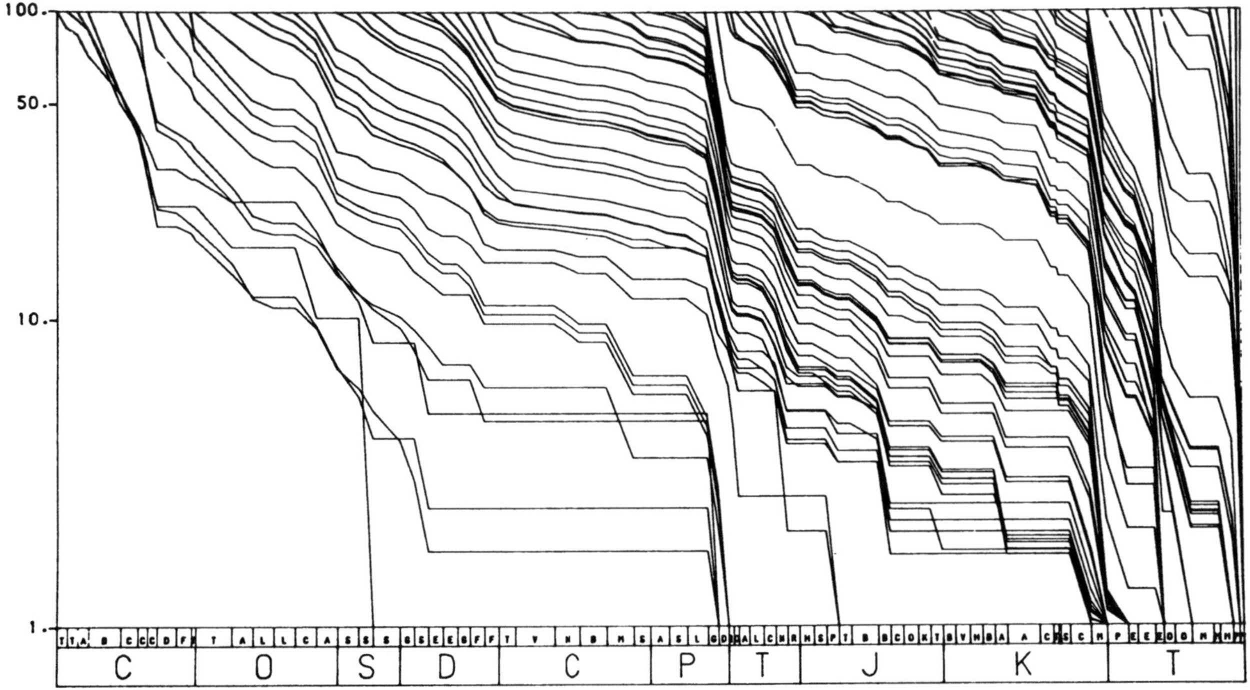

Fig. 2. Polycohort survivorship of 2316 extinct marine families during the Phanerozoic. Each curve traces the survivorship of families that were extant (the polycohort) during a geologic stage. Steep portions of the curve represent intervals of elevated extinction rates. Low slopes record intervals of low extinction rates. Because the same family may belong to several polycohorts (for example, a family present in both the Middle and Upper Devonian will belong to both polycohorts), the extinction of a single family may be reflected in more than one survivorship curve. This tends to exaggerate the intensity of episodes of mass extinction

chian-Toarcian, the Tithonian, and the Cenomannian-Turonian in an effort to assess their importance. Knowledge of the geographic extent of an extinction event may provide important clues as to its cause.

Extinctions During Historic Time

We have no knowledge of recent "natural" extinctions of entire species, at least those on continents and in the marine environment. Patterns and levels of extinction on contemporary islands can be instructive, but the extrapolation of the extinction of local populations to the global extinctions of species must be done with great care. The demise of species during historic time results almost exclusively from the direct and indirect consequences of human activities. The destructiveness of *Homo sapiens* may parallel some of the environmental catastrophes of the geologic past – estimates of the extinction of species resulting from the cutting of the South American rain forests approach values typical of the mass extinctions of the Phanerozoic. If the destructiveness of humans is comparable to the destructiveness and effects of purported meteorite impacts or other events that precipitate mass extinctions, we might be able to generalize about or predict the relative vulnerability of species of biotas to such prehistoric events.

Among contemporary species, there are certain life-history traits and extrinsic factors that appear to alter or affect the probability of extinction. These factors, endogenous or exogenous, deterministic or random, have been reviewed by many authors, including Simberloff (this volume). Many of these characteristics tend to occur as constellations within species, i.e., they are inter-correlated (see LaBarbera, this volume, for a review of those factors related to body size). To summarize these studies briefly, the best predictors of susceptibility to extinction based on our very limited knowledge of contemporary, anthropogenic extinctions of animals are a) large body size; b) low r (intrinsic rate of population increase); c) low N (population size); d) high variance of r or N; e) dependence on rare, dispersed, or ephemeral resources; f) poor or low vagility; and g) local endemism.

Exacerbating these factors would be a reduction in geographic range or effective population size. Should a species or one of its popuplations be reduced in numbers to the range 10^1 to 10^3 individuals, this will lead to inbreeding and a loss of genetic variation.

Recent reviews of the subject [1, 22] are approaching a consensus on the issue of heterozygosity: a significant loss of genetic variation by a population is frequently associated with a loss of immediate or short-term fitness of its individuals, expressed as decreased growth rate, metabolic efficiency, developmental homeostasis (e.g., bilateral symmetry), viability, and longevity. Inbreeding in species that do not normally undergo selfing or inbreeding is virtually always deleterious. The relevance of such consequences of reduced population size to understanding the causes of fossil extinction is not clear. The processes that result in the extinction of species having but 10^1 to 10^3 individuals (such species might be virtually "invisible" in the fossil record) may be quite different than the processes that initiate the decline in population size to such a precarious level.

Extinctions during historic time can be attributed to a) competition, b) predation, c) random population fluctuation, and d) habitat change. With the exception of island settings and direct competition with *Homo sapiens*, competition does not appear to play a major role in contemporary extinctions. Again, with the possible exceptions of islands and the hunting activities of historic and prehistoric humans, predation seems to be an uncommon cause of species extinctions. Random population fluctuations, while often the most proximal causes of species extinctions, probably play but a minor role in continent and continental shelf settings. Habitat alteration, either directly or indirectly induced by humans, seems to be the major cause of extinctions during historic time. In addition, the extinction of a key species within a community can lead to damaging effects on other species, perhaps resulting in an ecologically generated cascade of species extinctions.

Much of our data on prehistoric extinctions comes from the stratigraphic record of marine habitats, but we know virtually nothing about contemporary extinctions in such environments. Therefore, a major challenge to paleobiology lies in the prediction of the vulnerability of fossil marine taxa to changes in habitat size, temperature, oxygen tension, turbidity, seawater chemistry, and light. The characteristics of organisms that determine their susceptibility to those exogenous changes are metabolic rates, capacity to become dormant, resistance to sudden, variable, or prolonged changes in temperature, the ability to tolerate prolonged periods of starvation and to find and exploit isolated patches of resources, and the production of resistant and persistent dispersing offspring.

As Sousa's [39] review of the role of natural disturbance in extant communities shows, organisms have evolved mechanisms for coping with a wide variety of natural physical disturbances. Indeed, some species (i.e., fugitives, *sensu* Hutchinson [17]) have come to depend on them for their very persistence. However, the scales of disturbance that result in species extinction appear to be quite distinct from those to which these species have evolved.

Insight into the processes of extinction among fossil species might be gained from study of contemporary examples of large-scale habitat alteration, e.g., the ambitious experiments of the World Wildlife Fund in Brazil [23]. In addition, we expect that much can be gained from a systematic review of the literature concerning large-scale habitat alteration in such marine settings as the Baltic and the Mediterranean Seas.

The Record of Extinctions: Issues of Quality and Temporal Pattern

The past few years have seen considerable progress in our understanding of the character of extinctions, and the plausibility and persuasiveness of many extinction models is now considerable. We should not lose sight of the fact that our evidence on the precise timing, duration, and faunal effects of many extinctions is still remarkably limited. Consider, for example, that our knowledge of changes in the marine macrofossil record in the immediate vicinity of the Cretaceous-Tertiary boundary is based on very few sections. In addition to this limited degree of geographic documentation we should recognize that the stratigraphic and chronostratigraphic resolution of our data is very uneven. While the range terminations of many taxa are known with confidence, other precise age assignments are precarious, ambiguous, and almost always disputable. For example, the precise stratigraphic level of the final extinction of the ammonites remains a controversial issue. This situation requires an intensification of our empirical work, with detailed work on new as well as previously studied stratigraphic sections that contain important extinction horizons. Such careful stratigraphic work may provide important constraints and tests for many extinction models.

Our ability to discern the abruptness of an extinction in the fossil record is dependent on stratigraphic and paleontologic resolution. Stratigraphic incompleteness, diagenesis, facies changes, and biogeographic changes can cause gradual extinctions to appear as if they were sudden events of great magnitude. Two factors can conspire to *reduce* the observed magnitude of an event: reworking of fossils up a stratigraphic section and "backward smearing." Bioturbation, especially in deep-sea sediments sampled in cores, tends to cause the diffusion of fossils upward over scales of centimeters to decimeters, and together with resedimentation can cause microfossils to occur at horizons slightly above their actual level of extinction. Geochemical signatures within the tests of these microfossils may provide a means for the recognition of such upward mixing. More serious is the backward smearing of observed extinction resulting from incomplete sampling of the fossil record. In most paleontologic analyses, extinction events are assessed from the observed stratigraphic ranges of species or higher taxa; these ranges are simply the time range between the first and last observations of a taxon either in a local stratigraphic section or in a compilation based on the correlation of numerous sections. However, observed ranges are always minimum estimates of actual ranges, and the failure to sample a taxon over all of its true range will cause truncation at both the bottom and top of the actual range. Signor and Lipps [37] have shown by means of statistical models that the failure to sample taxa at the precise time of their last appearance can quickly alter the appearance of an abrupt extinction event in stratigraphic range charts, causing observed numbers of extinctions to increase gradually well before (sometimes several stages before, in the case of very large extinction events) and then accelerate just before the actual horizon of the mass extinction. This pattern will be further exacerbated if there are any facies changes below the extinction horizon. Jablonski [18] and Waterhouse and Bonham-Carter [48] have shown empirically that this backward smearing does indeed occur in the fossil record. In Jablonski's analysis of the "Lazarus effect," he showed that taxa that actually survived an extinction event "disappeared" well below the extinction event, only to "reappear" above it. The artificial last occurrences were smeared backward in the record as predicted (see also Jablonski, this volume). Once recognized, such artificial extinctions can be used to place confidence limits on the observed pattern of final extinction in critical time intervals [18].

Because an important current issue in the study of extinctions involves their temporal distribution, the quality of the geologic time scale will affect our effort to detect any periodic pattern in the record of extinctions. Inasmuch as the search for periodicity in the record of extinctions has focused on the record of the past 250 myr, the quality of the time scale for the Cenozoic and Mesozoic is especially important. Although the Cenozoic time scale seems well established, the Mesozoic time scale prior to the mid-Cretaceous is currently rather poor. This is because of an insufficient number of biostratigraphically well located, reliable radiometric dates and a lack of correlation with seafloor magnetic anomalies. In consequence, the range in age of some stages among several recently proposed time scales is on the order of five million years and is as much as 14 million years for the Jurassic-Cretaceous boundary. It is clear that there is an urgent need to obtain many more radiometric dates and to refine methods of establishing time scales through the use of ammonite chrons as the minimum unit of stratigraphic subdivision.

Taking into account reasonable estimates of sedimentation rates in deep-sea cores, the age of the Cretaceous-Tertiary Boundary event perhaps can be estimated on the basis of sedimentation rates to within a few thousand years or even less. But nowhere else in the Phanerozoic has a major extinction event been dated to such a degree of precision. Magnetic stratigraphy has proven useful, especially for correlation between marine and continental sections, but here the time resolution is commonly on the order of 0.5 myr. Thus, it remains an article of faith that the dinosaur and calcareous plankton extinctions of the Cretaceous-Tertiary were precisely synchronous, as required by certain extinction scenarios.

Our discussions of patterns in the record of extinctions focused almost exclusively on the detection and significance of patterns in the timing of mass extinctions and extinction events during the past 250 million years. Little attention was paid to potential patterns in the variation of extinction intensity through time. In part this may be due to the difficulty in providing confident estimates of the intensities of extinctions, in part to the allure of the issue of periodicity.

Are extinction events uniformly spaced (periodic) in geologic time? Claims of regular periodicity in the marine fossil record have been published by Fischer and Arthur [6], Raup and Sepkoski [32], Rampino and Stothers [27], and Kitchell and Pena [21] (a pseudo-periodicity, in their analysis) with periods of 32, 26, 30, and 31 myr, respectively. Discussion of these claims raises complex questions of statistical procedure and of uncertainties in the empirical data [13]. The more fundamental question, however, is whether the distribution of extinctions in time is a consequence of a) many independent causes operating in an unpredictable fashion, or b) a single driving mechanism or ultimate cause. If the former is true, and this has been the conventional wisdom, one can predict that the extinction events should show a random spacing in time; but it can be shown that the major extinctions have a distinctly nonrandom distribution. That is, they are more evenly spaced than is typical of random distributions. For extinctions from Permian time to Present, the nonrandomness appears statistically highly significant. Furthermore, stationary periodicity yields excellent fits to the extinction data at both generic and family levels for fossil marine animals. This does not prove that the simple periodicity is the best description of the extinction pattern because it is logically impossible to test all nonrandom patterns that could describe the actual distribution. However, given a choice between randomness and periodicity, tests of the data suggest periodicity, and periodicity thus emerges as the hypothesis to investigate.

As discussed above, the quality and temporal precision of the data base is a persistent issue. In the face of uncertainties over the data, two approaches are possible: a) the use of all available data in the hope that if the pattern of periodicity is strong enough, the signal will be apparent, even in the degraded data at hand, and b) the use of high-quality subset of the whole data base, culled so as to remove the most untrustworthy data. It is impressive that both approaches reveal a more or less uniform spacing between extinctions [35]. Uncertainties in the timing of extinctions persist, but errors in their age assignments would tend to decrease the chances of detecting a periodic signal and are not likely to generate or enhance a periodic signal. By the very nature of statistical tests, it is an easier task to reject the null hypothesis of randomness than it is to accept the conclusion of periodicity. Considering the work on this issue to date, either the patterns detected in the record of extinctions of the past 250 myr are robust or the statistical tests are not very sensitive (see Connor, this volume, for a review of analytical techniques useful for time-series data).

The periodicity issue is very important. This is not because it favors one causal hypothesis or another. Rather, it is important because it would suggest that most extinctions have similar causes and that the causes of one extinction, if discovered, would inform us about the causes of other extinctions.

Three research programs are likely to prove especially important in the study of temporal patterns of extinction.

Refinement of Taxonomic and Stratigraphic Data

Detailed local studies in the vicinity of extinction events and careful global compilations of such local studies will reveal crucial data on the timing and intensity of extinction. The need for such study is especially acute in the Paleozoic, where a periodic signal has yet to be detected.

Detailed Examination of the Fabric of Periodicity in the Marine Record

Efforts to detect periodicity within subsets of the marine fossil record may prove instructive. If the periodic pattern is particularly characteristic of certain clades or groups of clades, the paleobiology of those groups may suggest a cause of the pattern.

Extinctions in the Terrestrial Realm

The record of extinctions of terrestial plants and animals should be examined for evidence of synchronicity with the marine record and for periodicity. The presence or absence of periodicity in the terrestrial realm will have profound consequences for our understanding of the causes of global extinctions.

Victims and Survivors: The Selectivity of Extinction

Why do some groups survive extinction events while others perish? As suggested in studies of contemporary extinctions, the biological properties of species may affect their probability of extinction. The fossil record also provides some information on this issue. Inasmuch as we can document the occurrence (if not often the cause) of more extinctions in the fossil record than have been documented in historic time, the fossil record may also provide information on the susceptibility of living species to extinction.

Some fossil groups, such as the ammonites and some Cambrian trilobites, show a "boom and bust" pattern of diversity. In these volatile clades high extinction rates are accompanied by high origination rates at most times. The paleobiology of these groups might suggest characteristics that result in high extinction rates.

The features that we identify below seem important in determining a taxon's probability of extinction. Few of these features can be said to be well established predictors of evolutionary survival. Nevertheless, each has some evidence in its favor, and while all deserve greater scrutiny and additional study, all support the proposition that extinction is not random with respect to a species' biology.

Geographic Range

It does seem well established that the geographic range of a taxon is correlated with its geologic duration (e.g., [3, 4]). Widespread taxa tend to be geologically long-lived, while geographically restricted, endemic taxa have relatively brief geologic durations. Recent evidence suggests, and it is reasonable to assume, that geographically restricted species (species on islands are good examples) tend to be more susceptible to the ravages of disease, accidents, predation, competition, and regional environmental change.

Among the factors that tend to result in a broad geographic distribution is dispersal ability. Because the larval shell morphology of bivalves and gastropods is often preserved on the hard parts of the adult, dispersal abilities can often be inferred directly from the fossil specimens themselves [19]. That dispersal ability, as expressed by larval shell type, is correlated with geographic range among extinct species of Cretaceous and Tertiary molluscs [12, 18] suggests that dispersal ability limits geographic range in evolutionary as well as ecological time. Yet a broad geographic distribution does not invariably result in a low probability of extinction. Mesozoic ammonites and bivalves such as *Monotis* and the inoceramids typically have broad geographic ranges, yet these groups are characterized by high extinction and high origination rates [9, 10].

Although a broad geographic distribution appears to confer extinction resistance on species during times of background extinction, analysis of species longevities and geographic distributions of Cretaceous-Tertiary bivalves and gastropods of the southern U.S. fails to reveal any correlation (Jablonski, this volume). In this case, the role of geographic distribution may express itself at a higher taxonomic level. Bivalve and gastropod genera with representatives in more than one province tend to survive the end-Cretaceous event, whereas those genera with species restricted to but one province tend to perish. This suggests that selection during mass extinction regimes may differ and may be expressed at higher hierarchical levels than selection during background times (Jablonski, this volume).

Body Size

Species of large-bodied individuals appear to become extinct with greater frequency than species characterized by small-bodied individuals. This pattern has been suggested in studies of both terrestrial vertebrates [44] and marine invertebrates [8]. Explanations for this pattern are varied, but the lower population densities, lower birth rates, and/or greater nutritional requirements of larger organisms are the most likely proximal reasons for greater susceptibility to extinction. That large body size could tend to increase the probability of extinction is an especially interesting hypothesis inasmuch as the prevalence of Cope's Rule (see LaBarbera, this volume) has suggested to many that there are advantages to the evolution of large body size. If Cope's Rule is generated through selective pressures for a larger body size (however, see [40] for a cogent, nonadaptive explanation for Cope's Rule), a correlation between body size and the probability of extinction would suggest that what may benefit individuals (size increase) may be harmful to the species. Thus, selection could be seen as operating in opposing directions at different hierarchical levels. Another apparent paradox that presents itself in this regard is the correlation between body size and geographic range among terrestrial mammals of North America [5]. It would appear that whatever extinction resistance is conferred by a broad geographic range, it is outweighed by the disadvantages of lower population densities associated with large size.

Tropical Setting

Although little systematic work has been done, it often appears that species inhabiting tropical habitats (especially those in reefs) are more extinction-prone than those in extratropical habitats. This effect is most apparent during times of mass extinction [36] but has not been well documented for intervals of background extinction. It is not clear if this association of high extinction rates with tropical habitats is due to a) the sensitivity of reef-building organisms to environmental change (with the species dependent on the reef-builders for their habitat becoming extinct as a cascading effect), b) the sensitivity of tropical species in general, c) the smaller geographic ranges of tropical species (McCoy and Connor [24] document smaller geographic ranges among tropical species of North American mammals), or d) is a simple consequence of the greater number of species in the tropics (i.e., the "selectivity" may be more apparent than real).

Productivity

Areas of low productivity might contain biotas that would be prone to environmental crises that affect primary productivity. Vermeij [46] and Vermeij and Petuch [47] note different extinction susceptibilities among Tertiary molluscs on the west and east sides of the Isthmus of Panama. They find that those species that inhabit high productivity waters are the ones most likely to persist, perhaps because their high fecundity enabled them to recover quickly.

Species-Richness within Clades

All other things being equal, clades characterized by many species will tend to persist for longer intervals of geologic time than will clades having only few species. This pattern is apparent at long time scales with more diverse classes persisting longer than less diverse ones [34] and at shorter time scales among genera of Gulf Coast (USA) bivalves and gastropods during times characterized by background extinction rates (though not during the Cretaceous-Tertiary mass extinction) (Jablonski, this volume). Increasing species richness within families through geologic time has also been offered as an explanation [7] for the decline in famililal extinction rates noted by Raup and Sepkoski [31] and Van Valen [45] (but see also Sepkoski, this volume).

In the group discussions Bambach reported on some preliminary results of a study he is conducting with Gilinsky. Early results suggest that among clades of Phanerozoic marine invertebrates, the probability of family extinction is higher during mass extinctions that occur late in the history of the clade than during mass extinctions that occur early in the history of the clade. These results deserve far greater scrutiny and study, especially because they suggest that a family's probability of extinction during mass extinction times is dependent on how long its clade has been in existence.

Two research programs for exploring the issue of selectivity of extinctions were suggested.

Comparison of Extinction Intensities in Marine and Terrestrial Habitats

Terrestial habitats may be characterized by higher extinction rates than marine habitats because of the greater likelihood of restricted geographic distributions in the terrestrial realm. Marine habitats, because of their greater absolute extent and because of the interconnection of the oceans, may house species with greater geographic ranges. Furthermore, the terrestrial realm may be more frequently disturbed by global environmental changes whose effects are buffered in the aquatic environment.

Comparison of the Victims and Survivors of Both Background and Mass Extinctions

In light of the patterns discussed above, how do victims and survivors differ with respect to their geographic distribution, body size, feeding adaptations, trophic positions, substrate adaptations, biogeographic affinities, species richness, and other features? Is there any systematic variation in the intensity of extinction with respect to these categories? Are the patterns of selection different during background and mass extinction times? Are the patterns of selectivity different at different mass extinctions?

Causes of Mass Extinction in the Geologic Past

Physical factors, even though they can provoke a series of biotic changes in the environment, seem the most likely proximate causes for extinction.

The biotic changes that are associated with extinctions in the fossil record, with the possible exception of some floral changes (Niklas, this volume), are difficult to attribute to the direct effects of competition.

Changes in the physical environment must have global consequences if they are to precipitate a mass extinction. Raup [29] has shown that extinctions of the magnitude seen during mass extinctions could not be produced through the extirpation of one or a few provincial biotas.

It is important to note here that our review of the mechanisms of extinction focuses not on the ultimate causes (companion stars disturbing the paths of comets, variation in rates of seafloor spreading, and so forth) but rather on the more proximate causes of extinction, for at least two reasons: a) the same proximate cause may have different ultimate causes [25], and b) it is the proximate causes that are most likely to leave independent physical or geochemical evidence of their effect. We also note that we recognize that each mass extinction could, in principle, be caused by a different set of environmental changes, and that a particular extinction event could be the result of the interplay of several of the potential causes that we enumerate here.

Impact of Extraterrestrial Objects

While offered as an explanation for many Phanerozoic extinctions, this hypothesis has received the greatest attention (and the most supporting evidence) with regard to the terminal Cretaceous extinctions. Recent hypotheses of the impact of extraterrestial objects for the terminal Cretaceous extinctions were proposed on the basis of two lines of non-biotic evidence: sedimentological indicators (shocked quartz, boundary clays) of environmental disturbances, and enrichment of heavy metals such as iridium in deposits at the Cretaceous-Tertiary Boundary. These anomalies took on added significance because high-precision stratigraphic studies suggested a very short duration (less than 4.7×10^5 years, perhaps on the order of 10^4 years for many taxa [15]), and because an important component of marine life, the calcareous plankton, were almost completely wiped out. The mass extinctions constituted the corpse, the geochemical anomalies provided the circumstantial evidence, and the iridium anomaly in sediments around the world is the "smoking gun" that indicated an extraterrestrial bolide. The Snowbird Conference [38] presented several computer-based scenarios of the effects of the impact of a large body on the Earth. Assuming the impact of a 10 km diameter meteorite (an asteroid or a comet), the resulting disturbance of the physical environment would be sufficient to cause large-scale and profound environmental change and mass mortality among many groups. The theoretically predicted and partially confirmed (or at least not contradicted) scenarios entail a) global darkness (10^{-1} to 10^{0} years duration); b) atmospheric pollution (10^{-1} to 10^{0} years duration); c) destruction of stratospheric ozone (10^{0} to 10^{1} years duration); and d) chemical pollution of the ocean, in particular a lowering of the pH of seawater (10^{2} to 10^{4} years duration) (see Hsü, this volume). The biological consequences of such environmental changes include massive reduction in population sizes, temporary suppression of primary production on land (10^{-1} to 10^{0} years duration), and drastic suppression of primary productivity in the sea (10^{2} to 10^{4} years duration). These predicted changes (as well as others similar to many produced in nuclear winter scenarios) could account for the extinctions at the Cretaceous-Tertiary Boundary.

Most would agree that a good case, though perhaps not an overwhelming one, has been made for the impact of an extraterrestial body at the Cretaceous-Tertiary Boundary. This is a remarkable state of affairs considering that ten years ago such a hypothesis would not have been tolerated in scientific circles. This acceptability is due in no small measure to the value of independent, physical evidence for this particular hypothesis. Without the geochemical and sedimentological evidence, it would be just another wild idea. Of course problems persist, especially with regard to scenarios that predict environmental changes that seem to call for total extinction rather than the substantial (but obviously less than total) and selective extinctions that are recorded in the rocks. The paleontological evidence for selectivity at this boundary needs to constrain the models and scenarios that the impact hypothesis has generated. Further constraints on such extinction scenarios can be provided by a consideration of the role of primary productivity in the Recent ocean. Survival of even detritus-feeding benthos would be unlikely if primary productivity had indeed ceased for 10^{4} years (J. S. Levinton, personal communication).

Reduction of Available Marine Habitat as a Result of Change in Sea Level

That this is such a long-standing hypothesis [26] is testimony to at least circumstantial evidence in its favor. Many extinctions of marine organisms occurred at times of relative low stand of sea level (see Fig. 3 for the timing and extent of Phanerozoic sea level changes). Despite this association in timing of extinctions and marine regressions, problems plague this hypothesis as well. Not every regression is accompanied by an extinction, Pleistocene sea level fluctuations did not cause substantial extinctions of marine organisms, and the existence of island as possible refuges for families of the

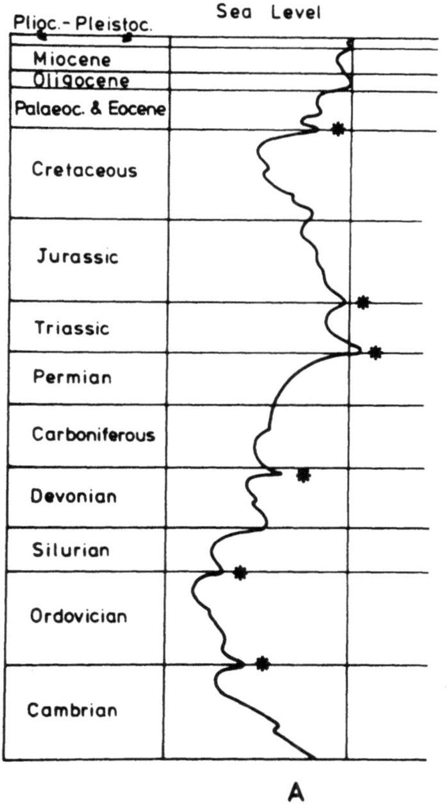

Fig. 3. Record of Phanerozoic sea level change. Relative low stands of sea level are shown as excursions of the curve to the right, high stands to the left. Times of mass extinctions are shown by * [11]

marine benthos could have served to repopulate the continental shelves during the following transgression [18]. These criticisms might suggest that, first, paleogeographic setting is important. A sea level change on the relatively emergent continents of the Pleistocence would not result in as large a change in area of shallow sea per meter of sea level change as would a change in sea level during times of low continental relief and low continental freeboard. Under paleogeographic conditions more typical of the Earth's history marine organisms might have become "perched" [20] – stenotopic and restricted to epeiric seas and thus prone to extinction during rapid regressions. Second, the duration of a sea level low stand may be important; the relatively brief Pleistocene low stands may have been insufficient for an equilibration of marine diversity to some lower value. Third, distinctive habitats, found only on continental shelves or in the broad, shallow inland seas characteristic of much of the Paleozoic might have harbored many of the faunas that suffered extinction. Finally, the withdrawal of the seas might prompt significant changes in patterns of nutrient cycling, entraining environmental changes far from the continents themselves.

Climatic Change

Climatic change, especially the effects of "refrigeration," has been recently championed as a significant agent of mass extinction by Stanley [41]. Global cooling is indeed associated with the extinctions of the late Eocene and the late Ordovician, and some evidence for a change in climate can be found at the Cretaceous-Tertiary Boundary. Climatic cooling is a plausible agent of extinction given the often narrow thermal tolerances of many marine species and the effect of climate on oceanic circulation. Nevertheless, the lack of precise chronological correlation of many major extinctions (the Permo-Triassic, the Late Triassic, the Late Devonian) with major episodes of climatic cooling and the lack of major extinctions associated with major episodes of climatic cooling (the Pleistocene, the Permo-Carboniferous) tends to reduce the attractiveness of climatic cooling as an agent of mass extinction.

Oceanic Anoxia

Episodes of oceanic bottom water stagnation can be recorded in the stratigraphic record by the widespread deposition of such organic-rich sediments as black shales. Such anoxic events may be prompted by sea level change [10, 11] and their onset appears to be sudden. The extinctions of the Cenomanian-Turonian (Middle Cretaceous), Toarcian (Early Jurassic), and Frasnian-Famennian (Late Devonian) occur at the same time as episodses of oceanic anoxia. Less convincing evidence for oceanic anoxia is seen at the Cretaceous-Tertiary Boundary, the Late Triassic, the Permo-Triassic, and the basal Silurian (shortly after the Late Ordovician extinctions). While anoxia might be implicated in some marine extinctions, it is clear that anoxia are not invariably associated with every Phanerozoic extinction, but a link between marine and terrestrial events leading to oceanic anoxia is not clear. Furthermore, the global extent of many episodes of oceanic anoxia is questionable – the Toarcian anoxic event, for example, appears limited in its geographic extent.

Declining Provinciality

Global diversity is a function not only of the degree of species packing within habitats but of the degree of biotic provinciality. Valentine and Moores [42] suggested that the changing geographies resulting from plate tectonic processes contributed to the regulation of Phanerozoic faunal diversity in the marine realm. And both Schopf [33] and Valentine and others [43] argue that changing patterns and levels of marine provinciality could explain much of the variation in familial diversity during Phanerozic time. The Late Permian extinction event seems to be the best candidate for a biogeographically induced extinction. In addition, origination rates might be lowered as a consequence of the paleogeographic setting of the Late Permian [16], thus exacerbating the environmental and biogeographic effects of continental assembly. The role of declining provinciality in other Phanerozoic extinction events is less clear.

Increased Volcanic Activity

The association in time between extensive outpouring of plateau basalts and the terminal Cretaceous extinctions should direct our attention to this hypothesis. This is particularly because extensive volcanism can produce not only geochemical anomalies much like those at the Cretaceous-Tertiary Boundary but environmental conditions much like those found in many nuclear winter and impact scenarios.

Magnetic Reversals

The weak but intriguing correlation between reversal frequency and extinction rate [30] suggests further inquiry into the possible links between magnetic reversals and extinctions. We simply know too little about the direct effects of zero magnetic fields on organisms. Our ignorance on this matter should promote rather than preclude research on the topic.

Two research programs that are directed towards understanding the causes of mass extinction were suggested.

Fine-scale Studies of Extinctions and Environmental Changes

Detailed examinations of changes in species occurrence and abundance in the vicinity of extinction events should be undertaken. Detailed environmental analyses, including geochemical studies, should accompany the paleontological work in an effort to relate changes in local environmental changes to the magnitudes and timing of extinction seen in the local stratigraphic sections.

Integration of the Physical and Paleobiological Record of Extinction

Efforts should be made to match the timing, duration, and magnitude of extinctions to independent (physical and geochemical) evidence for the timing, duration, and magnitude of environmental changes such as those recorded by iridium anomalies, isotopic fluctuations, sea level change, climatic cooling, oceanic anoxia, bursts of volcanism, and others. The pattern of extinction and survival in the face of each of these environmental changes may be predictable, given knowledge of the paleobiology of the species involved. We may be able not only to match the timing, duration, and magnitude of an environmental change with an extinction event, but also to compare the predicted outcome with the actual effect.

Evolutionary Effects of Mass Extinctions

To a large degree the immediate, and perhaps the subsequent, effects of mass extinctions depend on the selectivity and cause(s) of the extinction itself. For example, if extinctions selectively remove large-bodied species, perhaps we should not be too surprised at the prevalence of small-bodied forms as ancestors. We expect that the species that are likely to survive mass extinctions are probably opportunistic, "weedy" species – ones that are capable of survival in disturbed habitats. Such opportunistic species are often the most vagile forms and may have been present in refugia during the extinction episodes. We note that opportunistic species can often be morphologically simple species – the typical stock from which many groups diversify.

The proposition that clade-level properties, such as the distribution of genera among several provinces, are subject to selection during mass extinctions suggests the possibility that some species-level characteristics may persist simply by virtue of their fortuitous association with a higher level trait (Jablonski, this volume). Such properties, "carried through" the extinction by other traits, may affect the range of morphologies and adaptations seen in post-diversification biotas.

Mass extinctions may have biogeographic effects. If mass extinctions selectively remove geographically restricted taxa, the post-extinction biota will contain a smaller porportion of endemics than the initial fauna. The surviving biota will be characterized by geographically widespread forms, and global provinciality will be low. This line of reasoning suggests that the decline in provinciality that has often associated with mass extinctions may actually be an effect of rather than a cause or a contributing agent to the extinction event.

Of related biogeographic interest is the notion of refugia and their roles as "extinction shelters" and as source areas for rediversification. The geographic extent of the environmental catastrophe that precipitates a mass extinction may determine whether refugia are actually particular geographic areas (the high latitudes, for example) or somewhat species-specific or habitat-specific (nearshore settings, for example), and thus different for different species.

A common feature of post-extinction time is rediversification. Such rediversification is often concentrated in particular clades and may represent the replacement of previously dominant groups by species derived from heretofore subordinate forms. The replacement of the dinosaurs by the mammals following the Cretaceous-Tertiary extinctions is the most often cited example. There is little evidence to suggest that such replacements represent competitive displacement. Rather, the patterns of diversity change through time of potentially competing groups seems more consistent with the phenomenon of a preemptive occupation of adaptive zones. It appears that whichever clade first diversifies within an adaptive zone is the clade most likely to persist and dominate that zone (see [2] for an example of this among therapsids and dinosaurs). Thus, high speciation rates may be as important as "adaptive superiority" in determining which clade fills the newly vacated adaptive space.

The following research program on the evolutionary effects of mass extinctions was identified.

Detailed Examination of Instances of Ecological Replacements after Mass Extinctions

More examples with greater temporal precision are needed of this phenomenon. Studies which integrate our knowledge of the paleoecology, functional morphology, and temporal variation in diversity among clades should shed light on those features most important in shaping the course of post-extinction evolution. Particular attention should be paid to the marine fossil record, where few well documented examples of replacement exist, yet where the data are most amenable to such study.

References

1. Allendorf F, Leary RF (1986) Heterozygosity and fitness in natural populations of animals. In: Conservation biology: Science of diversity, ed Soulé ME. Sunderland, MA: Sinauer

2. Benton MJ (1983) Dinosaur success in the Triassic: a noncompetitive ecological model. Q Rev Biol 58:29–55
3. Boucot AJ (1975) Evolution and extinction rate controls. Amsterdam: Elsevier
4. Bretsky PW (1973) Evolutionary patterns in the Paleozoic bivalvia: documentation and some theoretical consideration. Geol Soc Am Bull 84:2079–2096
5. Brown JH (1981) Two decades of homage to Santa Rosalia: toward a general theory of diversity. Am Zool 21:877–888
6. Fischer AG, Arthur MA (1977) Secular variations in the pelagic realm. SEPM Spec Pub 25:19–50
7. Flessa KW, Jablonski D (1985) Declining Phanerozoic background extinction rates: effect of taxonomic structure? Nature 313:216–218
8. Hallam A (1975) Evolutionary size increase and longevity in Jurassic bivalves and ammonites. Nature 258:439–496
9. Hallam A (1976) Stratigraphic distribution and ecology of European Jurassic bivalves. Lethaia 9:245–260
10. Hallam A (1981) The end-Triassic bivalve extinction event. Palaeogeog Palaeoclimatol Palaeoecol 35:1–44
11. Hallam A (1984) Pre-Quaternary sea-level changes. Ann Rev Earth Planet Sci 12:205–244
12. Hansen TA (1980) Influence of larval dispersal and geographic distribution on species longevity in neogastropods. Paleobiology 6:193–207
13. Hoffman A (1985) Patterns of family extinction depend on definition and geological timescale. Nature 315:659–662
14. Hoffman A, Kitchell JA (1984) Evolution in a pelagic system: a paleobiologic test of models of multispecies evolution. Paleobiology 10:9–33
15. Hsü KJ, He Q, McKenzie JA, Weissert H, Perch-Nielsen K, Oberhänsli H, Kelts K, LaBrecque J, Tauxe L, Kränenbühl U, Percival SF Jr, Wright R, Karpoff AM, Petersen N, Tucker P, Poore NZ, Gombos AM, Pisciotto K, Carman MF Jr, Schreiber E (1982) Mass mortality and its environmental and evolutionary consequences. Science 216:249–256
16. Hüssner H (1983) Die Faunenwende Perm/Trias. Geologische Rundschau 72:1–22
17. Hutchinson GE (1951) Copepodology for the ornithologist. Ecology 32:571–577
18. Jablonski D (1986) Causes and consequences of mass extinctions: a comparative approach. In: Dynamics of extinction, ed Elliott EK. New York: Wiley, pp 183–229
19. Jablonski D, Lutz RA (1980) Molluscan larval shell morphology: ecological and paleontological applications. In: Skeletal growth of aquatic organics, eds Rhoads DC, Lutz RA, pp 323–378. New York: Plenum Press
20. Johnson JG (1974) Extinction of perched faunas. Geology 2:479–482
21. Kitchell JA, Pena D (1984) Periodicity of extinctions in the geologic past: deterministic versus stochastic explanations. Science 226:689–692
22. Ledig FT (1986) Heterozygosity, heterosis, and fitness in outcrossed plants. In: Conservation biology: Science of diversity, ed Soulé ME. Sunderland, MA: Sinauer
23. Lovejoy TE, Bierregaard RO, Rankin JM, Schubart HOR (1983) Ecological dynamics of forest fragments. In: Tropical rainforest: Ecology and management, eds Sutton SL, Whitmore TC, Chadwick AC. London: Blackwell
24. McCoy ED, Connor EF (1980) Latitudinal gradients in the species diversity of North American mammals. Evolution 34:193–203
25. McLaren DJ (1983) Bolides and biostratigraphy. Geol Soc Am Bull 94:313–324

26. Newell ND (1967) Revolutions in the history of life. Geol Soc Am Spec Paper 89:63–91
27. Rampino MR, Stothers RB (1984) Terrestrial mass extinctions, cometary impacts and the Sun's motion perpendicular to the galactic plane. Nature 308:709–712
28. Raup DM (1978) Cohort analysis of generic survivorship. Paleobiology 4:1–15
29. Raup DM (1982) Biogeographic extinction: a feasibility test. Geol Soc Am Spec Paper 190:277–282
30. Raup DM (1985) Magnetic reversals and mass extinctions. Nature 314:341–343
31. Raup DM, Sepkoski JJ Jr (1982) Mass extinctions in the marine fossil record. Science 215:1501–1503
32. Raup DM, Sepkoski JJ Jr (1984) Periodicity of extinctions in the geologic past. Proc Natl Acad Sci USA 81:801–805
33. Schopf TJM (1976) The role of biogeographic provinces in regulating marine faunal diversity through geologic time. In: Historical biogeography, plate tectonics, and the changing environment, eds Gray J, Boucot AJ, pp 449–458. Corvallis, OR: Oregon State University Press
34. Sepkoski JJ Jr (1981) A factor analytic description of the Phanerozoic marine fossil record. Paleobiology 7:36–53
35. Sepkoski JJ Jr, Raup DM (1985) Periodicity in marine mass extinctions. In: Dynamics of extinction, ed Elliott DK. New York: Wiley
36. Sheehan PM (1985) Reefs are not so different – they follow the evolutionary pattern of level-bottom communities. Geology 13:46–49
37. Signor PW III, Lipps JH (1982) Sampling bias, gradual extinction patterns and catastrophes in the fossil record. Geol Soc Am Spec Paper 190:291–296
38. Silver LT, Schultz PH (eds) (1982) Geological implications of impacts of large asteroids and comets on the earth. Geol Soc Am Spec Paper 190
39. Sousa WP (1984) The role of disturbance in natural communities. Ann Rev Ecol Syst 15:353–391
40. Stanley SM (1973) An explanation for Cope's Rule. Evolution 27:1–26
41. Stanley SM (1984) Marine mass extinctions: a dominant role for temperature. In: Extinctions, ed Nitecki MH, pp 69–118. Chicago: University of Chicago Press
42. Valentine JW, Moores EM (1972) Global tectonics and the fossil record. J Geol 80:167–184
43. Valentine JW, Foin TC, Peart D (1978) A provincial model of Phanerozoic diversity. Paleobiology 4:55–66
44. Van Valen LM (1975) Group selection, sex, and fossils. Evolution 29:87–94
45. Van Valen LM (1984) A resetting of Phanerozoic community evolution. Nature 307:50–52
46. Vermeij GJ (1986) Survival during biotic crises: the properties and evolutionary significance of refuges. In: Dynamics of extinction, ed Elliott DK. New York: Wiley, pp 231–246
47. Vermeij GJ, Petuch EJ (1986) Differential extinction in tropical American molluscs: endemism, architecture, and the Panama Land Bridge. Malacologia 27:29–42
48. Waterhouse JB, Bonham-Carter G (81976) Range, proportionate representation, and demise of brachiopod families through Permian Period. Geol Mag 113:401–428

Patterns and Processes in the History of Life,
eds. D. M. Raup and D. Jablonski, pp. 259–276. Dahlem Konferenzen 1986
Springer-Verlag Berlin, Heidelberg

The Proximate Causes of Extinction

D. Simberloff

Dept. of Biological Science
Florida State University
Tallahassee, FL 32306, USA

Well, ther' ain't no sense in it. A body might stump his toe, and take pison, and fall down the well, and break his neck, and bust his brains out, and somebody come along and ask what killed him, and some numskull up and say, "Why, he stumped his *toe*." Would ther' be any sense in that? *No.*

– Mark Twain
The Adventures of Huckleberry Finn

Abstract. The vast majority of contemporary extinctions can be viewed as anthropogenous in the sense that human activity greatly reduced population sizes and extinction would not likely have occurred now without the human activity. However, one would still wish to know why small populations, even when protected from further human interference, appear to be unusually prone to extinction. Empirical data on the last gasp of such declining species are almost nonexistent but there is evidence that four forces conspire to put small populations at increased risk: demographic stochasticity, genetic deterioration, social dysfunction, and extrinsic forces. There are presently no models that accurately apportion the threat of extinction among these forces and even the available guidelines for indicating which species are especially at risk are very imprecise.

Introduction

It is easy to find what is causing extinction these days – browsing through library shelves turns up such titles as *Extinction – The Causes and Consequences of the Disappearance of Species* [17] and lists of the causes of recent extinctions (e.g. [19, 56]). Other notable compendia of this sort include those by Fitter [20] and Simon and Géroudet [46]. These works tell us an enormous amount about the *ultimate* causes of extinction and unanimously indict human activity as the direct or indirect reason for virtually every historic extinction. For example, Ziswiler [56] provides information on the extinction of 116 birds and 97 mammals that are already extinct, plus 10 reptiles, 149 birds, and 158 mammals that are already extinct, plus 10 reptiles, 149 birds, and 158 mammals that are gravely threatened. Table 1 summarizes this information, and one sees immediately that, except for the few unknown cases, all the damage is anthropogenous. Soulé [48] believes that there are no documented examples of continental species extinguished by nonhuman agencies.

However, Ziswiler does not really answer all the questions. For example, he lists the heath hen (*Tympanuchus cupido cupido*) as having been eliminated in 1932 from the eastern United States because of anthropogenous habitat alteration plus hunting. Yet hunting and habitat alteration ceased well before 1932, and the extinction of the heath hen is a very complicated story [7]. Originally heath hens were found from Maine southward to at least Virginia and were very common in favorable habitat (sandy scrub-oak plains). Because they were easily killed, they were quickly exterminated from accessible areas and disappeared from Connecticut and mainland Massachusetts soon after 1840. A few persisted in Long Island, the plains of New Jersey, and the foothills of the Pocono Mountains in Pennsylvania, but by 1870 the last individuals were restricted to Martha's Vineyard, an island southeast of Massachusetts. By 1890 there were 200 birds left, and by 1896, fewer than 100.

Extinction would doubtless have ensued immediately but for the establishment in 1908 of a well guarded refuge of 1600 acres for the last 50 birds. The refuge habitat was systematically improved and by 1915 heath hens could be found all over the island: 300 or more birds could easily be flushed from the corn and clover plots planted on the reservation, and the population was estimated at 2,000. Then began a series of disasters that ended in extinction. In spite of unusual precautions to prevent fires from spreading, a conflagration during a gale in 1916 swept through the breeding area destroying birds, nests, eggs, cover, and food. A hard winter followed, and an

Table 1. Ultimate causes of extinction of 116 bird and 97 mammal species or subspecies, plus causes threatening extinction of 10 reptile, 149 bird, and 158 mammal species and subspecies. Data from Ziswiler (56)

Cause	Extinct Birds	Extinct Mammals	Threatened Reptiles	Threatened Birds	Threatened Mammals
A	14	32	6	31	82
B	2	1	1	3	24
C				1	6
D	3	6		2	6
E	2		2	5	
F	3		2	7	1
G		1			20
H	6	31	1	7	13
I	28	13		80	41
J				8	
K	40	1	1	29	13
L	6		1		4
M	1			2	4
N	1	1	2	3	1
O	33	31	3	37	7
P	3			2	
Q	14	6	1	42	
R		8		5	14
S	9	29		29	
T	1			9	
U	3			8	4
V	4	1		2	3

A = hunted for meat or fat
B = hunted for hides or feathers
C = hunted for trophies or souvenirs
D = hunted for sport or pleasure
E = eggs and young collected
F = live-animal trade
G = persecuted because of superstitious beliefs
H = combatted as alleged pest
I = habitat altered through deforestation
J = habitat altered through drainage
K = habitat altered through civilization or monoculture
L = denaturalized fauna through goats or sheep
M = denaturalized fauna through rabbits
N = denaturalized fauna through feral dogs
O = denaturalized fauna through feral cats
P = denaturalized fauna through feral pigs
O = denaturalized fauna through rats
R = denaturalized fauna through foxes
S = denaturalized fauna through mongooses
T = denaturalized fauna through mustelids
U = destroyed by introduced animal diseases
V = unknown

unprecedented flight of goshawks in its midst reduced the population to fewer than 150, mostly males.

A slight rally in the next few years was insufficient to compensate for this triple blow, and extensive inbreeding, declining sexual vigor, and the excess of males brought the birds to the brink of extinction. Worst of all, in 1920 the poultry disease, blackhead, brought to the island with domestic turkeys, killed many heath hens. By 1927 there were only 13 birds, of which 11 were males. In 1928 there were only 2 birds, and after December 8 of that year, only one survived. Until the last sighting on March 11, 1932, this individual male dutifully appeared, to be observed by many ornithologists and bird-lovers who travelled to the island to see the last heath hen.

In other words, even with deliberate habitat change minimized and hunting virtually stopped (though there was evidence of poaching as late as the 1920s [25]), several forces conspired to eliminate this species. So even though it is true that the ultimate cause of death was human activity, once the population size was sufficiently reduced, extinction was assured even without human activity. The problem is analogous to assigning the exact cause of death in certain human diseases. For example, the cause of many cancer deaths is listed as "congestive heart failure," but the patient was brought to this pass by the ravages of the cancer.

We would like to know the nature of the proximate causes of extinction. By "proximate" causes I mean the reasons why the last few die. Such causes are in contrast to the "ultimate" causes, the events that may have occurred much earlier that led inexorably to a situation in which there would be a small, terminal population. Soulé [48] suggests that proximate causes are not very interesting: "The extinction problem has little to do with the death rattle of the final actor. The curtain in the last act is but a punctuation mark – it is not interesting in itself. What biologists want to know about is the process of decline in range and numbers." I would argue that both questions are important, both from a conservation standpoint and from a more purely academic one. In many cases we do not know why range and numbers have declined, and it behooves us to find out. But in others (e.g., those cited by Ziswiler [56]), we do know why the declines occurred but not why the extinctions happened. Yet such knowledge is imperative if we are to plan reserves effectively. As imposing a personage as the President of the United States has asked how many individuals of a species we need [41], epitomizing a perception that simply providing a little piece of appropriate habitat is enough to maintain a population. *Why* exactly is it that very small populations cannot persist? Is the heath hen case typical?

The Minimum Viable Population Concept

Broadly speaking, there are four reasons why very small populations are unlikely to persist [43, 48]. Consideration of these has led to the concept of the "minimum viable population" (MVP), that number of individuals such that when populations fall below this point they are doomed to quick extinction. As originally conceived by MacArthur and Wilson [32], this point was a clear inflection in a curve of expected time to extinction. Populations above this point were virtually immune to extinction, while those below this point were likely to go extinct very quickly. However, MacArthur and Wilson [32] were considering only one of the four reasons for increased probability of extinction for small populations, and their simple model has been progressively made more realistic [40, 44] for the one force that they included, with the result that for this force one might expect a more gradually increasing expected persistence time as population size increases. For the other forces the exact shape of the curve has not even been predicted, and there are no empirical data. However, we are concerned not so much with the exact shape of the curve as with the proposition that swift extinction is ensured for very small populations, so the MVP can serve as a metaphor for the proposition, even if we cannot specify a specific number as the critical size.

The four forces are:

1 – Demographic stochasticity, the random variation in population variables such as sex ratio, birth and death rates, or the distribution of individuals among age classes [33]. This was the force treated by MacArthur and Wilson [32] and Richter-Dyn and Goel [40] and is more threatening to a small population. For example, if N is the expected number of offspring in any generation, the probability that they will all be one sex is 2^{1-N}. Thus extinction from this source alone is vastly greater for small than for large populations. There is also typically an increasing skewness towards males from a source discussed below, inbreeding depression [42]. Furthermore, the skewed sex ratio increases the rate of inbreeding. The highly skewed sex ratio of the declining heath hen population could have been due to either or both sources and was likely one of the contributing proximate causes of extinction.

2 – Genetic deterioration: small populations are liable to lose alleles by genetic drift and, in addition, are likely to inbreed. The threat of extinction is thus exacerbated for several reasons.

Over the long term, the resulting loss of genetic variation is at a rate roughly proportional to $1/2\ N_e$, where N_e is effective population size (see below). This loss may limit a population's ability to respond to environmental change through natural selection. There is dispute about just how small a population must be in order for drift to be a significant force in removing alleles in the face of normal natural selection, with estimates ranging from $N_e = 50$ [8] to $N_e = 500$ [23]. The concept of effective population size entails a hypothetical population of unchanging size in which all members have equal expectation of being parents of any progeny individual [27]. For a given number of individuals, genetic drift is lowest in such an ideal population, so it is natural to define for any real population an ideal population with the same amount of genetic drift and to call the size of this ideal population the "effective population size" (N_e) of the real population. The effective population size is often vastly smaller than the census population size [27], so even if the danger point to small populations from this direction were $N_e = 50$, this force would seem to be important. However, extinction of small populations seems frequently to take place so quickly that one might not have expected evolution to have solved the problem, even if there were a wealth of genetic variability. In other words, it is likely that other factors will eliminate a small population before failure to evolve endangers it.

In addition to causing alleles to be lost from a population, drift and inbreeding cause an increased number of individuals to be homozygous, which is likely to result in inbreeding depression. For species that normally outbreed, a sudden imposition of inbreeding with consequent production of homozygous offspring leads to loss of fitness as deleterious alleles are unmasked [39, 45]. Many fitness traits can be affected, including viability and fertility. Although the large effects of homozygosity for very deleterious genes, such as those causing death or gross deformity, are the most obvious aspects of inbreeding, even more important is the cumulative effect of increasing homozygosity for many loci. In some instances even small amounts of inbreeding can be surprisingly harmful in primates, ungulates, and small mammals [39].

At exactly what effective population size inbreeding depression becomes a threat to population survival depends very much on the species – its evolutionary history, normal breeding system, etc. If a species typically inbreeds, it is less likely to be greatly affected by additional inbreeding imposed by a decreased population. Those alleles that would be severely debilitating when homozygous are more likely to have already been selected out of the population. The near universality of at least some effect [37] suggests that increased inbreeding is to be avoided. Animal breeders, who typically work with already inbred lines, have found empirically that a per generation rate of inbreeding exceeding 2 or 3% allows deleterious alleles to be fixed more rapidly than selection can eliminate them [47]. Soulé [47] suggests that for conservation of wild, non-inbred animals, 1% would be a better limit, and this translates to $N_e = 50$. Breeders have also found empirically that there is usually a major effect on fecundity as the inbreeding coefficient approaches 0.5–0.6, and Soulé [47] shows that this suggests that the expected number of generations to a point where rapid extinction is insured is about $1.5 \times N_e$.

Bent [7] feels that inbreeding depression may have played a major role in the final elimination of the heath hen, through decreasing sexual vigor as well as possibly affecting sex ratio.

3 – Social dysfunction: Soulé [48] observes that some species have characteristic social behavior that renders them more liable to extinction. The increased risk arises in two ways. First, elaborate group mating displays, such as the lekking behavior of the heath hen [7], make it far easier for certain of the ultimate extinction forces, such as hunting, to make enormous inroads. Other kinds of social behavior such as group defense and herding or schooling can have a similar effect.

Second, once a population is already small, there may simply be too few individuals to stimulate or consummate social behavior. A good example is the "Fraser Darling" effect by which certain species satiate their predators by synchronous breeding. One proximate mechanism by which colonial birds achieve such synchrony is group stimulation of ovarian development of the colony's females [12, 55]. There are similar examples of social stimuli producing reproductive synchrony in mammals (including human females [34]), though there is no evidence that predator-satiation is the selective agent. One can then imagine, for birds at least, that even if the distribution of clutch sizes remained constant as population sizes decreased, mortality from predation would increase.

Asynchrony could also reduce mean clutch size. A species may require some sort of social facilitation in order for mating and offspring production to be vigorous [26]. Asynchrony between the sexes could prevent mating entirely. Allen [2] argues that "sex rhythm" is important for some bird species: female birds have short estrous cycles during which fertilization must occur, while males have short intervals during which they can fertilize a female. If there are too few individuals in a population, the probability that a fertile female encounters a capable male may fall precipitously.

4 – Extrinsic forces: temporal variation in habitat parameters and the population sizes of enemies, and random catastrophes such as fires, storms, and epizootics – there is a continuum of frequency and degree here, with

ordinary storms simply more frequent and less drastic than hurricanes. Mortality caused by these forces is density-independent. If the per capita probability of death is D and if deaths of individuals are independent, then the probability that all N individuals in the population die is D^N, an exponentially decreasing function of N. Probably the spatial extent of the population, or the number of separate populations of the species, is as critical a factor as the number of individuals for many extrinsic forces. For example, the fire that ravaged the last population of the heath hen would not have wrought such havoc had there been several populations, even if all of them were smaller than the single one on Martha's Vineyard [14]. Similarly, the epizootic of blackhead would have been restricted had there been discrete populations. Another extrinsic force that contributed to the heath hen's extinction, the severe winter of 1916/1917 was a regional effect, and several populations would not have mitigated this disaster, at least not if they were all in the Northeast. However, the extraordinary influx of goshawks might well have been restricted to one or a few populations, had there been discrete populations.

We thus see that all four classes of proximate causes of extinction may have contributed to the end of the heath hen. Demographic stochasticity may have skewed the sex ratio, and inbreeding depression may have skewed it also. Inbreeding depression may also have caused decreased sexual vigor. The bird's behavior undoubtedly made it an easier target for hunters and, in addition, the fact that it typically bred in a lek may have decreased its sexual vigor and fecundity when its numbers fell too low. Finally, a catastrophic extrinsic force, fire, combined with three more normal extrinsic forces – disease, predation, and severe weather – to deliver the coup de grâce.

Survey

My original intent was to scan at least several hundred cases like that of the heath hen – extinct or terminally declining species or populations – and see which proximate forces were important for which species, analogously to my scan of at least the suspected effects of several hundred introduced species [45]. Imagine my suprise when I found that the heath hen was by far the best studied species, and that for almost all those species whose demise or imminent demise can ultimately be attributed to anthropogenous insults, there have not even been guesses about the proximate causes.

One reason is that extinction frequently follows very quickly once populations are brought to the level where the proximate forces I have categorized above become important. In fact, sometimes the anthropogenous ultimate cause of extinction is the proximate cause as well, as for the famous Steven Island wren (*Xenicus lyelli*), whose only population was both discovered and extinguished in 1894 by the lighthouse keeper's cat [25]. Consequently, biologists are not always alerted in time to a situation that demands intensive study. Another reason is that, once a species is so rare that it is endangered by the proximate forces, it is often illegal, immoral, or ill-advised to do the experiments or even intensive observations that would be required to understand the cause for its decline. Removal of a few individuals for attempted captive propagation often elicits debate, and study simply for study's sake would be even less popular. This is the conservation analog of the American rationale for aerial bombardment during the Vietnam War: we had to destroy the village (species) to save it. Yet another reason for the paucity of information is that it is simply difficult to study very rare species intensively, even if we are allowed to do it. This is one reason why we advise our students to pick common species to study for their theses – at least they can find them consistently enough to learn something about them.

So I will not produce a list like mine for species introductions or like Ziswiler's for the ultimate causes of species extinction. However, there are some instructive examples of how the four classes of proximate factors have operated, and I will digress to discuss them.

As noted above, "natural extinction" on continents must be very rare, though the last gasp of a relict must happen occasionally for reasons beyond human control. Fjeldså [21] views the hooded grebe (*Podiceps gallardoi*) as such a doomed species. This Patagonian bird is an extreme food specialist (primarily on snails) and is restricted to a habitat (turbid lakes) that has been disappearing for millions of years because of a long-term drying trend in the region. Even on islands, however, extinctions that apparently are not anthropogenous seem, on very close examination, to be otherwise. For example, the most detailed account of a local mammalian extinction in Britain was of the house mouse (*Mus musculus muralis*) on St. Kilda [18]. This happened within 18 months of humans *leaving* the island. The mouse did not starve without man, since it exists on other islands without him. Berry and Tricker [9] cogently argue that the mouse is outcompeted in the absence of humans by the long-tailed field mouse (*Apodemus sylvaticus*), even though the two species do not appear *a priori* to be very similar ecologically, and even though one cannot prove at this time what they compete for. Similarly, the house mouse population on Brooks Island in San Francisco Bay declined from 12,000 to zero over 15 months after the invasion of a few California meadow voles (*Microtus californicus*) from a nearby island on which they existed by human agency. The two species have distinct ecologies, and there was no obvious reason for the extinction, which consisted of slow, steady excess of deaths over births. Yet Lidicker [31] feels that some subtle competition, possibly of the interference variety, is at the root of the disappearance.

Direct evidence that demographic stochasticity played a proximate role in any single extinction does not exist. There are examples such as the heath hen in which an excess of males appeared at the end. The last six observed dusky seaside sparrows (*Ammospiza maritima nigrescens*), for instance, were all males. However, there is strong evidence that by the time the population was all male, the local environment was so degraded that the subspecies would have gone extinct even had the sex ratio been 1/1 [52]. Of course, many species that went extinct gradually from non-catastrophic causes must at some point have reached a stage where the last few individuals were of one sex. By chance alone, for example, one would expect the last three individuals of any species to be of the same sex 1/4 of the time. At this point, unless there is parthenogenesis, the species will go extinct, and one could view this as resulting from demographic stochasticity. However, our interest is more in the range of the last 20–500 individuals or so, not the last three, and we do not have direct evidence that demographic stochasticity has been important in this range. In any event, I have noted above that inbreeding also may lead to a male-biased sex ratio.

There are models of specific endangered populations, based on field estimates of various population parameters, that suggest that demographic stochasticity is likely to be a major force in these instances. A model of the spotted owl (*Strix occidentalis caurina*) metapopulation, for example, suggests that demographic stochasticity is more likely than genetic factors to extinguish local subpopulations over the short term of decades (Salwasser, personal communication). Shaffer and Samson [44] describe a detailed model of demographic stochasticity for the grizzly bear (*Ursus arctos*) that estimates that this force alone will cause a population of $N_e = 50$ to go extinct, on average, in 114 years.

The role of genetic deterioration in extinction seems more substantiated. As for the heath hen, there is a suggestion of inbreeding depression in the end of the ivory-billed woodpecker [3]. The nearly universal effects of inbreeding on fitness traits is documented by Ralls and Ballou [39] for zoo animals, while Frankel and Soulé [22] and Beardmore [5] summarize much evidence for reduced fitness following inbreeding in many kinds of organisms. Even more germane are the many laboratory studies (cited in [22]) in which the great majority of replicate inbred lines went extinct. For example, of twenty lines of house mouse (*Mus musculus*) subjected to inbreeding, only ten survived to the fifth generation and only one to the twelfth generation [10]. Since many of the organisms, such as guinea pigs and chickens, subjected to such experiments were already inbred and thus might have been expected to have been purged of alleles that would cause major inbreeding depression, such experiments constitute strong evidence that inbreeding depression can play a proximate role in extinction. Many more informal breeding schemes for domestic animals that have run afoul of fertility and viability decrease after inbreeding [47] lead one to the same conclusion.

In the absence of much direct evidence, however, it is important to keep an open mind on the role of inbreeding depression in extinction. Even though inbreeding can lower fertility, lengthen age-at-maturity, reduce size, render organisms less symmetric, reduce sexual vigor, and wreak many other horrors that affect fitness, this does not in itself prove that the population will be endangered. In an abstract sense one can say that a species undergoing inbreeding with various of these traits is "poorer" than it was before, but one must remember that there *are* extant species that have low fertility, become mature late in life, have small size, etc., and we cannot automatically view them all as endangered (but see the discussion below on extinction-proneness).

To put it another way, any extant species is *ipso facto* successful, and one cannot claim that it is unsuccessful simply by positing a species with more desirable traits. "It is no use to point out that a *Drosophila* with a cerebrum, a vertebrate eye, and an opposable thumb would be [at] an advantage over other *Drosophila*" [30]. The "Red Queen Hypothesis" [54] would seem to indicate that any decrease in fitness in an existing species will be balanced by an increase in fitness in other species, and, over the long run, the first species' lineage is more likely to go extinct. Even if the hypothesis should prove to be true, this does not mean that the extinction will be anytime soon. After all, *all* species are ultimately doomed to extinction, taxonomic or phyletic, and we are concerned with the proximate, short-term causes.

The passenger pigeon and the ivory-billed woodpecker are examples of species in which social dysfunction may have played a proximate role in extinction. Halliday [26] argues that one reason why the colonially nesting passenger pigeon (*Ectopistes migratorius*) went extinct was that when hunting reduced group size below a certain level they were simply unable to facilitate one another's reproduction in some way. Allen and Kellogg [3] suggest that sex rhythm failure may have contributed to the final elimination of the ivory-billed woodpecker (*Campephilus principalis*). I know of no other species for which social dysfunction has been cited as a proximate factor in extinction, though the various forms of social behavior and the myriad ways in which it can be extremely important have not been studied exhaustively until recently, and such research is in an explosive growth phase [55]. There is certainly no dearth of reports from zoos in which an apparently well suited pair of individuals simply will not mate or produce no viable offspring. Usually one assumes, anthropocentrically, that they just do not like one another much. This may often be true, but there may also be instances in which some subtle social facilitation is required but cannot occur with just one pair.

The proximate role of extrinsic forces in population extinction is well documented in butterfly population studies [15]. For example, Ehrlich suggests that two observed extinctions of small checkerspot (*Euphydryas editha bayensis*) populations were the result of drought, which limited nectar resources which, in turn, limited egg production. It appears that genetic deterioration typically does not play a role in extinction of small *Euphydryas* populations, not because inbreeding depression does not exist, but because extrinsic forces prevent small populations from persisting long enough for inbreeding depression to be important. Populations either rebound to large size or are quickly extinguished. Similarly, Ehrlich et al. [16] found that a montane population of the silvery blue (*Glaucopsyche lygdamus*) was extinguished when a late season snowstorm destroyed the flowers that are the caterpillars' main resource.

For the butterflies, these population extinctions did not translate into species extinctions because the extrinsic forces were of limited geographic extent and there were always other populations to redress the extinctions. When there are few populations of a species and they occupy a small region, such events can threaten the very species. For example, a hurricane in 1963 reduced the single population of the Laysan teal (*Anas laysanensis*) to a fraction of its original size [19], a severe winter in 1906/1907 eliminated many herds of the pronghorn (*Antilocapra americana*) [51], and a substantial fraction of the remaining great auks (*Pinguinus impennis*) were eliminated by a series of eruptions in 1830 that destroyed one of their breeding and roosting grounds [6, 25].

Yet another proximate force threatens elimination of species, or other distinctive genetic entities (subspecies, races, etc.), though not by termination of the lineage. This force is quite analogous to the threat of elimination by assimilation of small enthnic groups, such as some of the Tungus people of Ussuria, who have interbred extensively with the more numerous Japanese. If any population is reduced to very small numbers and is interfertile with another, more common form, it may disappear as a distinct genetic entity simply through interbreeding. Assimilation is the likely end of the red wolf (*Canis rufus*), which hybridizes with the much more numerous coyote (*Canis latrans*) [24].

Such interfertility is especially common in birds, and Cade [11] lists a number of examples in which species or races are threatened with extinction by this route. Particularly notorious is the case of the dusky seaside sparrow, in which the plan to maintain the genes of this race by crossing the final six individuals (all male) with conspecific females from distinct races [37] elicited objection from the U.S. Fish and Wildlife Service on the grounds that the hybridization would extinguish the dusky seaside sparrow and thus contradict the intent of the Endangered Species Act [28]. This semantic delicacy was manifested in spite of little evidence that the dusky seaside sparrow was other than a local variant of a widely ranging species [29]. Fortunately, Disneyworld has proven not nearly so squeamish about taxonomic definition and has provided funding for the hybridization project that the Fish and Wildlife Service had denied.

Prospect

Since several forces frequently combine, as for the heath hen, to produce extinction, and since even the parameters required for realistic models of one force are extremely difficult to estimate, it is unlikely that a single model or algorithm or protocol will be concocted to apportion the threat of extinction among different causes, or even to tell us at exactly what population size extinction is imminent [49]. Even the less ambitious goal of determining which species are most at risk seems elusive. There are several lists [13, 22, 48, 53] of traits that seem, *ceteris paribus*, to place a species at higher risk. Mertz [35] adumbrated this approach and even cited many of the same traits in his study of California condor (*Gymnogyps californianus*) demography. Observing that large body size, long developmental time, low reproduction, and long life expectancy might all place a species at higher risk of extinction, he suggested that any species whose success depends on high survival rather than high reproduction is more vulnerable to any major new impact, be it anthropogenous, climatic, or whatever. Thus, he suggests, the late Pleistocene and Recent mammalian megafaunal extinctions might well have affected the same sorts of species whether hunting, climatic deterioration, or a combination of the two was the key cause.

Most of the traits said to make a species extinction-prone are eminently reasonable, but it is difficult to see how they can lead to further insights on exactly why particular species go extinct. The problem is that for every trait listed one can find species that have the trait and have *not* gone extinct. In this regard, the lists seem quite analogous to the lists of traits that make species "good colonizers" that appeared in many papers in *The Genetics of Colonizing Species* [4]. One was always able to find species with all the requisite traits that nevertheless were not adept at colonization, and other species with glaring deficiencies (in terms of the lists) that colonized quite well.

For extinction-proneness, consider the related traits "rarity" and "population size," which are viewed as most important in all four cited works. First of all, many of the data relating extinction to population size do not rest on observed extinctions – the extinctions are inferred from inferred presences of species on land-bridge islands during times of lowered sea level. Doubtless some currently missing species were originally present, but there is reason for skepticism about our ability to determine how many and which ones [1]. Second, even when there is good fossil evidence that an extinction actually occurred (e.g., in some of the data marshalled by Diamond [13]), there is no direct evidence on population size, so one must rely on presumed correlates, such as body size or degree of habitat specialization. Finally, when there are direct data on both extinction and population size, as for the birds of Bardsey Island, the correlation between the two is not perfect, which is precisely what led Diamond [13] to suggest that there must be factors other than population size involved in extinction-proneness.

A closer consideration of the meaning of "rarity" suggests why rarity *per se* is not predictably associated with extinction-proneness. Rabinowitz [38], following Drury [14], has observed that there are three distinct aspects of a species' situation that have all been used as synonyms of "rarity": geographic range, habitat breadth, and local population size. A species found in only one locality is often viewed as rare, no matter how common it is and how many habitats is occupies in that locality. Or a species may be found over a broad geographic region, and even in a variety of habitats, but one never finds more than a few individuals in any one place. Finally, a species may require a particular habitat that itself is rare, such as a peculiar soil type, even if the habitat is found (rarely) over a wide area, and even if the species is common where the habitat is found.

Concatenating these three traits, Rabinowitz [38] finds that there are seven distinct categories of rare species. When she looks for plants of each kind, she finds that very few fall into two categories, narrow geographic range and broad habitat range, whether or not population size is large. Whether such species never evolve in the first place of whether they are very likely to be estinguished cannot be deduced from available data, though Rabinowitz [38] hypothesizes that demographic stochasticity might frequently extinguish populations in some habitats, thus producing a narrow-habitat endemic. It seems counter-intuitive that species occupying several habitats should be more prone to extinction than those occupying one, all other things being equal – especially when population sizes are typically large. Nevertheless, the data exist. My guess is that this is an evolutionary phenomenon rather than a question of extinction. Perhaps gene flow prevents species with narrow geographic ranges from evolving sufficient adaptations to several distinct habitats.

The striking observation from Rabinowitz's classification is that lots of species are common nowhere. If they are in addition restricted to rare habitats, they are certainly in danger of anthropogenous extinction, especially if the geographic range is also narrow. For example, *Gambusia amistadensis*, a poeciliid fish found only at Goodenough Spring and the creek that flowed 1.3 km from the spring to the Rio Grande River, was eliminated in nature when a reservoir was constructed that inundated the spring, and the laboratory populations are not likely to survive because of the species' very precise habitat requirements [36]. Not only could humans destroy the species, but a natural extrinsic event such as a lava flow or fire could eliminate a species that is sufficiently restricted, though I know of no examples.

However, if they are found over a wide area in many habitats, or in a particularly common kind of habitat, there is perhaps no reason to think that rare species are particularly extinction-prone. Rabinowitz [38] gives examples of rare grasses that seem to be adapted to rarity and grow better when rare. For that matter, many species that we would call weeds seem adapted *as species* to local population extinction. They are adept at getting to newly available sites before other species but inevitably are outcompeted once the other species get there. However, by then they have moved on to yet other sites, *ad infinitum*. Drury [14] objects that it is teleological to view rarity *per se* as an adaptive strategy, on the grounds that selection acts on individuals, not groups. Perhaps strategy *is* automatically a teleological word, but one can still hypothesize that certain species have survived partly because of adaptations such as typically being sparse – this is classical "species selection" [50]. By the same token, it is not teleological to hypothesize that species have persisted because they occupy many sites but undergo rapid extinction at any of them, no matter what the selection – individual or group – that originally selected for these traits. Drury [14] himself stresses the insurance against extinction provided by having disparate populations, even to the extent of advising for conservation the dispersion of individuals from small relictual populations.

References

1. Abele LG, Connor EF (1979) Application of island biogeography theory to refuge design: Making the right decision for the wrong reasons. In: Proceedings of the first conference on scientific research in the National Parks, ed Linn RM, vol I. Washington, DC: U.S. Department of the Interior, pp 89–94
2. Allen AA (1934) Sex rhythm in the ruffed grouse (*Bonasa umbellus Linn.*) and other birds. Auk 51:180–199
3. Allen AA, Kellogg PP (1937) Recent observations on the ivory-billed woodpecker. Auk 54:164–184
4. Baker HG, Stebbins GL (1965) The genetics of colonizing species. New York: Academic
5. Beardmore JA (1983) Extinction, survival, and genetic variation. In: Genetics and conservation, eds Schonewald-Cox CM, Chambers SM, MacBryde B, Thomas L. Menlo Park, CA: Benjamin/Cummings, pp 125–151
6. Bengston S-A (1984) Breeding ecology and extinction of the great auk (*Pinguinus impennis*): Anecdotal evidence and conjectures. Auk 101:1–12
7. Bent AC (1932) Life histories of north american gallinaceous birds. Smithsonian Institution United States National Museum Bulletin 162. Washington, DC: U.S. National Museum
8. Berry RJ (1971) Conservation aspects of the genetical constitution of populations. In: The scientific management of animal and plant communities for conservation, eds Duffy E, Watt AS. Oxford: Blackwell, pp 177–206
9. Berry RJ, Tricker JK (1969) Competition and extinction: The mice of Foula, with notes on those of Fair Isle and St. Kilda. J Zool Lond 158:247–265
10. Bowman JC, Falconer DS (1960) Inbreeding depression and heterosis of litter sizes in mice. Genet Res 1:262–274
11. Cade TJ (1983) Hybridization and gene exchange among birds in relation to conservation. In: Genetics and conservation, eds Schonewald-Cox CM, Chambers SM, MacBryde B, Thomas L. Menlo Park, CA: Benjamin/Cummings, pp 288–310
12. Daly M, Wilson M (1978) Sex, evolution, and behavior. North Scituate, MA: Duxbury
13. Diamond JM (1984) "Normal" extinctions of isolated populations. In: Extinctions, ed Nitecki MH. Chicago: University of Chicago Press, pp 191–246
14. Drury WH (1974) Rare species. Biol Conserv 6:162–169
15. Ehrlich PR (1983) Genetics and the extinction of butterfly populations. In: Genetics and conservation, eds Schonewald-Cox CM, Chambers SM, MacBryde B, Thomas L. Menlo Park, CA: Benjamin/Cummings, pp 152–163

16. Ehrlich PR, Breedlove DE, Brussard PF, Sharp MA (1972) Weather and the "regulation" of subalpine populations. Ecology 53:243–247
17. Ehrlich PR, Ehrlich AH (1981) Extinction: The causes and consequences of the disappearance of species. New York: Random House
18. Fisher J (1948) St. Kilda, a natural experiment. New Nat J 1:91–108
19. Fisher J, Simon N, Vincent J (1969) Wildlife in danger. New York: Viking
20. Fitter R (1968) Vanishing wild animals of the world. New York: Franklin Watts
21. Fjeldså J (1984) Three endangered South American grebes (*Podiceps*): Case histories and the ethics of saving species by human intervention. Ann Zool Fennici 21:411–416
22. Frankel OH, Soulé ME (1981) Conservation and evolution. Cambridge: Cambridge University Press
23. Franklin IR (1980) Evolutionary change in small populations. In: Conservation biology: an evolutionary-ecological perspective, eds Soulé ME, Wilcox BA. Sunderland, MA: Sinauer, pp 135–150
24. Grainger D (1978) Animals in peril. Toronto: Pagurian
25. Greenwood JC (1967) Extinct and vanishing birds of the world, 2nd ed New York: Dover
26. Halliday T (1978) Vanishing birds. New York: Holt, Rinehard, and Winston
27. Hedrick PW (1983) Genetics of populations. Boston: Science Books International
28. Hillinger C (1980) Proposed bird cross-breeding stirs furor. *Los Angeles Times*, Oct 19, 1980, part 1, pp 14–15
29. James FC (1980) Miscegenation in the seaside sparrow? BioScience 30:800–801
30. Lewontin RC (1965) Selection for colonizing ability. In: The genetics of colonizing species, eds Baker HG, Stebbins GL. New York: Academic Press, pp 79–94
31. Lidicker WZ (1966) Ecological observations on a feral house mouse population declining to extinction. Ecol Monogr 36:27–50
32. MacArthur RH, Wilson EO (1967) The theory of island biogeography. Princeton, NJ: Princeton University Press
33. May RM (1973) Stability and complexity in model ecosystems. Princeton, NJ: Princeton University Press
34. McClintock MK (1971) Menstrual synchrony and suppression. Nature 229:244–245
35. Mertz DB (1971) The mathematical demography of the California condor population. Am Nat 105:437–454
36. Peden AE (1973) Virtual extinction of *Gambusia amistadensis* n.sp., a poeciliid fish from Texas. Copeia 1973(2):210–221
37. Post W, Antonio FB (1981) Breeding and rearing of seaside sparrows (*Ammospiza maritima*) in captivity. Intl Zoo Yrbk 21:123–128
38. Rabinowitz D (1981) Seven forms of rarity. In: The biological aspects of rare plant conservation, ed Synge H. London: Wiley, pp 205–217
39. Ralls K, Ballou J (1983) Extinction: Lessons from zoos. In: Genetics and conservation, eds Schonewald-Cox CM, Chambers SM, McBryde B, Thomas L. Menlo Park, CA: Benjamin/Cummings, pp 164–184
40. Richter-Dyn N, Goel NS (1972) On the extinction of a colonizing species. Theoret Pop Biol 3:406–433
41. Segerberg O Jr (1971) Where have all the flowers, fishes, birds, trees, water, and air gone? New York: David McKay

42. Senner JW (1980) Inbreeding depression and the survival of zoo populations. In: Conservation biology: an evolutionary-ecological perspective, eds Soulé ME, Wilcox BA. Sunderland, MA: Sinauer, pp 209–224
43. Shaffer ML (1981) Minimum population sizes for species conservation. BioScience 31:131–134
44. Shaffer ML, Samson FB (1985) Population size and extinction: a note on determining critical population sizes. Am Nat 125:144–152
45. Simberloff D (1981) Community effects of introduced species. In: Biotic crises in ecological and evolutionary time, ed Nitecki MH. New York: Academic Press, pp 53–82
46. Simon N, Géroudet P (1970) Last survivors. New York: World Publishing Co
47. Soulé ME (1980) Thresholds for survival: maintaining fitness and evolutionary potential. In: Conservation biology: an evolutionary-ecological perspective, eds Soulé ME, Wilcox BA. Sunderland, MA: Sinauer, pp 151–170
48. Soulé ME (1983) What do we really know about extinction? In: Genetics and conservation, eds Schonewald-Cox CM, Chambers SM, MacBryde B, Thomas L. Menlo Park, CA: Benjamin/Cummings, pp 111–124
49. Soulé ME, Simberloff D (1986) What do genetics and ecology tell us about the design of nature reserves? Biol Conserv 35:19–40
50. Stanley SM (1979) Macroevolution, pattern and process. San Francisco: WH Freeman
51. Stewart D (1978) From the edge of extinction. New York: Methuen
52. Sykes PW Jr (1980) Decline and disappearance of the dusky seaside sparrow from Merritt Island, Florida. Am Birds 34:728–737
53. Terborgh J, Winter B (1980) Some causes of extinction. In: Conservation biology: an evolutionary-ecological perspective, eds Soulé ME, Wilcox BA. Sunderland, MA: Sinauer, pp 119–134
54. Van Valen L (1973) A new evolutionary law. Evol Theory 1:1–30
55. Wittenberger JF (1981) Animal social behavior. Boston: Duxbury Press
56. Ziswiler V (1967) Extinct and vanishing animals. New York: Springer-Verlag

Patterns and Processes in the History of Life,
eds. D. M. Raup and D. Jablonski, pp. 277–295. Dahlem Konferenzen 1986
Springer-Verlag Berlin, Heidelberg

Phanerozoic Overview of Mass Extinction

J. J. Sepkoski, Jr.

Dept. of Geophysical Sciences
University of Chicago
Chicago, IL 60637, USA

Abstract. Mass extinctions are episodes of accelerated extinction of variable magnitude that affect widespread taxa and cause at least temporary declines in their diversity. Although such episodes are often difficult to identify and characterize precisely in the fossil record, it is clear that they have been frequent throughout the history of complex life. In this paper, I briefly summarize 29 definite and potential events of mass extinction that can be recognized in a new compilation of data on fossil marine genera.

Introduction

Mass extinctions are the most dramatic events in the history of life. The largest eliminated huge numbers of species and higher taxa, changed patterns of adaptation, and shaped the evolutionary pathways that generated the variety of taxa and morphologies seen in the succeeding quiet intervals of geologic time. There has been a renewed interest in the patterns, causes, and evolutionary consequences of mass extinctions in recent years, following the publication by L. W. Alvarez et al. [1] of the extraterrestrial impact hypothesis for the terminal Cretaceous mass extinction. To date, at least 25 purported mass extinctions have been identified in the marine and terrestrial fossil records. In this contribution, I will briefly review some of what is known (and not known) about these events as they are manifested in the Phanerozoic marine fossil record. This review in many respects will be an update of my previous summaries [41, 42]. But whereas those papers concentrated on extinction patterns among taxonomic families, this contribution will summarize patterns exhibited in a new (and still preliminary) data set on marine genera. Below, I first discuss problems encountered in defining and identifying mass extinctions; I then describe the nature of the generic data set; and finally I present brief descriptions of the mass extinctions, and events that look like mass extinctions, evident among fossil marine genera.

Difficulties in Identifying Mass Extinctions

General Definition

A mass extinction is any substantial increase in the amount of extinction (i.e., lineage termination) suffered by more than one geographically widespread higher taxon during a relatively short interval of geologic time, resulting in an at least temporary decline in their standing diversity. This is a general definition that is designed purposely to be somewhat vague, especially with respect to the meanings of "substantial increase" and "relatively short interval." We will have only limited knowledge of the frequency, magnitudes, durations, selectivity, and macroevolutionary consequences of events of accelerated extinction. Five events, occurring in the Late Ordovician (Ashgillian), Late Devonian (Frasnian), Late Permian, Late Triassic, and Late Cretaceous (Maestrichtian), are of such magnitude that they clearly stand out as distinct from the evolutionary turnover of normal geologic intervals [35, 41]. Other extinction events, such as that in the Late Eocene, are much smaller and not so clearly distinct from background extinction. In the absence of precise, high-resolution data on such smaller events, it is not clear whether they represent discrete episodes of mass extinction (that is, scaled-down versions of the five largest events) or simply slight accelerations of normal, background extinction over finite, but imprecisely known, time intervals.

Biostratigraphic Resolution

There are two basic and complementary empirical means by which mass extinctions can be investigated: detailed biostratigraphic studies and global taxonomic analyses. Local biostratigraphic studies are the best means for determining the durations and internal patterns of extinction events. Detailed range charts of species sampled bed-by-bed or decimeter-by-decimeter in outcrops or cores can reveal whether such events represent abrupt truncations of species' ranges over several centimeters to meters of section (such as seen in the several Cambrian biomere events; see [29–31]) or if they represent more protracted range terminations without concomitant replacement. However, because of the time-consuming labor involved, detailed biostratigraphic analyses of presumed mass extinctions are still very limited. Available range charts generally have restricted taxonomic and stratigraphic coverage and generally have been assembled for only a few, geographically scattered localities. Thus, for most extinction events, published biostratigraphic range charts do not provide sufficient resolution to determine how many higher taxa were affected and whether detailed patterns of extinction differ from those of background intervals. Furthermore, because biostratigraphic studies are generally local, they often do not provide information on how widespread the events are and how much of the range termination might be induced by local hiatuses, nonconformities, or facies changes (such as seen in the regressions that frequently accompany extinction events). Finally, even when available from widespread localities, problems of detailed correlation often preclude assessment of true durations of extinction events in range charts.

Taxonomic Compilations

Synoptic compilations of the stratigraphic ranges of taxa over the entire globe (or even continents) can overcome some of the problems inherent in local biostratigraphic studies. Compilations have the advantages that a) they represent the summed effort of numerous systematic and biostratigraphic studies and therefore reflect a much larger sample of the biota; b) they often encompass many or all taxonomic groups in a major ecosystem and therefore permit assessment of the overall effects of extinction events; c) because they may be global, they are presumably less influenced by local hiatuses and facies changes; and d) they usually cover longer intervals of geologic time and therefore more readily permit comparative assessment of the frequency, magnitudes, and changes in diversity associated with events.

Despite these positive properties, taxonomic compilations do not provide the detailed resolution on specific events afforded by local biostratigraphic studies. Compilations are normally executed at taxonomic ranks above the species level (frequently at the family level; e.g., [8, 28, 41, 52]) and register last occurrences of taxa only to globally or regionally recognizable stratigraphic intervals (zones, stages, or series) which may have durations of millions of years as well as imprecisely defined or correlated boundaries.

These last two properties induce a dampening of the signal of extinction events in taxonomic compilations: polytypic higher taxa reflect an event only if all constituent species become extinct (cf. [34]), and data summed over finite stratigraphic intervals often reflect an averaging of intensities of both short-term extinction events and longer background phases. Dampening may be further exacerbated by imperfect sampling of the fossil record, involving both the failure to find rare taxa and the backward smearing of last appearances of long-ranging taxa that have not been found in their last interval of existence [45]. Because of this dampening, taxonomic compilations can provide only minimum estimates of the magnitudes of well documented extinction events and maximum estimates of their durations. Thus, it is impossible in many cases to determine from a compilation whether an interval of increased extinction represents a discrete event or merely a protracted acceleration of background extinction.

There are also situations, however, in which taxonomic compilations with fine stratigraphic resolution might enhance the magnitudes of extinction events or even create nonexistent events. Failure to sample a substantial portion of the biota in one stratigraphic interval will cause truncation of the observed ranges of some taxa and can thereby artificially concentrate apparent extinctions in a preceding, better-sampled interval. This situation can occur when monographic efforts or geographic coverage of sampling varies considerably from one stage or substage to another or when facies changes, such as accompany sea level fluctuations, make the modal strata of the one stage considerably less fossiliferous than the preceding (cf. discussion of the middle Caradocian, below). Both situations can induce artificial maxima in time series for extinction intensity which, without additional knowledge, might be interpreted as apparent extinction events.

The solution to these various sampling problems is to compile global range data at lower taxonomic levels and finer stratigraphic intervals and to study fine-scale patterns of mass extinction in directed biostratigraphic analyses. In this paper, I provide a preliminary report on the first endeavor, an attempt to compile global data on the stratigraphic ranges of genera resolved to the level of substages.

Generic Data

The effort to compile data on generic extinctions at the substage level represents an extension of my previous compilation of familial data [40]. This new work was initiated in 1984 and is still very incomplete (and thus the preliminary results reported here must be interpreted with care). The data base consists of all information on fossil marine invertebrates and protozoans (but not vertebrates) listed in the *Treatise on Invertebrate Paleontology* [26]. This basic information has been augmented with taxonomic and stratigraphic data from approximately 500 other literature sources, including various secondary compilations (e.g., [13, 18]), monographs, and taxonomic papers. To date, approximately 25,000 genera have been incorporated into the data set, of which more than 20,000 are extinct. This represents nearly an order of magnitude more data than afforded by the previous familial compilation.

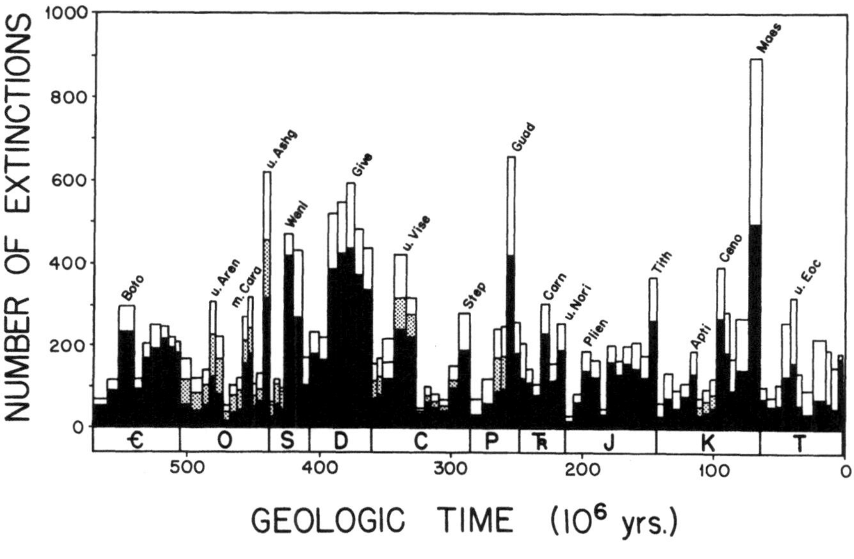

Fig. 1. Numbers of extinctions of marine animal genera (excluding vertebrates) in 94 stages and substages through the Phanerozoic. Black bars indicate high-resolution data; stippled bars indicate data at the stage level of resolution in cases where substages are shown; open bars indicate low-resolution series- and system-level data spread proportionally among the stages and substages. Local maxima of extinction are labelled here and in Fig. 2 with the following abbrevations: "l" = lower, "m" = middle, "u" = upper, "Boto" = Botomian, "Dres" = Dresbachian, "Trep" = Trempealeauan, "Aren" = Arenigian, "Cara" = Caradocian, "Ashg" = Ashgillian, "Wenl" = Wenlockian, "Give" = Givetian, "Fras" = Frasnian, "Vise" = Visean, "Serp" = Serpukhovian, "Step" = Stephanian, "Guad" = Guadalupian, "Olen" = Olenekian, "Carn" = Carnian, "Nori" = Norian, "Plien" = Pliensbachian, "Tith" = Tithonian, "Apti" = Aptian, "Ceno" = Cenomanian, "Maes" = Maestrichtian, "Eoc" = Eocene, "Mio" = Miocene, "Pl" = Pliocene

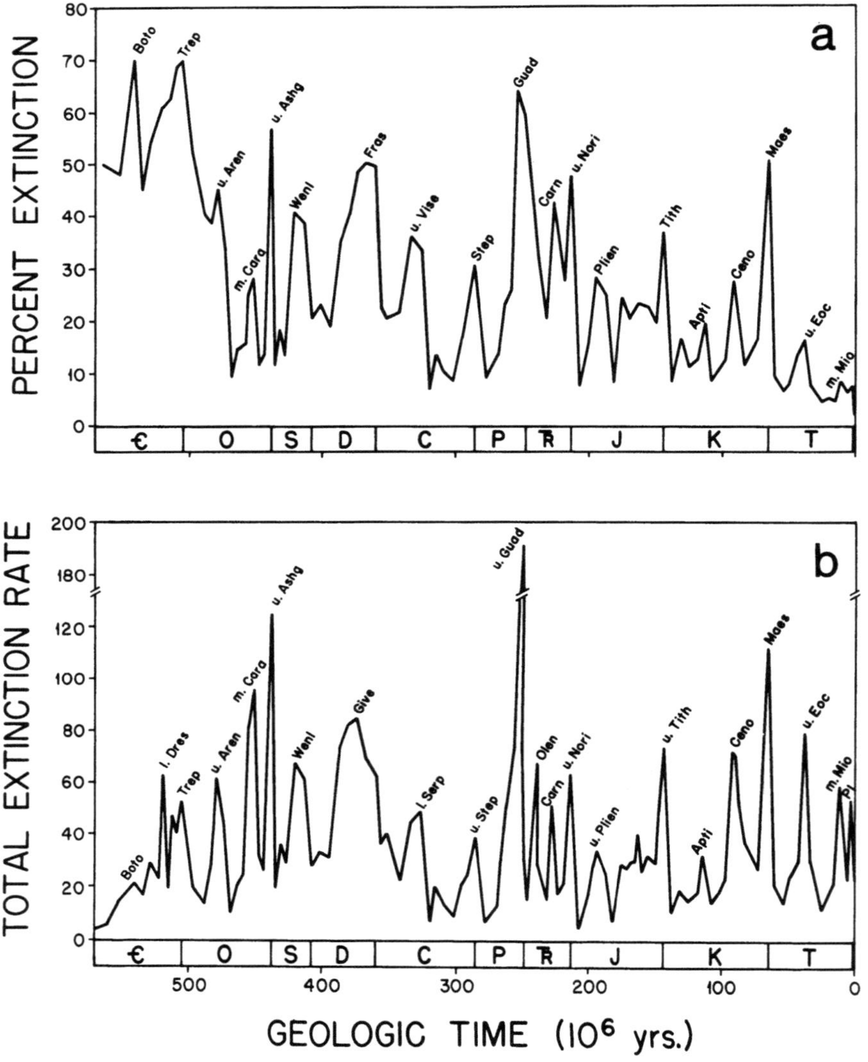

Fig. 2 a, b. Two metrics of generic extinction intensity through the Phanerozoic. **a** Percent extinction, calculated as the number of extinctions divided by standing generic diversity. Approximately the same stratigraphic units as in Fig. 1 are used. **b** Total rate of extinction, calculated as the number of extinctions divided by the estimated durations of the stratigraphic intervals. 111 intervals are used. Abbreviations for designated extinction maxima are listed in the caption to Fig. 1

The stratigraphic framework being employed is almost twice as detailed as that used in the familial compilation. Using the correlation charts presented by Harland et al. [14] as a primary base (but including numerous corrections and additions), the 82 "stages" of the previous compilation have been divided into 161 "substages." However, only a small proportion of the data have yet been resolved to this level. Of the generic extinctions, 33.5% are resolved to substage and 36% to stage; most of the remaining extinctions are listed to the series level, and 8% are known only to the system level.

The number of generic extinctions per stratigraphic stage are illustrated in Fig. 1. (Actually, several stages have been split or lumped in this figure to make the histogram widths more even.) Data resolved to the stage and/or substage level are represented by black bars, and series- and system-level data are shown by open bars. These low-resolution data have been distributed among the constituent stages in proportion to the amount of higher resolution data.

Figure 2a illustrates the raw data recalculated as percent extinction by dividing numbers of extinctions by estimated generic diversity, thus normalizing extinction numbers for the number of genera at risk. Again, stages have been variously split and lumped as in Fig. 1 in order to equalize their durations as much as possible. (Such manipulation is necessary so that the average amount of generic turnover within a sampling interval is not affected by systematic variations in interval duration through time.) Fifty-two intervals are shown for the Paleozoic, providing an average resolution of 6.3 myr, and 43 for the post-Paleozoic, giving an average of 5.8 myr.

Figure 2b illustrates a time series for the total rate of generic extinction, calculated by dividing raw numbers of extinctions by the stage durations as estimated by Harland et al. [14]. All stages that contained mostly substage-level data have been split into their constituent subdivisions (each assumed to have equal duration) and lower-resolution data have been distributed proportionally. This manipulation resulted in 111 stratigraphic intervals, providing an average resolution of 5.5 myr in the Paleozoic (50% better than in the familial data) and 4.8 myr in the Mesozoic-Cenozoic (25% better).

Maxima in Marine Extinctions

The curves in Fig. 2 exhibit a pattern of widely fluctuating extinction intensities through the Phanerozoic with few long-term trends. The only consistent trend is exhibited in Fig. 2a, where the extinction intensity is normalized for standing diversity. This metric shows a persistent decline in average background intensity from an extraordinary high of around 50% generic extinction per stratigraphic unit in the Cambrian (average unit duration = 7.8 myr) to a low of around 5% in the Neogene (average duration = 5 myr) (cf. [35]). Superimposed upon this trend are numerous local maxima, or "peaks," of extinction, most of which are evident in all graphs in Figs. 1 and 2. The majority of the larger peaks in the Mesozoic and Cenozoic are known from more detailed biostratigraphic information to reflect events that can be considered true extinction events. Those in the Paleozoic are less thoroughly studied and their identity is consequently less certain.

In the text below, I briefly describe each of the major extinction peaks evident in Fig. 2. I begin with the Late Permian event, which is the largest mass extinction of the Phanerozoic, then consider the Mesozoic-Cenozoic peaks in sequence, and finally end with the Paleozoic peaks. Although I will treat most of these extinction maxima as if they were mass extinctions, it must be understood that I am not strictly adhering to the definition presented earlier. Because most events are still imprecisely characterized in terms of magnitude, abruptness, and taxonomic and geographic selectivity, I use an operational definition that any peak in extinction intensity that appears, at some scale, like a peak known to represent a documented extinction event should be considered at least a potential candidate for a mass extinction.

Late Permian

The greatest mass extinction of the Phanerozoic occurred within the Late Permian [28, 41, 52]. Using data on familial and ordinal extinctions, Raup [34] estimated by means of rarefaction curves that the Late Permian event may have eliminated 96% of species in the oceans. The generic data in Fig. 2a support this estimate; they indicate that 83% of marine genera became extinct over the last two "stages" of the Permian (Guadalupian + Tatarian [= Djhulfian + Dorashamian or Changhsingian]). Using Raup's rarefaction curves, this provides an estimate of 94 to 96% extinction at the species level (as opposed to Raup's estimate of 88% from cruder generic data). Virtually all taxonomic groups were affected by this event, with members of the "Paleozoic evolutionary fauna" [39], including articulate brachiopods, crinoids, cephalopods, corals, and bryozoans, suffering the greatest extinction.

Although the great magnitude of the Late Permian event is clear, its precise timing is not. Usually it has been assumed that this was a protracted mass extinction, encompassing at minimum 10 myr of the Permian (cf. [38]). The broad peaks for Late Permian generic extinctions in Fig. 2 might appear to support this contention. The peaks begin in the Leonardian, increase to a maximum in the Guadalupian (upper Guadalupian in the high-resolution data of Fig. 2b), and then decline somewhat in the Tatarian. The peak actually continues into the Lower and lower Middle Triassic, reflecting high rates of evolutionary turnover among groups, especially ceratite ammonoids, radiating after the extinction event ([51]; cf. [53]). The sharp peak in generic extinction rate over the Olenekian (upper Lower Triassic) in Fig. 2b seems to be a reflection of this rapid turnover among ceratites and not a distinct extinction event; it is balanced by equally high origination and results in no net decline in ceratite diversity at the zonal level.

Even though it is not demonstrable at this time, it is possible that much of the breadth of the Late Permian extinction peak reflects inadequate sampling resulting from the worldwide paucity of latest Permian stratigraphic sections. Complete sections across the Permo-Triassic Boundary are known from only three major areas [44], and paleontologic exploration of these sections is far from complete. Thus, many taxa known to range across the boundary have not yet been found in latest Permian rocks (e.g., [3, 27]; see also Jablonski, this volume). Among groups that did suffer major extinction, the timing of disappearance may be controlled more by available facies than by actual extinction. This may be particularly true for crinoids, which are well-known in diverse late Guadalupian faunas on Timor but are virtually unknown from younger Permian strata; whether this represents extinction 5 myr before the Permo-Triassic Boundary or simply lack of preservation of appropriate facies is not certian. Recent study of Chinese sections [44] suggests that most of the extinction may indeed have been concentrated in the very latest Permian, but further paleontologic exploration of all Permo-Triassic Boundary sections is needed to resolve this problem.

Mesozoic-Cenozoic Extinction Maxima

Excluding the Olenekian peak, the Triassic-to-Recent interval in Fig. 2 contains at least ten maxima of generic extinction, of which eight correspond to independently recognized extinction events. Each of these is briefly summarized below.

1. Upper Triassic – Two peaks of extinction intensity occur in the Upper Triassic in Fig. 2. The larger is in the upper Norian (= Rhaetian in part) and reflects the well-known "end Triassic extinction event" [12], which is one of the larger mass extinctions of the Phanerozoic [41]. At the generic

level, this event has a magnitude of 48% extinction, contributed largely by cephalopods, bivalves, gastropods, and brachiopods.

The preceding peak in the Carnian Stage is somewhat enigmatic and may not reflect an extinction event. The peak is produced mainly by large numbers of extinctions among the rapidly evolving ceratites and, to a lesser extent, gastropods, for which information of Triassic occurrences is rather poor. Thus, it is possible that the Carnian peak reflects both high turnover and backward smearing of extinction records from the latest Triassic event. However, detailed biostratigraphic analysis should still be directed toward the Carnian to determine the precise nature of extinction.

2. Pliensbachian – Both time series in Fig. 2 exhibit a low peak centered on the Pliensbachian Stage of the Lower Jurassic. This local maximum appears to correspond to an event Hallam [10, 11] recognized as a minor mass extinction affecting European bivalves in the lower third of the succeeding Toarcian Stage. In the generic data the Pliensbachian point, which is contributed mostly by bivalves and brachiopods, only slightly exceeds the magnitude of the Toarcian point, contributed largely by better-resolved cephalopods. Thus, it is likely that both points reflect the same event with apparent extinctions smeared backward into the Pliensbachian (especially upper Pliensbachian) as a result of incomplete sampling.

The Pliensbachian-Toarcian peak is followed in the Middle Jurassic by high rates of evolutionary turnover, expecially among ammonoid cephalopods. No extinction events can be identified in this interval, even with the generic data resolved to the level of substage (see also [10, 11]). The distinct minimum of extinction in the Aalenian (lower Middle Jurassic) may be artifactual, since the Aalenian is not distinguished from the lower Bajocian in many parts of the world.

3. Upper Tithonian – The Tithonian Stage of the latest Jurassic has been recognized as a time of marked extinctions among marine animal families [36, 42] as well as bivalve genera [11] and dinoflagellate species [50]. The generic data indicate that this may be the largest extinction event in the 150 myr interval between the ends of the Triassic and Cretaceous Periods. Approximately 37% of marine genera became extinct across the Tithonian, which is nearly twice the magnitude in the preceding Upper Jurassic and succeeding Neocomian. Most of the extinctions, which are concentrated in the upper half of the Tithonian, are contributed by ammonoids and, to a lesser extent, bivalves and corals.

4. Aptian – A small peak in extinction intensity, corresponding to 19% generic extinction, is located in the Aptian Stage of the Lower Cretaceous. The scattered substage data suggest that much of this extinction may have

occurred in the upper part of the stage. To my knowledge, no distinct extinction event has been recognized here on the basis of detailed biostratigraphic analyses (although see [19]). However, an Aptian event was predicted by Sepkoski and Raup [42] in their analysis of periodicitiy in post-Paleozoic extinctions. Thus, the small Aptian peak in Fig. 1 is intriguing and should be subjected to greater scrutiny.

5. Cenomanian – The Cenomanian Stage has recently been recognized as containing a significant and fairly short-term (< 1 myr) extinction event at its top [20]. The generic data indicate that about 28% of existing genera disappeared across the Cenomanian, as contrasted to 9 to 19% in surrounding stages. Major taxa contributing to this extinction are ammonoids, bivalves, echinoids, foraminifers, malacostracans, and ostracodes. A major extinction among planktonic dinoflagellates has also been noted within the Cenomanian [21, 50].

6. Maestrichtian – The last half of the Maestrichtian Stage contains one of the largest mass extinctions of the Phanerozoic, affecting both marine and terrestrial taxa [2, 9, 37, 41]. The data in Fig. 2 corroborate Russell's [37] estimate of 50% generic extinction at the end of the Cretaceous. This magnitude makes the Maestrichtian the largest of post-Paleozoic extinction events and comparable to the Paleozoic Ashgillian and Devonian events which occurred at times of much greater background extinction. Virtually all major taxonomic groups were affected by the Maestrichtian event, with foraminifers, bivalves, bryozoans, ammonoids, gastropods, sponges, echinoids, and ostracodes all suffering more than 20 generic extinctions.

7. Upper Eocene – The Upper Eocene (Priabonian Stage) is usually considered to be an interval of major extinction, although some workers (e.g., [46, 53]) have recently questioned its magnitude and abruptness. Figure 2 indicates that the Upper Eocene does indeed contain a fairly substantial maximum in generic extinction, although it may be spread over up to 4 myr. Approximately 16% of marine genera disappear in this stage as compared to the surrounding low background intensities of 5 to 9%. In addition to previously documented extinctions among planktonic dinoflagellates, coccolithophorids, ebridians, and silicoflagellates [21, 50], the generic data indicate declines among foraminifers, echinoids, gastropods, bryozoans, and malacostracans.

8. Middle Miocene and Pliocene – The generic data exhibit two comparatively low peaks of extinction in the Neogene. The presumed Middle Miocene (Serravallian) event has not been widely recognized, although it was noted for marine animal families [36, 42] and appears as a sharp decline in survivorship among tropical planktonic foraminifers [15]. In addition to

foraminifers, the generic data suggest that bivalves, gastropods, and echinoids were involved.

Significant extinction among marine bivalves in the Pliocene has been documented in detailed analyses of subtropical and temperate faunas by Stanley [47, 48] and Stanley and Campbell [49]. The generic data also indicate accelerated extinction among foraminifers and echinoids, although neither group exhibits the magnitude seen in the Middle Miocene.

Paleozoic Extinction Maxima

The frequency of extinction events in the Paleozoic Era has appeared in the past to be considerably less than in the Mesozoic and Cenozoic [41]. Other than major events in the Ashgillian, Upper Devonian, and Upper Permian, only a few minor mass extinctions, largely confined to the Cambrian, had been recognized. The new generic data, coupled with the recent analysis of House [17], suggest that this impression is incorrect and that the frequency of events in the 325 myr of the Paleozoic was at least equivalent to that of the post-Paleozoic.

1. Cambrian – The Cambrian was a period of exceptionally high evolutionary turnover and frequent extinction events. The data on percent extinction for genera in Fig. 2a indicate that background extinction in the Cambrian through Early Ordovician was, on average, higher than that in most post-Paleozoic stages containing discrete extinction events. Most of this high extinction intensity was contributed by trilobites, which dominated the Cambrian fauna. Trilobites also suffered at least five extinction events (the so-called "biomere events") during the Cambrian, best documented in North America by Palmer [29–31] and others. Three of these events, which presumably are the largest of the five, appear as peaks in Figs. 1 and 2; these occur in the Botomian (upper Lower Cambrian), lower Dresbachian (lower Upper Cambrian), and Trempealeauan (end of the Cambrian). Each represents 15 to 20% extinction added above the background. Other biomere events that are not so evident in the generic data occur in the lower Franconian (middle Upper Cambrian) and probably the mid-Middle Cambrian.

2. Upper Arenigian – A low peak of generic extinction occurs in the upper Arenigian at the top of the Lower Ordovician. This peak represents an addition of only 6% to apparent background and is located near the end of the early Paleozoic phase of rapid evolutionary turnover. However, it corresponds in time to a biotic crisis noted by Boucot [5] in his synthesis of benthic community change through the Phanerozoic. The principal con-

tributors to the increased generic extinctions are cephalopods, trilobites, and articulate brachipods.

3. Middle Caradocian – The middle Caradocian contains a major peak in the total rate of generic extinction (Fig. 2b), contributed largely by bryozoans, brachiopods, and ostracodes. It is not clear, however, whether this peak represents a true extinction event or merley an artifact of sampling. In North America especially, the middle Caradocian ("Trentonian") is very fossiliferous and is succeeded by less fossiliferous, or at least less intensely studied, strata of Edenian and Maysvillian age; thus, the apparent extinction could largely represent truncations of ranges resulting form insufficient sampling. Alternatively, Brenchley [6] has suggested that there was a real decline in marine diversity caused by a reduction in biogeographic provinciality following the middle Caradocian. Comparative analyses of biostratigraphic data from different regions will be needed to evaluate these alternatives.

4. Upper Ashgillian – The end of the Ordovician was marked by a major mass extinction which may have been the second largest in Phanerozoic history [6, 41, 43]. Most of the extinction in the generic data is concentrated in the upper Ashgillian (Cautleyan through Hirnantian), over which approximately 57% of marine genera disappeared (Fig. 2a). Groups particularly affected include cephalopods, brachiopods, corals, bryozoans, and crinoids; trilobites also suffered major extinction and never again regained their high diversity of the early Paleozoic. Brenchley's [6] more detailed analysis of this event indicates that most of the decline in trilobite diversity occurred during the Rawtheyan interval before the terminal Hirnantian; brachiopods, on the other hand, suffered their greatest decline in the Hirnantian. Thus, the Ashgillian event may have been spread over several million years at the end of the Ordovician.

5. Silurian – No event of mass extinction has previously been recognized within the Silurian, although some data, such as Boucot's [4] synthesis of articulate brachiopods, show substantial declines in diversity late in the Silurian. The generic data in Fig. 2 exhibit strong peaks in extinction centered on the Wenlockian and Ludlovian. This suggests that there may have been one or more extinction events somewhere in the last half of the Silurian. Precisely where these are, however, is not clear. The generic data consistently show a maximum in the Wenlockian (Middle Silurian), primarily contributed by nautiloid cephalopods, articulate brachiopods, ostracodes, and crinoids. These same groups along with corals continue to exhibit high rates of extinction into the succeeding Ludlovian, closer to where Boucot's data show the decline in diversity. It is possible that these apparently pro-

tracted extinctions reflect backward smearing of last appearances from an event in the Ludlovian (or even Pridolian), where strata are often regressive.

6. Devonian – The Devonian has generally been considered to contain one of the five major mass extinctions of the Phanerozoic, centered on the Frasnian-Famennian Boundary within the Upper Devonian [16, 23]. McLaren [24, 25] has argued that this event was particularly extreme in terms of biomass and was associated with a worldwide disappearance of tropical reefs and concomitant extinction of most reef-building corals and stromatoporoids (see also [7, 32]). House [17], however, has shown that the Frasnian-Famennian ("Kellwasser") event was but one of a number of extinction events that affected the early ammonoid cephalopods during the Devonian. He has documented a pattern, reminiscent of the Cambrian biomere pattern, of eight rather closely spaced extinction events in the Emsian through Famennian Stages. Three of these events appear substantial: those at or near the Givetian-Frasnian, Frasnian-Famennian, and Famennian-Tournaisian (lowest Carboniferous) Boundaries. McGhee [22] has also shown that brachiopod extinctions in North America were not confined to the Frasnian-Famennian Boundary but rather spread out, probably in several pulses, between the upper Givetian and upper Frasnian.

The data in Fig. 2 are consistent with this emerging picture of multiple events within the Devonian. Both percent extinction and total extinction rate exhibit high intensities from the Emsian (upper Lower Devonian) through the Famennian (upper Upper Devonian) with indistinct maxima in either the Givetian or Frasnian. Virtually all diverse higher taxa, especially the brachiopods, cephalopods, trilobites, corals, and crinoids, exhibit high extinction rates across this interval without any clear pattern of one group suffering major extinction before another. Unfortunately, the generic data cannot yet be resolved to the substage level within the Devonian in order to determine directly whether these data show House's multiple events.

7. Upper Visean-lower Serpukhovian – The Carboniferous, like the Silurian, has generally not been recognized as containing major mass extinctions. However, the generic data suggest that two fairly substantial events may have occurred within this period, one in the upper Visean and/or lower Serpukhovian (=lower Namurian A) and the other in the upper Stephanian (end of the Carboniferous). The magnitude of the first event is rather low (about 35% generic extinction) and is comparable to that of many of the post-Paleozoic events. However, it is set against the low background magnitude that characterizes most of the Carboniferous and there-

fore appears significant. Groups exhibiting excess extinction around this time include foraminifers, brachiopods, and corals in the upper Visean and cephalopods, crinoids, and bryozoans in the lower Serpukhovian. (Ramsbottom [33] has also noted a significant extinction among ammonoids in the lower Serpukhovian.) It is not clear whether the extended extinction through the upper Visean and lower Serpukhovian reflects a protracted event of up to 15 myr duration or, more probably, backward smearing of extinction records resulting from inadequate sampling of frequently regressive Serpukhovian strata.

8. Upper Stephanian – The end of the Carboniferous seems to be marked by a small extinction event that eliminated about 30% of marine genera. A wide variety of higher taxa show slightly increased extinction rates at this time, especially the crinoids, cephalopods, foraminifers, and

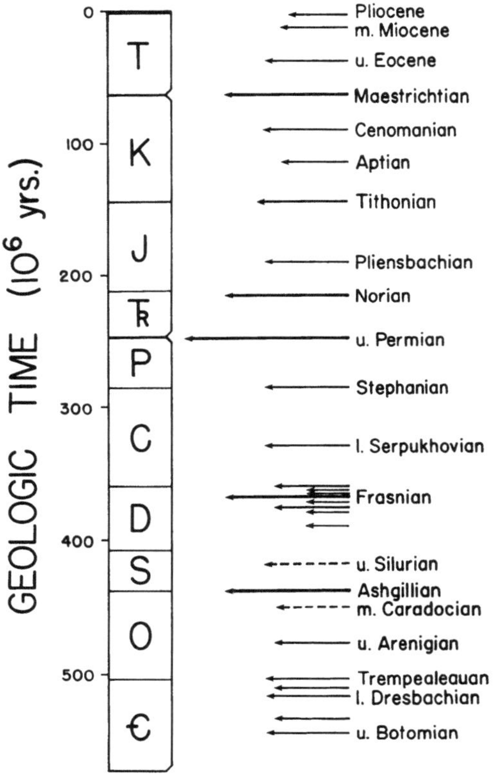

Fig. 3. Summary of the temporal distribution of extinction events through the Phanerozoic marine fossil record. The lengths of the arrows, which indicate the events, show very approximately their relative magnitudes

gastropods. This event is also evident at the familial level as a low but statistically significant maximum in extinction intensity [42].

Conclusions

Only a few events of mass extinction are well documented in the fossil record. However, the preliminary analysis of extinction intensity at the generic level indicates that mass extinction may have been far more frequent than generally realized (Fig. 3). I have briefly summarized 29 local maxima in extinction intensity that may represent anything from protracted intervals of high evolutionary turnover to catastrophic events of mass extinction. Viewed against the entire panorama of Phanerozoic evolution, this frequency reinforces the emerging notion that the history of the marine biota has not been one of gradually developing evolutionary pathways. Rather, it suggests that longterm evolution may have been fairly chaotic, especially at lower taxonomic levels, with numerous perturbations that were important in shaping the adaptations and ecological patterns seen in the fossil record.

Acknowledgements. Compilation and analysis of generic level data have been supported by NASA grant NAG 2-282. I thank K. W. Flessa for his helpful criticism of the draft manuscript.

References

1. Alvarez LW, Alvarez W, Asaro F, Michel HV (1980) Extraterrestrial cause for the Cretaceous-Tertiary extinction. Science 208:1095–1108
2. Alvarez W, Kauffman EG, Surlyk F, Alvarez LW, Asaro F, Michel HV (1984) Impact theory of mass extinctions and the invertebrate fossil record. Science 223:1135–1141
3. Batten RL (1973) The vicissitudes of the gastropods during the interval of Guadalupian-Ladinian time. In: The Permian and Triassic Systems and their mutual boundaries, eds Logan A, Hills LV. Calgary: Canadian Society of Petroleum Geologists, pp 596–607
4. Boucot AJ (1975) Evolution and extinction rate controls. Amsterdam: Elsevier
5. Boucot AJ (1983) Does evolution take place in an ecological vacuum? II. J Paleontol 57:1–30
6. Brenchley PJ (1984) Late Ordovician extinctions and their relationship to the Gondwana glaciation. In: Fossils and climate, ed Brenchley PJ. New York: Wiley
7. Copper P (1977) Paleolatitudes in the Devonian of Brazil and the Frasnian-Famennian mass extinction. Palaeogeog Palaeoclimatol Palaeoecol 21:165–207

8. Cutbill JL, Funnell BM (1967) Computer analysis of *The Fossil Record*. In: The Fossil Record, eds Harland WB et al. London: Geological Society of London, pp 791–820
9. Emiliani C, Kraus EB, Shoemaker EM (1981) Sudden death at the end of the Mesozoic. Earth Planet Sci Lett 55:317–334
10. Hallam A (1976) Stratigraphic distribution and ecology of European Jurassic bivalves. Lethaia 9:245–259
11. Hallam A (1977) Jurassic bivalve biogeography. Paleobiology 3:58–73
12. Hallam A (1981) The end-Triassic bivalve extinction event. Palaeogeog Palaeoclimatol Palaeoecol 35:1–44
13. Harland WB et al. (eds) (1967) The fossil record. London: Geological Society of London
14. Harland WB, Cox AV, Llewellyn PG, Pickton CAG, Smith AG, Walters R (1982) A geologic time scale. Cambridge: Cambridge University Press
15. Hoffman A, Kitchell JA (1984) Evolution in a pelagic planktic system: a paleobiologic test of models of multispecies evolution. Paleobiology 10:9–33
16. House MR (1967) Fluctuations in the volution of Palaeozoic invertebrates. In: The fossil record, eds Harland WB et al. London: Geological Society of London, pp 41–54
17. House MR (1985) Correlation of mid-Palaeozoic ammonoid evolutionary events with global sedimentary perturbations. Nature 313:17–22
18. House MR, Senior JR (eds) (1981) The Ammonoidea. New York: Academic Press
19. Kauffman EG (1979) Cretaceous. In: Treatise on invertebrate paleontology, Part A, eds Robison RA, Teichert C, pp A418–A487. Lawrence, KS: Geological Society of America and University of Kansas Press
20. Kauffman EG (1983) Mass extinction within the Cretaceous: earthbound events for calibration of extraterrestrial effects. Geol Soc Am Abstr Prog 15:608
21. Lipps JH (1970) Plankton evolution. Evolution 24:1–22
22. McGhee GR Jr (1982) The Frasnian-Famennian extinction event: a preliminary analysis of Appalachian marine ecosystems. In: Geological implications of impacts of large asteroids and comets on the earth, eds Silver LT, Schultz PH. Geol Soc Am Spec Paper 190:491–500
23. McLaren DJ (1970) Time, life and boundaries. J Paleontol 44:801–815
24. McLaren DJ (1982) Frasnian-Famennian extinctions. In: Geological implications of impacts of large asteroids and comets on the earth, eds Silver LT, Schultz PH. Geol Soc Am Spec Paper 190:477–484
25. McLaren DJ (1983) Bolides and biostratigraphy. Geol Soc Am Bull 94:313–324
26. Moore RC, Teichert C, Robison RA (eds) (1953–1984) Treatise on invertebrate Paleontology. Lawrence, KS: Geological Society of America and University of Kansas Press
27. Nakazawa K, Runnegar B (1973) The Permian-Triassic Boundary: a crisis for bivalves? In: The Permian and Triassic Systems and their mutual boundary, eds Logan A, Hills LV. Calgary: Canadian Society of Petroleum Geologists, pp 608–621
28. Newell ND (1967) Revolutions in the history of life. In: Uniformity and simplicity: a symposion on the principle of the uniformity of nature, ed Albritton CC Jr. Geol Soc Am Spec Paper 89:63–91
29. Palmer AR (1979) Biomere boundaries re-examined. Alcheringa 3:33–41
30. Palmer AR (1982) Biomere boundaries: a possible test for extraterrestrial perturbation of the biosphere. In: Geological implications of impacts of large asteroids and

comets on the earth, eds Silver LT, Schultz PH. Geol Soc Am Spec Paper 190:469–476
31. Palmer AR (1984) The biomere problem: evolution of an idea. J Paleontol 58:599–611
32. Pedder AEH (1982) The rugose coral record across the Frasnian-Famennian Boundary. In: Geological implications of impacts of large asteroids and comets on the earth, eds Silver LT, Schultz PH. Geol Soc Am Spec Paper 190:485–490
33. Ramsbottom WHC (1981) Eustatic control in Carboniferous ammonoid biostratigraphy. In: The Ammonoidea, eds House MR, Senior JR. New York: Academic Press, pp 369–388
34. Raup DM (1979) Size of the Permo-Triassic bottleneck and its evolutionary implications. Science 206:217–218
35. Raup DM, Sepkoski JJ Jr (1982) Mass extinctions in the marine fossil record. Science 215:1501–1503
36. Raup DM, Sepkoski JJ Jr (1984) Periodicity of extinctions in the geologic past. Proc Natl Acad Sci USA 81:801–805
37. Russell DA (1977) The biotic crisis at the end of the Cretaceous Period. In: K-TEC, Cretaceous-Tertiary extinctions and possible terrestrial and extraterrestrial causes, eds Béland P et al. Ottawa: National Museum of Natural Sciences Syllogeus No 12, pp 11–24
38. Schopf TJM (1974) Permo-Triassic extinctions: relation to sea-floor spreading. J Geol 82:129–143
39. Sepkoski JJ Jr (1981) A factor analytic description of the Phanerozoic marine fossil record. Paleobiology 7:36–53
40. Sepkoski JJ Jr (1982) A compendium of fossil marine families. Milwaukee Public Museum Contr Bio Geol 51:1–125
41. Sepkoski JJ Jr (1982) Mass extinctions in the Phanerozoic oceans: a review. In: Geological implications of impacts of large asteroids and comets on the earth, eds Silver LT, Schultz PH. Geol Soc Am Spec Paper 190:283–289
42. Sepkoski JJ Jr, Raup DM (1986) Periodicity in marine mass extinctions. In: Dynamics of extinction, ed Elliott D. New York: Wiley, pp 3–36
43. Sheehan PM (1975) Brachiopod synecology in a time of crisis (Late Ordovician – Early Silurian). Paleobiology 1:205–212
44. Sheng J-Z, Chen C-Z, Wang Y-G, Rui L, Liao Z-T, Bando Y, Ishi K, Nakazawa K, Nakamura K (1984) Permian-Triassic Boundary in middle and eastern Tethys. J Fac Sci, Hokkaido University Series IV, 21:133–181
45. Signor PW III, Lipps JH (1982) Sampling bias, gradual extinction patterns, and catastrophes in the fossil record. In: Geological implications of impacts of large asteroids and comets on the earth, eds Silver LT, Schultz PH. Geol Soc Am Spec Paper 190:291–296
46. Snyder SW, Müller C, Miller KG (1984) Eocene-Oligocene Boundary: biostratigraphic recognition and gradual paleoceanographic change in DSDP Site 549. Geology 12:112–115
47. Stanley SM (1979) Macroevolution: pattern and process. San Francisco: Freeman
48. Stanley SM (1984) Marine mass extinctions: a dominant role for temperature. In: Extinctions, ed Nitecki MH. Chicago: University of Chicago Press, pp 69–118
49. Stanley SM, Campbell LD (1981) Neogene mass extinction of western Atlantic molluscs. Nature 293:457–459

50. Tappan H, Loeblich AR Jr (1971) Geobiologic implications of phytoplankton evolution and time-space distribution. Geol Soc Am Spec Paper 127:247–340
51. Tozer ET (1981) Triassic Ammonoidea: classification, evolution and relationship with Permian and Jurassic forms. In: The Ammonoidea, eds House MR, Senior JR. New York: Academic Press, pp 65–100
52. Valentine JW (1969) Patterns of taxonomic and ecological structure of the shelf benthos during Phanerozoic time. Palaeontology 12:684–709
53. Van Valen L (1984) A resetting of Phanerozoic community evolution. Nature 307:50–52

Patterns and Processes in the History of Life,
eds. D. M. Raup and D. Jablonski, pp. 297–312. Dahlem Konferenzen 1986
Springer-Verlag Berlin, Heidelberg

Environmental Changes in Times of Biotic Crisis

K. J. Hsü

Geologisches Institut, ETH-Zentrum
8092 Zürich, Switzerland

Abstract. The fossil record indicates accelerated rates of extinction and of evolution at era transitions or times of biotic crisis. Geochemical anomalies suggest that the dissolved CO_2 content in ocean surface waters was then abnormally high, probably because the ocean was almost devoid of plankton. The oxygen-minimum zone in the oceans expanded, and the surface waters were unusually corrosive. Meanwhile the sea bottom underlying the expanded oxygen-minimum zone became locally anoxic. The environmental changes curtailed drastically the fertility of marine organisms and were probably the cause of mass extinctions. On land, pollen evidence indicates serious disruption of the terrestrial plant ecosystem at the end of Cretaceous; there was widespread destruction of the vegetation, although only a few plant species became extinct. We have no positive indications of comparable changes at other era boundaries, although a total elimination of the upiquitous algal community at the end of the Precambrian era has been speculated upon. Oxygen istotope data suggest that temperature may have dropped and/or risen abruptly at times of biotic crisis, but systematic trend has not yet been established. Of the numerous ideas advanced to explain the environmental catastrophes ending geological eras, I find the theory of large-body impact a most attractive working hypothesis.

Introduction

Geological eras were first defined in the last century on the basis of major faunal changes. Mass extinctions were thus known to Darwin, but he did not accept the concept of biotic crisis. Convinced of slow and gradual biologic evolution, he postulated that large time intervals were represented by era boundaries, and "in these intervals there may have been much slow extermination" [7]. Geological studies, using radiometric dating and magnetostratigraphical techniques, have proved Darwin wrong: the fossil record is demonstrably continuous across the era boundaries at many localities, and sudden mass extinctions did take place. The extinction rate at the end of Cretaceous of some groups of fossil organisms, for example, is at least 1,000 times faster than the extinction rate of Cenozoic molluscs [12], and similarly fast rates for the beginning of Paleozoic and Mesozoic are possible. Those were times of extraordinarily accelerated evolution or times of biotic crisis. Darwin made a second mistake when he viewed biotic interactions – the struggle for existence – as being the main cause of evolution. As cited by Hallam recently [10], Darwin wrote in his masterpiece: "As species are produced and exterminated by slowly acting causes and not by miraculous acts of creation, and as most important of all causes of organic change is one which is almost independent of altered and perhaps suddenly altered physical condition, namely, the mutual relation of organism to organism – the improvement of one organism entailing the improvement or the extermination of others." We now believe that the survival of organisms depends upon the viability of environment. Mass extinction implies drastic environmental changes and biotic crisis implies their suddenness. Those environmental catastrophes left their imprint as geochemical anomalies. This article reviews our current knowledge and discusses the probable cause of those changes.

Strangelove Ocean

The main parameters characterizing the chemistry of oceans are oxygen content, dissolved CO_2, pH, carbon-isotope composition, etc. Systematic patterns are illustrated by Fig. 1.

Oxygen comes from the air and is dissolved in surface waters of the oceans. When surface waters cool and sink to become bottom waters, the bottom is oxygenated. Intermediate depth bypassed by bottom circulation remains, however, relatively stagnat and the oxygen content is minimal there [4, 17]. Disseminated particles of organic matter impart to sediments of this oxygen-minimum zone a gray, dark gray, or black color.

The oceans get much of their dissolved CO_2 from land, but a more important source is recycled CO_2. Dead organisms falling on the seafloor are oxidized and release biogenic CO_2 to seawater. A descending ocean water

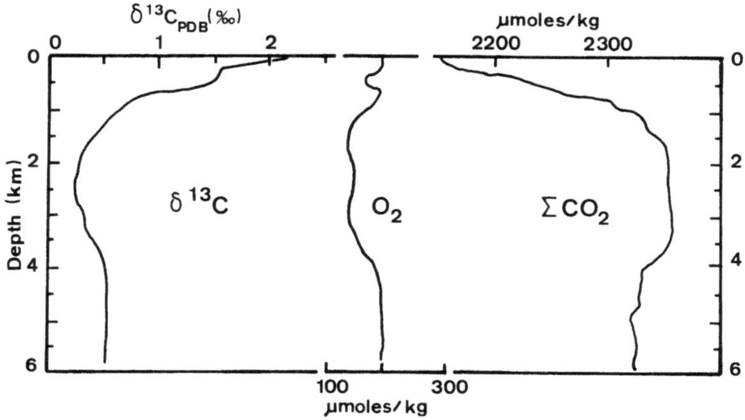

Fig. 1. Geochemical gradients in the oceans. The dissolved carbonate dioxide is depleted in the surface waters because of the utilization of CO_2 by plankton in their photosynthesis. At the same time the number of heavy carbon atoms ^{13}C (expressed as $\delta^{13}C_{pdb}$‰) in surface waters is increased because of the preferences of cell tissues for the light carbon atoms (after [17])

mass collects CO_2 as it sweeps past decaying organisms and becomes CO_2-rich bottom current. The oxygen-minimum zone outside of the current paths is stagnant, and there CO_2 accumulates to a maximum. In constrast the surface waters have the least amount of dissolved CO_2 because the CO_2-depleted surface water is alkaline with pH of about 8.1 to 8.3, and the CO_2-enriched intermediate and bottom waters are more acidic with pH values of about 7.5 [15].

The carbon isotope composition of dissolved carbonate also varies as a function of ocean depth. The variations are expressed in delta ^{13}C per mil referring to the difference between the number of carbon-13 atoms in a sample compared to that in a standard.

The vertical gradient in carbon isotopes is caused by a natural fractionation and is linked directly to the growth of plankton in surface waters [4, 17]. A living cell does not care about isotopic equilibrium. The tissues prefer a much lower level of carbon-13 atoms than that of a PDB standard, and they will take up less (20 or 30 per mil less), no matter what the isotopic composition of the dissolved carbonate in the surrounding seawater is. Since phytoplankton and zooplankton live in surface waters and their cell tissues prefer carbon-12 atoms, the surface seawater is normally impoverished in the light carbon and becomes enriched in carbon-13 atoms. On the other hand, the biomass of bottom dwellers is very small and carbon-isotope fractionation of the bottom waters is insignificant, so that their dissolved carbonate has an isotope composition not much different from that in average ocean waters. Figure 1 shows, for example, that the Pacific bottom water has a $\delta^{13}C$ value of about 0.5‰, whereas the surface water has 2‰ more carbon-13 atoms.

The effect of plankton production on water chemistry has been clarified by a study of chalk sedimentation in Swiss lakes [20]. Lake Greifen is a eutrophic lake in northwestern Switzerland. The peak season of algal growth occurs in July and August. The preferential utilization of carbon-12 (or the selective rejection of carbon-13) atoms leads to a maximum fractionation of carbon isotope during the summer. The dissolved carbonates in surface waters then have a maximum delta carbon-13 value 4.5‰ more than that of the bottom water (Fig. 2). As autumn sets in, the biologic productivity is reduced till organic growths reaches a dormant state in deep winter. The mid-December water samples have dissolved carbonates with the same isotopic composition from top to bottom. When spring comes, the growing season starts, and the early May sample already shows a slight gradient again, having resulted from the fractionation by the earlierst blooms of algae. The carbon-13 cycle of Lake Greifen is thus repeated annually.

The oceans have a much higher reservoir of dissolved carbonates. The annual fertility cycle could have little influence on the isotopic composition of dissolved carbonate, but long-term and sustained changes in productivity could have an effect. A hypothetical ocean of zero growth, nicknamed Strangelove ocean (after a fictitious character who wanted to start a nuclear holocaust), can be compared to the dormant winter season of Lake Greifen. If the carbon isotope fractionation is no longer affected by continued plankton production, a turnover of the ocean – nameley, descent of surface waters and upwelling of bottom waters, should homogenize the ocean chemistry within one or two thousand years. The delta carbon-13 values of the dissolved carbonates in a Strangelove ocean should thus be the same from top to bottom as they are during the winters in Lake Greifen.

We can conclude, therefore, that sudden changes in biologic productivity can produce two types of carbon isotope anomaly in planktic sediments. Sudden fertility increase produces a positive anomaly, which may be called a *fertility perturbation*. A nearly sterile ocean produces a negative anomaly, or a *Strangelove perturbation* (Fig. 3).

Brennecke and Anderson first discovered in 1977 a negative perturbation of 1.5 to 3.0‰ in $\delta^{13}C$ value across the Cretaceous/Tertiary boundary [3]. This change may seem insignificant but is about ten times the back-

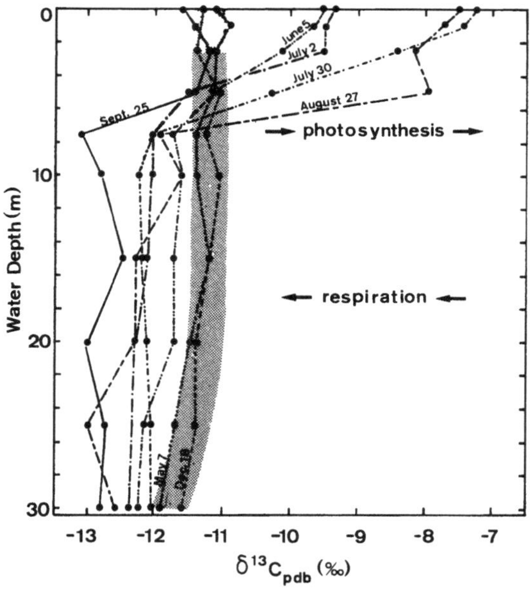

Fig. 2. Biogenic fractionation of carbon isotopes in Lake Greifen. During the early Spring months (May 7), the carbon isotope composition of the dissolved carbonate ions in the lake water is about the same from top to bottom. With the bloom of photosynthesizing algae which prefer light carbon atoms, relatively more heavy carbon atoms are left in surface waters, causing a maximum enrichment in mid-Summer (July 30). With the decrease of biologic activities during the Autumn (September 25), the enrichment decreases rapidly until the lake water again has the same isotopic composition from top to bottom when Winter comes (December 18). The cyclic change repeats itself every year

ground variation. Applying the concept of the "Strangelove perturbation" to explain the carbon isotope anomaly, Hsü and others concluded that the oceans after the terminal Cretaceous event became almost devoid of plankton [14]. The earliest Tertiary Ocean was thus a Strangelove ocean. At that time the normal process of carbon isotope fractionation by plankton growth was inhibited. Homogenization by oceanic mixing resulted in identical carbon isotopic compositions from top to bottom. The δ^{13}C values in the benthic foraminifers remain about the same across the boundary because the isotopic composition of the bottom waters did not change much. The temporary equalization of the δ^{13}C values in the benthic and planktic foraminifers is manifested in the minus 2‰ Strangelove perturbation in the boundary sediment across the Cretaceous/Tertiary transition (see Fig. 4).

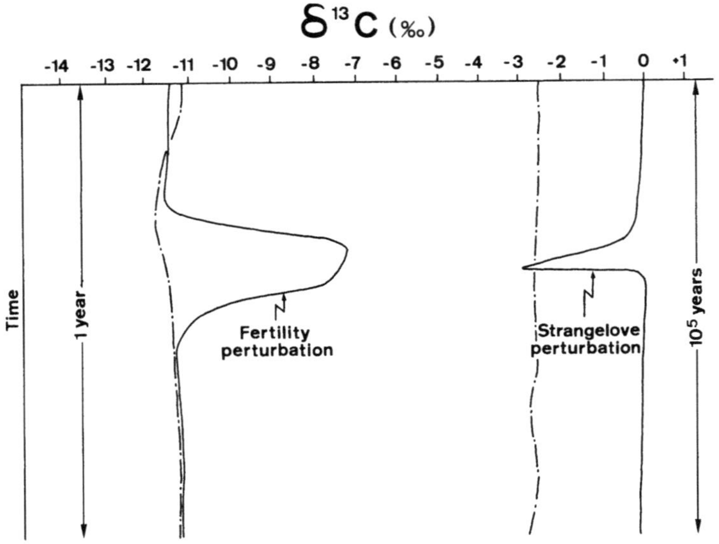

Fig. 3. Carbon isotope anomaly related to plankton productivity. The left side is the annual variation of carbon isotope values in dissolved carbonates of surface water (*solid line*) and of bottom water (*dashed line*) in Lake Greifen, Switzerland. Time at the bottom is January and at the top is December. A positive perturbation in surface water is observed in July–August when the plankton fertility reaches the maximum. The right side of the figure is the short-term (in geological terms) perturbation of carbon isotope values across the Cretaceous/Tertiary boundary at DSDP Site 524 in the South Atlantic Ocean, showing the changes in the planktic (*solid line*) and in the benthic (*dashed line*) feraminifers. The base of the negative Strangelove perturbation coincides with the C/T boundary, marked by an iridium anomaly; the perturbation is believed to indicate that the biomass of ocean plankton had been drastically reduced

An ocean can become temporarily sterile because of mass mortality, but ocean biomass is easily restored by reproductive regeneration. The normal life cycles of unicellular phytoplankton and zooplankton last for weeks or months only. To produce a Strangelove perturbation that lasted 10^4 years required that plankton reproductivity be curtailed for a sustained period of time. The biomass must then have remained low because many species of plankton ceased to multiply. The cause of this reduced fertility was probably ocean pollution, and the result was mass extinction of ocean plankton.

Calcareous plankton, such as foraminifera and nannoplankton, first evolved during the late Mesozoic. Yet Precambrian, Paleozoic, and early Mesozoic pelagic or hemipelagic sediments contain carbonates. The cal-

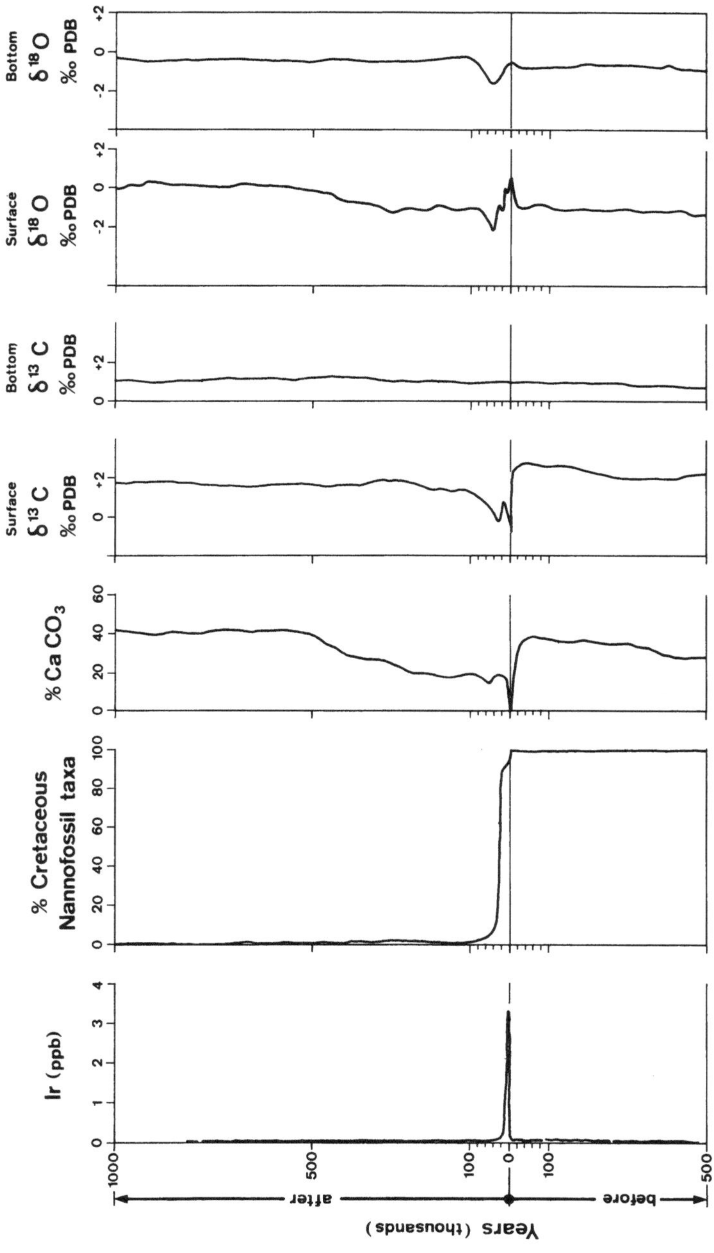

Fig. 4. Geochemical perturbations across the Cretaceous/Tertiary boundary, DSDP Site 524, South Atlantic. Note the carbon-13 and oxygen-18 anomalies in the sediments deposited during th 10^4 years after the boundary event marked by the iridium anomaly. Nannofossils became almost extinct during this interval of environmental perturbation

cium carbonate was probably produced through the biologic activity of planktic organisms such as phytoplankton or green algae. Those left no fossil skeletons, but their photosynthesis withdrew CO_2 from surface water to cause calcium carbonate oversaturation and precipitation. This process of calcite production is thus comparable to chalk sedimentation in Swiss lakes. If plankton production was suppressed in Precambrian or Paleozoic oceans, the biologic fractionation of carbon isotopes should also have ceased. We should thus find a "Strangelove perturbation" of the carbon isotope value in the sediments deposited during the earlier biotic crises. Indeed, a sharp excursion of $\delta^{13}C$ values of minus 2 to 5 per mil has been detected across the Permo/Triassic boundary in a boundary clay characterized by an iridium anomaly [5]. The pattern of geochemical changes is thus similar to that observed across the Cretaceous/Tertiary boundary.

No geochemical anomalies have been found across the normally defined Precambrian/Cambrian boundary [21], but an iridium anomaly has been found in a clay at a horizon called the China-C marker which defines the base of a formation containing the first trilobite [16]. We sampled across this boundary at two localities in southwest China and also found the "Strangelove perturbation" at the predicted horizon [15]. The perturbation is minus 4‰ compared to the average of underlying samples and minus 2‰ compared to the average of the immediately overlying samples. The duration of the perturbation, estimated on the basis of sedimentation rate, is on the order of 10^3 to 10^4 years. Again the pattern is comparable to that across the C/T boundary. Some Chinese scientists proposed, therefore, that the China-C datum be adopted as the level marking the Precambrian/Cambrian boundary [19].

I might emphasize that many thousands of samples have been analyzed for $\delta^{13}C$ and many sequences have been studied in detail, but the "Strangelove perturbation" has been found only across the three era boundaries. This is unmistakable evidence that the biotic crisis defining era boundaries was accompanied by drastic perturbations of ocean environments which all but completely suppressed the plankton production in the oceans.

Expansion of Oxygen-Minimum Zone

If plankton production were all but stopped, the dissolved CO_2 would have become the same in the surface and the bottom waters after a turnover of the ocean [4]. The dissolving power of seawater lies in its carbonic acid. The bottom waters rich in CO_2 are undersaturated with respect to $CaCO_3$ so that lime oozes on the deep-sea bottom undergo dissolution. The surface waters are normally oversaturated because the dissolved CO_2 content is reduced by plankton during their photosynthesis. But a Strangelove ocean devoid of plankton has a surface water as corrosive as the bottom water. The high dissolution rate, combined with the low production in such an ocean, should result in the deposition of a pleagic sediment low in $CaCO_3$. We expect, therefore, that the sediments deposited during biotic crises be a clay or a marl with reduced carbonate content. Indeed, a pelagic clay has been found just above the Cretaceous/Tertiary boundary at many outcrop localities [13]. Diligent searches have led to the discovery of a boundary clay across the Permo/Triassic and the proposed Precambrian/Cambrian transitions at localities where the sedimentary record is demonstrably continuous [5, 15].

The degree of undersaturation with respect to $CaCO_3$ commonly increases with depth in the oceans. The outcropping Cretaceous/Tertiary sequences on land were deposited on ancient shallow-sea bottom. The presence of a boundary clay at shallower sites of deposition leads one to expect an even more extensive dissolution at depth. Yet the boundary sediment at several deep-sea drilling sites is a marl or even a calcareous ooze [13]. There was apparently a chemical inversion: less corrosive water was then present at greater depth [8]. This fact is not surprising if we recall that an oxygen-minimum zone is present at some intermediate depth where the dissolved CO_2 and the corrosive power are maximal. An unusually high input of organic carbon to the oceans should have led to an expansion of this oxygen minimum for hundreds of meters upward to the photic zone and hundreds of meters downward to bathyal or abyssal depth [13]. Below this expanded oxygen minimum and above the calcite-compensation depth the ocean water should have been less corrosive and thus more favorable for the preservation of calcareous sediments. This apparent inversion of the calcite-dissolution gradient at the earlier Tertiary time is thus evidence for an expansion of the oxygen-minimum zone during biotic crisis.

Oxygen Deficiency

The dissolved oxygen content in ocean waters depends upon supply and consumption. Vigorous circulation accelerates the supply of oxygen, and a superabundance of decaying matter increases consumption. Since few organisms could be produced in a sterile ocean, one might instinctively suppose a poverty of organic carbon, or well oxygenated water, for a Strangelove ocean. Geological observations of boundary sediments point, however, to the contrary.

The oxidation-reduction state of ancient environment is manifested in the coloration of sediments. An oxygen-rich water will oxidize iron compounds and produce ferric minerals with brown or red color, whereas sediments deposited under reducing conditions are rich in green ferrous compounds as well as dark gray carbonaceous materials. A color change from red to green or black is, therefore, an indication of a change from well oxygenated to oxygen-deficient conditions. We indeed see a change from red- or white-colored sediments to green- or black-colored sediments across the C/T boundary, and this change takes place exactly at the horizon where the dual $\delta^{13}C$ and iridium anomalies are present. The famous outcrop at Stevns Klint serves as an illustration: the Cretaceous chalk is white and the Tertiary chalk is white, but the boundary clay is black [8]. One also sees paper-thin laminations in the clay. Both the color and the laminations indicate that the bottom water was devoid of oxygen at the time of boundary clay deposition. The black color owes its origin to very fine disseminated organic matter. The laminations owe their origin to the settling of sedimentary particles on a bottom where no life existed to disturb the delicate sedimentary structure. There was no life because there was no oxygen, and there was no oxygen because it had all been used up, all consumed to oxidize the extra decaying matter that was in the sediment. The finding of a black and laminated boundary clay implies, therefore, that a great amount of organic material was deposited on the bottom of the Strangelove ocean at the beginning of the Tertiary. Where dit it all come from?

The living biomass in the ocean today has about 30 billion tons of organic carbon. This should be recycled in a steady-state ocean. If this biomass were destroyed by a catastrophe and not recycled by normal fertility in a Strangelove ocean, the supply of organic carbon to the sea bottom would have been 1,000 times more than the normal annual supply of 30 million tons from river input. The biomass on land, mainly in the form of tropical forests, has 300 billion tons of carbon. If this biomass were also destroyed, the available material for delivery would contain 10,000 times the normal annual supply of organic carbon to the oceans. A still greater reservoir are the 2.6 trillion tons of carbon in the organic debris of forest soil, etc. A global deforestation and the consequent acceleration of erosion would set free a portion of this carbon for transport.

How Much of All that Organic Carbon was Delivered?

We could make lan estimate on the basis of carbon isotope data. All organic materials are enriched in carbon-12 and deficient in carbon-13 atoms, with delta carbon-13 values of minus 20 or 30 per mil. If all the organic carbon on land got into the oceans, the carbon isotope composition of the whole ocean would have been changed, by a few parts per thousand at least. This did not happen, although the delta carbon-13 value of the ocean surface waters did become slightly more negative, by one per mil or so. An arithmetic exercise indicates that only a small fraction, probably less than 10%, of the available organic carbon on land did get into the Strangelove ocean of the earliest Tertiary.

The influx of excess organic carbon from devastated continents could not have continued very long. In thousands of years, much of the dead organic debris should have been either removed or buried, and new trees made new forests to retain organic carbon in the soil again and to minimize erosion. Evidence from sediments confirms this deduction. Alvarez et al. described, for example, two layers of the boundary clay at Gubbio, Italy, "each about 0.5 cm thick, the upper being red in color and the lower gray" [1]. They estimated a time span of not more than 5,000 years for the clay deposition. The rate of pelagic clay sedimentation in the oceans is on the order of 1 or 2 millimeters per thousand years. Using the latter figure, we might conclude that the gray layer was laid down during the first 2,500 years of the Tertiary. The excessive influx of organic debris from land then caused a depletion of oxygen in seawater. The residual organic matter made the sediment gray. When the red layer was deposited during the second 2,500 years of the Tertiary, the organic carbon supply form rivers returned to normal. Dissolved oxygen in seawater was again in sufficient supply to oxidize the iron minerals to give the upper boundary clay its red coloration.

The varied nature of the first sediment above the Cretaceous/Tertiary boundary indicated differing severity, distribution, and duration of the oxygen-deficient event in different regions. A dark gray or black boundary clay was found mainly at outcrop section. The clay contains rich organic carbon, probably because the area was located on ancient continental margins, close to the source of land-derived debris. On abyssal plains of the central Pacific no color change has been observed at the boundary; little of the land-derived organic debris got out far enough to make a black layer there. Kyte and Wasson detected, nevertheless, a short episode of oxygen deficiency in abyssal waters through a study of mineralogy [18]. Pyrite and glauconite, two minerals formed typically under reducing conditions and common in the boundary sediment of many localities, are present in the Pacific boundary sediment. Unusual concentrations of arsenic and antimony compounds have also been detected; those metals could only have been precipitated from oxygen-poor bottom waters. The normal residence time of arsenic and antimony in the oceans is more or less 10,000 years. Kyte and Wasson [18] reasoned, therefore, that the episode of temporary oxygen deficiency for their precipitation lasted only thousands of years. This is in line with what I have just concluded.

The oxygen crisis may nevertheless have done its damage during the millenia. All living organisms require oxygen for respiration. Marine benthic organisms depend on dissolved oxygen in seawater. The anoxic part of the ocean could not support life. The animals which survived months of blackout and hunger were now to face death by asphyxia. The Strangelove ocean was thus not only devoid of plankton, it was also a nonviable or unhealthy environment for many bottom dwellers in coastal waters.

We have no information on the oxidation-reduction state of the boundary sediment across the Permo-Triassic transition. We do have indications that anoxic conditions were widespread in South China during the biotic crisis marking the proposed Precambrian/Cambrian boundary. The first sediment above the China-C marker is black shale, and the organic carbon content increases from less than 1% in sediments below to 20% above the marker [9].

Strangelove Continents

There was a catastrophic disruption of the terrestrial plant ecosystem after the C/T boundary event [25]. At a horizon marked by the iridium anomaly, there is a sharp perturbation in the pollen composition of sediments: the percent of fern spores increases from about 20% of almost 100% (Fig. 5). The data indicated a "massive destruction of the vegetation," although only a few plant species failed to survive the biotic crisis.

We do not know the cause of this deforestation. It could have been darkness. It could have been giant forest fires. It could have been freeze. It could have been atmospheric pollution by NO_x. It could have been acidic rains. Or it may have been a combination of several factors. We have to wait for a systematic study of the regional distribution and taxonomic selectivity of the damage before we can clarify the nature of this terminal Cretaceous

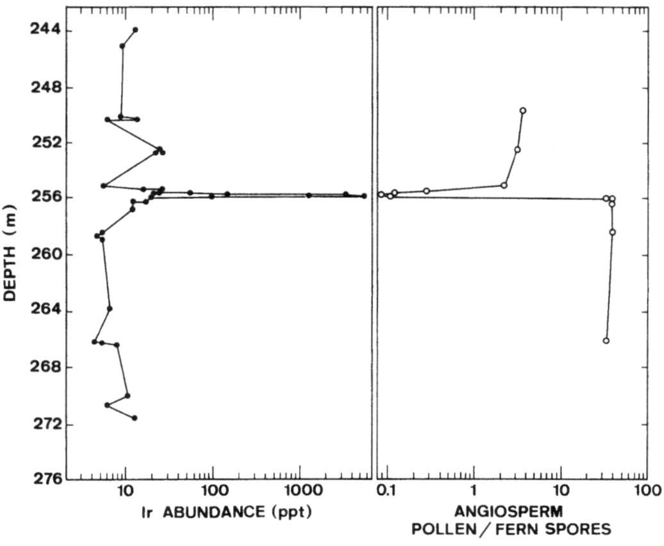

Fig. 5. Pollen/spore ratio in C/T boundary sediment. The perturbation in the Angiosperm pollen/fern spore ratio indicates widespread deforestation related to the terminal Cretaceous event [25]

catastrophe. Nevertheless, we are sure that something did happen, and it was a catastrophe.

Temperature Changes at Times of Biotic Crisis

Oxygen isotope analysis provides data for paleoclimatic interpretations. The first discovery by Shackleton that the ocean temperature increased 5 °C across the Cretaceous/Tertiary boundary inspired me to propose the theory of cometary impact [11]. Our subsequent work on South Atlantic cores revealed that the paleotemperature oscillated with an overall warming trend during the 40,000 years after the boundary event [14]. Oxygen isotope shifts at rates and magnitudes similar to that have been detected across the C/T boundary at other deep-sea drilling sites [2]. The preliminary data led to a working hypothesis that the ocean waters were warmed globally during the first 30 or 50 thousand years of the Tertiary [14]. Temperature increase at such a rate was most probably caused by the greenhouse effect of an increased CO_2 content in the Earth's atmosphere. Broecker pointed out that the CO_2 content in the air should be tripled if the atmosphere is equilibrated with a Strangelove ocean [4], and such a tripling should cause a temperature rise of 5° to 10 °C [26].

Temperature perturbations of very short duration (10^0–10^2 years) can hardly be expected to be recorded by sediments. Cooling has been postulated on a theoretical basis; insulation of solar radiation by ejecta dust in the stratosphere could, for example, cause a drastic chilling [24], a phenomenon called "nuclear winter" by the daily press. We found that the plankton skeletons in the very first sediment above the C/T boundary in our South Atlantic cores record a positive $\delta^{18}O$ anomaly of about 2‰ [14]. If this anomaly is a manifestation of the temperature effect, the cooling should be about 8 °C. Could that be evidence of a "nuclear winter" after the terminal Cretaceous event?

While ocean temperature changes after a "nuclear winter" may have been registered as a decrease of a few degrees, the cooling of the atmosphere could have been considerable. Temperatures of 40 °C below zero may have prevailed locally for several months. Under such circumstances, large reptiles should have been frozen to death even if they could survive the hunger.

It should be emphasized that we cannot place much reliance on the preliminary data on oxygen isotope. Numerous factors could influence the value of $\delta^{18}O$ of fossil skeletons in boundary sediments, and conflicting evidence is not uncommon. We should perhaps disregard all oxygen isotope data until a systematic trend is established.

We have no reliable data on sudden temperature changes across the Permo/Triassic or Precambrian/Cambrian boundaries. There is, however, unmistakable evidence that ocean waters cooled by a few degrees during the first 10^5 years of the Oligocene epoch [6, 14]. No mass extinctions occurred during this environmental crisis, some warm-loving species did become extinct, but the faunal turnover rate was not much different from the norm [6].

Cause of Environmental Crisis

The Strangelove perturbation, the calcite dissolution, the anoxic event, and the massive destruction of plant communities on land described above were all of very short duration, Estimates on the basis of sedimentation rate suggested that the era-ending catastrophes came suddenly without forewarning. Of the numerous ideas proposed to account for this sort of extraordinary catastrophes, the theory of meteorite impact seems most attractive to me.

I have reviewed the geochemical evidence for a large-body impact at the end of Cretaceous [14]. The discovery of iridium anomalies across the two other era boundaries [5, 9, 15, 16] indicates that the environmental perturbations at those times could also have been triggered by meteorite fall.

The physical, chemical, and biological consequences of large-body impacts have been investigated by computer modelling [22]. Four types of environmental changes could have resulted from the impact of a body with a mass of 10^{17} to 10^{18} g, namely: a) darkness because of an ejecta dust envelope in the stratosphere, b) chemical pollution because of atmospheric reactions in the ejecta cloud producing NO_x and other pollutants, c) increased ultraviolet radiation because of the total destruction of the stratospheric ozone layer, and d) short- and long-term temperature changes. Mass mortality after an impact event could have been the cause of catastrophic pollution. The environmental crisis in turn could have suppressed fecundity. According to this scenario, large impacts killed, but the environmental crises are the culprits directly responsible for mass extinctions.

Acknowledgements. I have benefited from discussions with my colleagues J. A. McKenzie, K. Perch-Nielsen, K. Kelts, and H. Oberhänsli. A. Hallam read the first draft critically and made valuable suggestions. B. Das Gupta helped in the preparation of the manuscript.

References

1. Alvarez LW, Alvarez W, Asaro F, Michel HY (1980) Extraterrestrial cause for the Cretaceous-Tertiary extinction. Science 208:1095–1108
2. Boersma A, Shackleton N, Hall M, Given Q (1979) Carbon and oxygen isotope records at DSDP site 384 (North Atlantic) and some Paleocene paleotemperatures and carbon isotope variations in the South Atlantic. Initial Rept DSDP 43:695–718
3. Brennecke JC, Anderson TF (1977) Carbon isotope variations in pelagic carbonates (abs): EOS (Am Geophys Union Trans) 58:415
4. Broecker W (1982) Glacial to interglacial changes in ocean chemistry. Progr Oceanogr 11:151–197
5. Chen J (1982) Carbon isotopic variation in carbonates at the boundary between Permian and Triassic in China. In: Developments in geoscience. Beijing: Science Press, pp 247–254
6. Corliss BH (1984) The Eocene/Oligocene boundary event in the deep sea. Science 226:806–810
7. Darwin C (1859) On the origin of species. London: John Murray

8. Ekdale AA, Bromley RG (1984) Sedimentology and icknology of the Cretaceous-Tertiary boundary in Denmark: Implications for the causes of the terminal Cretaceous extinction. J Sed Petrol 54:681–703
9. Fang D, Yang Z, Huang Z (1984) The Lower Cambrian black shale series and the iridium anomaly in South China. In: Developments in geoscience. Beijing: Science Press, pp 215–224
10. Hallam A (1983") Plate-tectonics and evolution. In: Evolution from molecules to men, ed Bendall DS. Cambridge: Cambridge University Press, pp 367–386
11. Hsü KJ (1980) Terrestrial catastrophe caused by cometary impact at the end of Cretaceous. Nature 285:201–203
12. Hsü KJ (1984) A scenario for the terminal Cretaceous event. Initial Rept DSDP 73:755–763
13. Hsü KJ (1984) Geochemical markers of impacts and of their effects on environments. In: Patterns of change in earth evolution, eds Holland HD, Trendall AF. Dahlem Konferenzen. Berlin, Heidelberg, New York, Tokyo: Springer-Verlag, pp 63–76
14. Hsü KJ, He Q, McKenzie J et al. (1982) Mass mortality and its environmental and evolutionary consequences. Science 216:249–256
15. Hsü KJ et al. (1985) Strangelove Ocean before Cambrian Explosion. Nature, in press
16. Krähenbühl lU (1984) Siderophile enrichment in boundary sediments. Chimia 38:107–113
17. Kropnick PM, Margolis SV, Wong CS (1977) Delta C-13 variations in marine carbonate sediments as indicators of the CO_2 balance between the atmosphere and oceans. In: The fate of fossil fuel CO_2 in the oceans, eds Anderson NR, Molahoff XA. New York: Plenum Press, pp 295–321
18. Kyte FT, Wasson JT (1985) The Cretaceous-Tertiary boundary in GPC-3, an abyssal clay section. Geochim Cosmochim Acta, in press
19. Luo H (1984) Sinian-Cambrian boundary stratotype section at Meishucun, Jinning, Yunnan, China. Beijing: People's Publishing House
20. McKenzie JA (1982) Carbon-13 cycle in Lake Greifen: A model for restricted ocean basins. In: Nature and origin of cretaceous carbon-rich facies, eds. Schlanger S, Cita M. London: Academic Press, pp 197–207
21. Nazarov MA et al. (1983) Iridium abundances in the Precambrian-Cambrian boundary deposits and sedimentary rocks of Russian Platform. Lunar Planet Sci 14:546–547
22. Silver LT, Schultz PH (eds) (1982) Geological implications of impacts of large asteroids and comets on the earth. Geol Soc Am Spec Paper 190
23. Stanley SM (1973) An ecological theory for the sudden origin of multicellular life in the late Precambrian. Proc Natl Acad Sci USA 70:1486–1489
24. Toon OB (1984) Sudden schanges in atmospheric composition and climate. In: Patterns of change in earth evolution, eds Holland HD, Trendall AF. Dahlem Konferenzen. Berlin, Heidelberg, New York, Tokyo: Springer-Verlag, pp 41–61
25. Tschudy RH, Pillmore CL, Orth CJ, Gilmore JS, Knight JD (1984) Disruption of the terrestrial plant ecosystem at the Cretaceous-Tertiary boundary, Western Interior. Science 225:1030–1032
26. Wigley TML, Jones PD (1981) Detecting CO_2-induced climatic change. Nature 292:205–207

Patterns and Processes in the History of Life,
eds. D. M. Raup and D. Jablonski, pp. 313–329. Dahlem Konferenzen 1986
Springer-Verlag Berlin, Heidelberg

Evolutionary Consequences of Mass Extinctions

D. Jablonski

Dept. of Geophysical Sciences
University of Chicago
Chicago, IL 60637, USA

Abstract. Mass extinctions are relatively brief, global excursions of extinction rates above normal, background levels for a number of higher taxa. At least five major mass extinctions have perturbed the Earth's biota over the past 600 million years, but current research has so completely emphasized causal mechanisms that the biological nature of the victims and survivors, and other evolutionary aspects of extinctions, have been neglected. Paleontological analyses suggest that selectivity during mass extinctions may be qualitatively different from patterns of extinction and survival during background times. This apparent indifference to background-regime adaptation means that mass extinctions can play a profoundly disruptive role in the evolutionary process. Taxa and morphologies can be lost not because they were poorly adapted by the standards of background processes (which constitute the bulk of evolutionary time), but because they occurred in lineages lacking the environmental tolerances or geographic distributions necessary to survive the mass extinction regime. For reasons that are still not clear, mass extinctions seem preferentially to remove taxa that are endemic, large-bodied, or tropical. Despite this disruption of background processes, some evolutionary trends persist across mass extinction boundaries; such long-term survival may require the chance occurrence within a single lineage of traits that enhance survivorship under both background and mass extinction regimes. By removing or reducing dominant groups, mass extinctions provide opportunities for diversification of taxa that had been minor constituents of the pre-extinction biota, channeling evolution in directions not predictable from situations established during background times.

Introduction

Much geological evidence has accumulated recently on the timing and triggers of mass extinctions in the fossil record ([20], and Sepkoski and Hsü, both this volume), but very little is known about the biological significance of these events and their relationship to processes that operate during times of normal, background levels of extinction. Mass extinctions would be important evolutionary phenomena if they simply accelerated biotic changes already under way during background times. They assume even greater significance as evolutionary forces if they alter the rules of extinction and survival in ways that have little or no correspondence to background patterns. In removing or decimating successful and seemingly well adapted clades (monophyletic evolutionary lineages), mass extinctions may set the stage for the origin or diversification of major groups, and thus fall outside of normal background processes in both their generative and destructive effects. However, we have little theoretical basis for understanding these biotic turnovers. We do not even know whether mass extinctions can be grouped as a single class of evolutionary phenomena: regardless of whether the major mass extinctions in the fossil record share a common cause, do they exhibit similar biological effects? We need to consider these biological effects in an explicitly evolutionary context, and this paper is an attempt to encourage investigation in this direction. After a brief discussion of the nature of the available data and of some pitfalls to the study of mass extinctions as evolutionary phenomena, I will focus on two themes:

1 – Patterns of survivorship, including the problems of a) selectivity – what factors determine which taxa and adaptations persist and which are lost in a mass extinction, and b) continuity – how do evolutionary trends begun before the mass extinction carry over into the post-extinction biota? For evolutionary purposes, emphasis should shift from the victims of mass extinctions to the survivors [11], and it is useful to recognize that survival of a given taxon or morphology brings no guarantee that evolutionary trends – net expansion or contraction of clades or directional shifts in the morphology of a clade – will continue in the new, post-extinction setting.

2 – Patterns of diversification in the wake of mass extinctions, which often contain not a rebound along earlier lines but an expansion of clades that had been minor constituents of the pre-extinction biota. Adaptive radiations are a common sequel to mass extinction, and these should be assessed not only in terms of the origin of new taxa but in terms of morphologies that arise or are lost. Thus, mass extinctions have shaped the history of life in both a creative and a destructive fashion – removing major taxa,

yet promoting the origin or diversification of new groups as the aftermath of the extinction event.

Extinctions in the Fossil Record

A number of major mass extinctions have perturbed the marine and terrestrial biota since the beginning of the Phanerozoic: the largest include the end-Ordovician, late Devonian, late Permian, late Triassic, and end-Cretaceous events (see Sepkoski, this volume). The rigorous definition of mass extinctions is still a problem: in theory they should involve relatively brief, global excursions of extinction rates above normal, background levels of extinction for a number of higher taxa (see [20] and Sepkoski, this volume), but background levels themselves have not been well quantified. Further, data vary in quality among regions, and temporal resolution is coarse for many portions of the stratigraphic record. The five major events mentioned above are sufficiently marked that few contest their classification as mass extinctions, but other events that would fit the above definition are probably to be found elsewhere in the fossil record as well; possible candidates include mid-Silurian, mid-Carboniferous, and mid-Cretaceous extinctions ([7, 39], Sepkoski, this volume). Enlarging our sample of well characterized extinction events, across a spectrum of extinction magnitudes, should greatly improve our ability to test hypotheses regarding their evolutionary significance.

Stratigraphic resolution is another major problem in assessing patterns of mass extinction. The fine temporal structure of major biotic change is difficult to decipher within a single environment or region, and establishing synchroneity on scales of decades, centuries, or millennia between – and among – marine and terrestrial habitats is a major undertaking and sometimes virtually impossible. Even in the relatively complete marine record, in which gaps are fewest and stratigraphic correlations are most precise, uncertainties range in the thousands to hundreds of thousands of years. Detailed local sections can provide more refined time scales, but local environmental changes almost inevitably confound global patterns. Moreover, small-scale stratigraphic breaks, which often accompany extinction events owing to the association of both breaks and extinctions with marine regressions, can artificially enhance abruptness of extinctions and their apparent simultaneity among higher taxa. On the other hand, a decline in the completeness of the fossil record approaching an extinction event, or even random sampling, can generate artificially gradual extinction patterns [44].

These two opposing biases are difficult to quantify, so that detailed analyses of extinction events must be undertaken with considerable caution.

Jablonski [20] suggested one approach that can help control for sampling and preservational biases in taxonomic patterns across critical time intervals. Those taxa that disappear from the record but return at some later point in time (exhibiting what he termed "the Lazarus effect") provide a measure of the quality of the fossil record across the interval in question. The Lazarus effect reflects purely artificial extinction owing to preservational and other sampling biases and gives a baseline against which to compare the observed overall extinction pattern. For example, Batten [5] documented a massive Lazaraus effect among the gastropods of the Permo-Triassic interval; more late Paleozoic families and genera occurred in the well sampled Late Triassic than in the latest, poorly represented stage of the Permian! Such a tremendous artificial decline suggests that the record is too poor to distinguish directly between a gradual or abrupt extinction in the Late Permian, despite data that initially suggested a protracted extinction process. The Lazarus effect can be used to set confidence limits for any horizon in which extinction patterns are of interest, and may permit a more rigorous assessment of intervals notorious for their uncertain record.

The fossil record is far more reliable at the higher taxonomic levels (see [33] and Jablonski et al., this volume) and, not surprisingly, most mass extinction analyses have relied on family-level taxonomic data, with fewer studies undertaken at the generic level, and fewer still at the species level. Clearly, an array of complementary data sets at different levels in the taxonomic hierarchy are needed in order to understand better the dynamics underlying mass extinctions. Backtracking from family-level data based on the taxonomic structure of present-day organisms has yielded some remarkable estimates on the magnitudes of mass extinctions (e.g., the 52% familial extinction at the end of the Permian translates into 91–96% extinction at the species level [34, 50], but this approach is of limited application and may even be deceptive. If species-family ratios have changed significantly over the Phanerozoic, the same magnitude of family-level extinction would reflect different scales of species-level extinction [12, 48].

Although an understanding of species-level processes is desirable, it is important to recognize that the extinction pattern of families and genera is significant in its own right. So long as higher taxa reflect true monophyletic evolutionary lineages (i.e., are genuine clades) and even roughly correspond to adaptive zones or unique fields in morphospace, then their survivorship patterns represent far more than mere epiphenomena. Instead, such patterns are indicators of the origin, maintenance, and loss of evolutionary novelties, and of the damping or acceleration of branching processes that have maintained life's diversity in the face of ongoing background and mass extinction. That generic- and family-level data are biologically meaningful in their own right is suggested by a) the maintenance of a diversity plateau at the level of 500 marine families for the 200 myr from the Late Ordovician to the mid-Permian, despite mass extinction perturbations [40] and the typically exponential rebound of family-richness after those extinction events ([40] and Sepkoski, this volume), and b) the inadequacy of species-level patterns as predictors of patterns at higher taxonomic levels during both background times (e.g., [21, 24]) and mass extinctions ([20, 22] and see below).

Patterns of Survivorship

By definition mass extinctions involve the extirpation of many evolutionary lineages. One important problem is whether the extinction event is selective or random in its removal of clades, and if selective whether the selectivity is consistent with survivorship and selective forces during background times [34, 36]. A related problem is the continuity of certain large-scale evolutionary patterns across major extinction events: we can see that the evolutionary clock is perturbed but not entirely reset during mass extinctions, but have little insight into why some trends persist but others do not.

Selectivity

How readily can the nature of the survivors of a mass extinction be predicted from a knowledge of evolutionary processes in the interval preceding the extinction event? We do not yet know. Neither patterns of background extinction nor of mass extinctions are sufficiently well characterized to permit a general comparison. However, fragmentary evidence from a variety of sources suggest that mass extinctions are qualitatively as well as quantitatively different from background extinctions.

From a comparison of patterns of end-Cretaceous survivorship among gastropods of the Gulf and Atlantic Coast of North America with evolutionary patterns in the preceding 16 myr, Jablonski [20, 22] concluded that mass extinctions were neither random nor consistent with background patterns. He found that in background times planktotrophic larval development, broad geographic range at the species level and high species richness enhanced species and clade survivorship. However, none of these traits affected clade survivorship across the end-Cretaceous event. Instead, clades with broad geographic ranges (regardless of the geographic range of constituent species) showed preferential survivorship. In the ensuing Tertiary Period, the traits that had promoted survivorship during Cretaceous background times regained their effectiveness.

Patterns of extinction and survival among other taxa and time intervals also suggest that background and mass extinctions represent qualitatively different regimes, although useful data are extremely sparse. Examples include Paleozoic bivalves, in which endemic genera suffer disproportionately during the end-Ordovician, late Devonian, end-Permian, and end-Triassic extinctions [8]; and Paleozoic bryozoans, in which morphologically complex genera (inferred specialists) suffer disproportionately during the end-Ordovician and late Devonian extinctions but have higher survivorship than simple genera during intervening background times [3]. Clearly, further comparative studies between background and mass extinctions, and among mass extinctions, would be extremely valuable.

If the results reported by Jablonski [20, 22] are indeed applicable to other taxa and other extinctions events, they suggest that mass extinctions are a potent macroevolutionary force, profoundly disrupting normal processes and removing taxa well adapted to background regimes. Taxa or morphologies could be lost not because they are maladaptive as measured under background conditions (which constitute the bulk of evolutionary time), but because they happened to lack the appropriate biogeographic deployment or other traits necessary to weather the mass extinction. Extinction patterns will be complicated by the concomitant removal of taxa that were already dwindling or "endangered" (cf. [25]), but even expanding clades can be lost if they are endemics or if their centers of distribution lie within regions that are particularly disrupted (e.g., tropical groups, discussed below). Survivorship thus appears to follow Raup's [36] paradigm of "nonconstructive selectivity" – determined by traits not tightly linked to adaptations honed during background times, and thus failing to promote long-term adaptation of the biota.

Such "nonconstructive" extinction may have shaped the fates of numerous major adaptations through the Phanerozoic. For example, the shell-drilling habit in carnivorous naticid gastropods was apparently lost soon after its initial appearance in the late Triassic, despite the undoubted expansion of available resources that this innovation entailed at the organismic level [13]. What should have been a valuable new adaptation originated in the wrong place at the wrong time and was extinguished by the end-Triassic mass extinction, only to re-originate in a related group some 120 myr later. Similarly, the ability to bore into hard substrata was first achieved among the Bivalvia near the end of the late Ordovician [30 a], and this habit surely made available new living space and refuges from disturbance and predation. However, this lineage did not survive the end-Ordovician mass extinction, and undoubtedly boring bivalves do appear in the fossil record until the Triassic, over 100 myr later. The strong differential in survivorship between marsupial and placental mammal lineages during the end-Cretaceous mass extinction [2] also probably does not reflect selectivity against the marsupial mode of reproduction, but differences in clade geography [9a] and environmental tolerance that happened to be correlated with the reproductive dichotomy.

The survivorship of widespread clades regardless of the geographic range of their constituent species supports a hierarchical view of evolution, in which evolutionary processes operate at a variety of focal levels, with consequences both upward and downward within a genealogical hierarchy from gene to clade [14, 22, 53]. Thus survivorship of end-Cretaceous molluscan genera (=clades) could not be predicted on the basis of species-level characteristics, but the emergent property, overall clade geographic range, exerts a significant influence on the fate of the suite of morphological and other traits unique to the lineage. This is not to suggest that natural selection at the level of the individual organism became inoperative at the close of the Cretaceous Period, but that selectivity at a higher focal level yielded a predictable pattern while processes at the lower level did not. Such dynamics again validate the use of supraspecific taxa in tracking the history of the Earth's biota but warn against assuming a perfect correspondence between patterns at higher taxa and at the species level (see also [12]).

Other patterns of differential survivorship associated with mass extinction events may also be consistent with the "nonconstructive selectivity" model. Many authors have argued that all five of the major mass extinctions are most severe in low latitudes, particularly among marine reef communities (see [20, 42, 46]). This generalization requires further testing: Hickey [18] suggests that tropical floras were *least* affected by the end-Cretaceous extinction, although the data are not very convincing [20, 26]. Furthermore, because the tropics are the most diverse regions of the globe, they would be expected to lose the highest number of taxa even in the absence of latitudinal trends in extinction magnitudes. Jablonski [20] did present evidence that families restricted to the tropics were disproportionately affected among Permo-Triassic brachiopods and Cretaceous-Tertiary bivalves and gastropods, but the biogeography of extinction deserves much more analysis in both marine and terrestrial settings (see [23]). For example, a number of rival hypotheses are available to explain the apparent vulnerability of low-latitude taxa to mass extinction: a) mass extinctions may be driven by global climatic changes, which might be particularly devastating to stenothermic tropical biotas. b) The tropical reef community may be such a tightly woven network of biological interactions that the initial removal of the same proportion of species as were lost at high latitudes could have far more disruptive consequences. c) The apparent vulnerability of the tropics might reflect the biogeographic structure of its biota, containing a high proportion of endemic and thus extinction-prone taxa, rather than recurrent climatic or other crises that focus directly upon low-latitude environments and the adaptations of their inhabitants.

Another remarkable pattern of differential extinction lies in the lack of correspondence between the major floral turnover events and the major mass extinctions among marine or terrestrial animals. Biogeographic changes rather than losses of higher plant taxa characterize the end-Cretaceous extinction, and the floral change associated with the end-Permian extinction occurs relatively rapidly but at different times within a 25 myr interval in different geographic areas [27]. Knoll [27] has suggested that the vascular plants exhibit a pattern of extinction and large-scale evolutionary change that is fundamentally different from the animal pattern. Major floral changes appear to be gradual and time-transgressive rather than sudden and globally synchronous, driven by large-scale competitive replacements rather than by extinctions (see diversifications discussion, below). These macroevolutionary differences between plants and animals have been attributed to differences in their basic biologies: plants are trophically homogenous and thus have few options for niche displacement with the advent of a new competitor, and they possess vegetative regeneration abilities that should make them relatively invulnerable to global, catastrophic mass mortalities. These may not be the precise explanations for the observed evolutionary differences (e.g., niche subdivision may be rampant in plants along other niche axes; mass extinctions among animals need not always involve catastrophic mass mortality), but such profound differences among major groups deserve further analysis. Knoll's reasoning might suggest that clonal marine invertebrate taxa, which can exhibit remarkable powers of vegetative regeneration (although lacking a precise analogy to a seed bank dormant in the soil), should exhibit histories more similar to plants than to other animals, for example. However, clonal groups such as graptolites, bryozoans, and corals exhibit a mass extinction pattern more similar to aclonal invertebrates than to plants.

Among many other parameters, size selectivity has been attributed to a number of mass extinctions. Few, if any, terrestrial vertrebrates with estimated masses exceeding 25 kg survived the end-Cretaceous extinction [38]; similarly, large marine predators (exceeding 10 m in length) vanished during each of the major post-Paleozoic mass extinctions [10]. There are many plausible explanations for this size-related pattern, stemming from such factors as characteristic population sizes, generation times, home ranges and position in trophic webs (see also LaBarbera, this volume), but neither cause nor evolutionary effect has been rigorously examined. It is also important to recognize that the pattern does not represent a simple threshold effect: many smaller co-occurring taxa, related or not, suffered extinction at the same time. The one exception may lie in the late Pleistocene large-mammal extinctions: approximately 10,000 years ago, North American species and genera of large-bodied mammals suffered extinction rates far in excess of the preceding 3 myr of background extinction, while small-bodied mammals did not [28]. This extinction, which may be unique in the strength of its bias towards large-bodied forms, has been attributed to the depredations of humans (see [29] for debate on this issue).

Certain taxa exhibit exceptionally high volatility during their histories, exhibiting high diversification rates but high extinction rates as well, such that they undergo repeated, rapid excursion in standing diversity (e.g., most ammonite lineages relative to the less volatile bivalves (see [45]). If mass extinctions involve multiplication of species-extinction rates, these volatile clades would be far more likely to be driven into extinction than would less volatile clades, in which an increase in the characteristically low extinction rates would bring only modest losses in standing diversity. Alternatively, perhaps volatile clades are more sensitive to perturbations that can drive them into the mass extinction regime while leaving the nonvolatile clades virtually undisturbed. We still do not know what biological attributes impart macroevolutionary volatility, but insight into this problem and to the influence of intrinsic biological traits on survivorship during both background and mass extinction might be gained through studies of the paleobiology of volatile clades and their evolutionary patterns (capture of major adaptations, interactions with other higher taxa, developmental variability within and among clades, etc.) relative to groups having lower speciation and extinction rates and less radical fluctuations in standing diversity.

These different patterns of selectivity – and I have mentioned only a few of those hypothesized over the years – do suggest a nonrandom component to mass extinctions. If the survivorship patterns are indeed distinct from those exhibited during background times, then it is important to explore the transition between the two macroevolutionary regimes. Again, comparative studies would be valuable here: are all mass extinction regimes alike, or is each a unique excursion away from the background regime? How large a perturbation is required to force the transition from background to mass extinction regime, and can that transition occur on a taxon-by-taxon basis, or is a true global threshold involved? We still are not certain that the magnitudes of mass extinction events constitute a discrete statistical population relative to background extinction [31, 37], and it would be of great interest to test whether survivorship patterns at some of the smaller peaks in the post-Paleozoic extinctions curve conform to the end-Cretaceous mass extinctions (and thus fall within the mass extinction regime), to the background extinction pattern (and thus do represent intensification of background processes), or even to an intermediate pattern along a continuum.

Dynamics: Heightened Extinction or Failure of Origination?

The evolutionary dynamics underlying extinction events are themselves still poorly understood. If mass extinctions are produced primarily by increase in the per-taxon extinction rate, which now appears to be the favored explanation (e.g., [36]), then all taxa should exhibit an acceleration in extinction rates but per-taxon origination rates should remain relatively constant. However, this prediction might be confounded by other factors that cause the answer to this question of dynamics to be somewhat elusive. For example, a) a mass extinction might enrich the biota with taxa that are speciation-resistant (see Jablonski et al., this volume, for examples), so that the overall origination rate of the biota would decrease even though the characteristically low, background-level speciation rate of the survivor clades never changed. It is not yet clear whether the post-extinction rebound rate is sufficiently rapid to rule out this possibility (see [9]). b) Presumably, a mass extinction driven by mass mortality over less than a century (the impact hypothesis predicts that the end-Cretaceous biotic change was accomplished in 1 to 10 years according to [1], p 1136; not all would agree) should have little effect on per-taxon speciation rates because normal processes should return so rapidly. At the same time, however, it might be argued that c) the unfavorable factors that heighten species extinction on any time scale might also act against new, vulnerable isolates and thus suppress speciation as well.

On the other hand, if mass extinctions are produced or exacerbated by suppression of origination (e.g., [4, 19, 45]), then the taxa with the highest per-taxon extinction rates should be the most vulnerable, as extinction in the absence of replacement takes its toll. The hectic, boom-and-bust history of the ammonites may reflect such processes. The higher diversification and extinction rates of Sepkoski's Paleozoic fauna relative to the post-Paleozoic fauna may foreshadow the greater vulnerability of the former to mass extinctions [40].

Persistence of Trends

Despite sweeping changes in the taxonomic composition of the biota – and, presumably, in its array of adaptations – certain evolutionary patterns persist across mass extinction boundaries. Clades that do not merely survive mass extinctions but actually manage to maintain continuity of evolutionary trends over the course of such events clearly enjoy a significant advantage over less resilient contemporaries. Favorable adaptations captured or maintained during the preceding interval of background extinction can form the basis of further diversification and elaboration once background conditions have returned – escaping the "two steps forward, one step back" syndrome implied by the contrast between background and mass extinction regimes. Does the maintenance of a pre-extinction trend into the post-extinction world simply represent the chance co-occurrence within a single clade of traits advantageous during background times with other, discrete traits that enhance survival during mass extinctions, or is there a single feature or suite of traits advantageous under both regimes?

One large-scale pattern that extends across mass extinction events is the replacement – and the change is more properly termed a biotic replacement rather than an evolutionary trend, since the units are not monophyletic – of the Paleozoic fauna dominated by articulate brachiopods, stalked crinoids, and rugose and tabulate corals, by the post-Paleozoic or "Modern" fauna dominated by bivalves, gastropods, and gymnolaemate bryozoans. Analysis of the long-term diversity trends of these faunas suggests that the rise of the post-Paleozoic fauna was already underway and would eventually have taken place even in absence of the end-Permian and end-Triassic mass extinctions [40]. Similarly, within the history of the post-Paleozoic fauna, the "Mesozoic revolution" in marine organisms, which saw the expansion of durophagous predators and heavily armored prey and the decline of predator-vulnerable taxa [52], continued unabated through the end-Cretaceous extinction. Indeed, since their Paleozoic beginnings the extratropical bivalves and gastropods show an exceptionally nonvolatile history of rather slow but steady diversification, weathering mass extinctions with considerable turnover but with only modest losses in total taxonomic richness.

As we begin to analyze the role of mass extinctions in shaping the global biota, it becomes clear that long-term patterns could actually represent a combination of both background and mass extinction processes. At the same time, as noted for the end-Permian event, mass extinctions can hasten or magnify trends even if persistence through the mass extinction is not determined by the triats being selected for during background time. The mid-Paleozoic increase in antipredatory morphologies in marine benthos, a "precursor to the Mesozoic marine revolution" [43] illustrates some of the complexities involved. The increase was accomplished in different ways in different taxa: bellerophontid mollusks exhibited preferential extinction of taxa lacking antipredatory morphologies, and the decline in unarmored bellerophontids is most pronounced near the time of the late Devonian mass extinction. One possible explanation would be preferential extinction of the less predator-resistant taxa during the event. However, in the late Devonian there were only four genera having shells with anti-predatory sculpture, and it is equally possible that they survived the mass extinction not because of their predator-resistant morphology but because they possessed other traits, such as broad geographic distributions, which favored survival under the mass extinction regime. Situations in which evolutionary trends continue across one or more extinction episode afford an opportunity to examine the interactions of background and mass extinction at their most complex.

Post-Event Diversification

The survivors of mass extinctions can give rise to much more than the continuation of evolutionary trends that had been under way before the extinction event. Rebound from mass extinctions commonly involve diversifications of taxa that had been minor components of the pre-extinction biota. Thus a number of adaptive radiations that had once been attributed to competitive exclusion and triumphant diversification by adaptively superior taxa now appear to have been preceded and presumably triggered by the extinction of the earlier group; examples from terrestrial vertebrates include a series of radiations among the mammal-like reptiles (therapsids) [26]; replacement of therapsids by dinosaurs [6]; replacement of dinosaurs by mammals, which themselves had originated at almost exactly the same time as the dinosaurs; and replacement of archaic carnivorous mammals by the modern Order Carnivora in the Tertiary [32]. A similar evolutionary pattern, in which an extinction event removes or severly depletes a dominant group and rapid evolution of other taxa ensues, is exhibited among marine invertebrates. For example, rapid evolution in the 3–5 myr after the end-Ordovician extinction "produced most of the important new brachiopod groups that dominated for the following 85 m.y." [41], an interval in turn terminated by the late Devonian event (see also [7]). The repeated changes in the composition of reef communities, mentioned above, were also mediated by mass extinctions rather than by progressive competitive exclusion of a dominant group by a new taxon.

On the basis of these patterns of faunal replacement, it appears that dominant groups, once in place, can suppress the origination or diversification of potential competitors; on a macroevolutionary scale, this sort of preemptive competitive exclusion appears to be a potent force (Valentine [49] models an adaptive landscape shaped by this preemptive effect). In contrast, displacement of a clade by a competitively superior one appears to be less frequent than generally believed. Several authors have speculated (e.g., [15, 36]) that the global evolutionary system would slow down, perhaps grind to a halt at higher taxonomic levels in the absence of external perturbations that unpredictably remove dominant taxa, and thus release other groups into adaptive radiation. On the other hand, some diverse clades such as the trilobites exhibit prolonged declines spanning several mass extinction events (another instance of continuity across extinction boundaries), suggesting that neither random background extinction nor truncation by mass extinction is sufficient to explain their histories (see [35]).

Given that mass extinctions affect a broad range of higher taxa, the composition of the post-extinction biota may depend in part on differential rediversification rates. For example, the articulate brachiopods suffered greater losses than the bivalves during the end-Permian event, but the differences in their post-Paleozoic histories was most clearly set by the far greater – and virtually uninterrupted – rate of rediversification of the now-dominant bivalves. What suppressed the rebound of the post-Paleozoic brachiopods while permitting substantial diversification of the bivalves? It is difficult to invoke direct competitive interaction between the two groups; other biotically mediated processes may have restricted the brachiopod rebound, such as increased shell-crushing predation [52] or bioturbation [47], ecological trends that persisted across several mass extinction events. Alternatively, the damage may have been done entirely during the end-Permian event: loss of major adaptations with the extinction of so many constituent taxa at the end of the Permian may have placed the articulate brachiopods at a disadvantage in the post-Paleozoic world. For example, the group may have lost all of its members having high-dispersal, planktotrophic larval development (a mode of development absent in the articulates today) [51]. This mode predominates in low-latitude marine communities, and the absence of such reproductive capabilities may have hampered brachiopod rediversification except in the lower-diversity, high-latitude settings in which the alternative mode most frequently occurs.

Two different processes may lead to post-extinction dominance of a previously low-diversity group. Firstly, a group might have suffered losses along with the rest of its contemporaries, but then achieved dominance by a more rapid, preemptive diversification than was achieved by other survivors, either because of a higher intrinsic rate of evolution or because of change capture of a major innovation early in the rebound phase. Thus, the race goes to the swiftest re-diversifier. Secondly, new adaptive radiations might arise from among the seemingly small number of taxa that are essentially unscathed during the extinction. For example, the placental mammals suffer remarkably little extinction at the end of the Cretaceous and then undergo exuberant diversification at the beginning of the Tertiary. We will have a better understanding of the evolutionary effects of mass extinctions once we place the groups that undergo radiations after each extinction within a theoretical framework.

Finally, the distinction between changes in numbers of taxa and in numbers of individuals, emphasized by McLaren [30] in terms of detecting the destructive effects of mass extinctions, should also be examined during rebound times. Although the Paleozoic fauna is heavily diminished in taxa and many post-Paleozoic taxa have originated by the mid-Triassic [40], many Triassic faunas contain elements of decidedly Paleozoic affinities, including large numbers of stalked crinoids, articulate brachiopods, and epifaunal, pterioid bivalves (e.g., [17]). Only in the Jurassic (and then in nearshore habitats [24]) does the already diverse post-Paleozoic fauna become the truly predominant biotic component.

Conclusion

Extinction has long been recognized as an evolutionary force: differential survival is as much a part of the neo-Darwinian equation as differential reproduction. However, the fossil record contains episodes of mass extinction that appear to fall outside of the operation of normal extinction processes. Instead of constituting an attritional, selective process that continually weeds out the less well adapted taxa, mass extinctions are relatively discrete events that affect a wide range of taxa and environments simultaneously and may exhibit little selectivity, or a selectivity unrelated to that operating in normal times. New geological and paleontological evidence suggests that mass extinctions may be more frequent (although still uncommon and brief relative to the 600 myr of the Phanerozoic) and may play a more pervasive role in shaping the history of life than is generally assumed. Much research remains to be done, but by shifting emphasis from the causes of mass extinctions to their effects and giving equal weight to the phenomena of survival and re-diversification as well as to extinction, evolutionary biologists and paleobiologists can greatly expand the data and theory of macroevolution.

Acknowledgements. I thank K. W. Flessa and S. M. Kidwell for discussions and criticism. Supported by NSF Grants 81-21212 and 84-17011.

References

1. Alvarez W, Kauffman EG, Surlyk F, Alvarez LW, Asaro F, Michel HV (1984) Impact theory of mass extinctions and the invertebrate fossil record. Science 223:1135–1141
2. Archibald JD, Clemens WA (1984) Mammal evolution near the Cretaceous-Tertiary boundary. In: Catastrophes and earth history, eds Berggren WA, Van Couvering JA. Princeton, NJ: Princeton University Press, pp 339–372
3. Anstey RL (1978) Taxonomic survivorship and morphologic complexity in Paleozoic bryozoan genera. Paleobiology 4:407–418
4. Bakker RT (1977) Tetrapod mass extinctions – A model of the regulation of speciation rates and immigration by cycles of topographic diversity. In: Patterns of evolution, ed Hallam A. Amsterdam: Elsevier, pp 439–468
5. Batten RL (1973) The vicissitudes of the gastropods during the interval of Guadalupian-Ladinian time. In: The Permian and Triassic systems and their mutual boundary, eds Logan A, Hills LV. Can Soc Petrol Geol Mem 2:596–607
6. Benton MJ (1983) Dinosaur success in the Triassic: A noncompetitive ecological model. Q Rev Biol 58:29–55
7. Boucot AJ (1983) Does evolution occur in an ecological vacuum? II. J Paleontol 57:1–30
8. Bretsky PW (1973) Evolutionary patterns in the Paleozoic Bivalvia: Documentation and some theoretical considerations. Geol Soc Am Bull 84:2079–2096
9. Carr TR, Kitchell JA (1980) Dynamics of taxonomic diversity. Paleobiology 6:427–443

9a. Clemens WA (1984) Evolution of marsupials during the Cretaceous-Tertiary transition. In: Third Symposium on Mesozoic Terrestrial Ecosystems, eds Reif W-E, Westphal F. Tübingen: ATTEMPTO Verlag, pp 47–52

10. Fischer AG (1981) Climatic oscillations in the biosphere. In: Biotic crises in ecological and evolutionary time, ed Nitecki MH. New York: Academic Press, pp 103–131

11. Flessa KW, Jablonski D (1984) Extinction is here to stay. Paleobiology 9:315–321
12. Flessa KW, Jablonski D (1985) Declining Phanerozoic background extinction rates. Effects of taxonomic structure? Nature 313:216–218
13. Fürsich FT, Jablonski D (1984) Late Triassic naticid drillholes: Carnivorous gastropods gain a major adaptation but fail to radiate. Science 224:78–80
14. Gould SJ (1982) Darwinism and the expansion of evolutionary theory. Science 216:380–387
15. Gould SJ (1984) The cosmic dance of Siva. Nat Hist 93(8):14–19
16. Gould SJ, Calloway CB (1980) Clams and brachiopods – ships that pass in the night. Paleobiology 6:383–396
17. Hagdorn H, Mundlos R (1982) Autochthonschille im Oberen Muschelkalk (Mitteltrias) Südwestdeutschlands. N Jb Geol Paläont Abh 162:332–351
18. Hickey LJ (1984) Changes in the angiosperm flora across the Cretaceous-Tertiary boundary. In: Catastrophes and earth history, eds Berggren WA, Van Couvering JA. Princeton, NJ: Princeton University Press, pp 279–313
19. Hüssner H (1983) Die Faunenwende Perm/Trias. Geol Rundsch 72:1–22
20. Jablonski D (1986) Causes and consequences of mass extinctions: A comparative approach. In: Dynamics of extinction, ed Elliott DK. New York: Wiley, pp 183–229
21. Jablonski D (1986) Larval ecology and macroevolution in marine invertebrates. Bull Mar Sci, in press
22. Jablonski D (1986) Mass and background extinctions: The alternation of macroevolutionary regimes. Science 231:129–133
23. Jablonski D, Flessa KW, Valentine JW (1985) Biogeography and paleobiology. Paleobiology 11:75–90
24. Jablonski D, Sepkoski JJ Jr, Bottjer DJ, Sheehan PM (1983) Onshore-offshore patterns in the evolution of Phanerozoic shelf communities. Science 222:1123–1125
25. Kauffman EG (1984) The fabric of Cretaceous marine extinctions. In: Catastrophes in earth history, eds Berggren WA, Van Couverin JA. Princeton, NJ: Princeton University Press, pp 151–246
26. Kemp TS (1982) Mammal-like reptiles and the origin of mammals. London: Academic Press
27. Knoll AH (1984) Patterns of extinction in the fossil record of vascular plants. In: Extinctions, ed Nitecki MH. Chicago: University of Chicago Press, pp 22–68
28. Martin PS (1984) Catastrophic extinctions and late Pleistocene Blitzkreig: Two radiocarbon tests. In: Extinctions, ed Nitecki MH. Chicago: University of Chicago Press, pp 153–189
29. Martin PS, Klein RG (eds) (1984) Quaternary extinctions: a prehistoric revolution. Tucson: University of Arizona Press
30. McLaren DJ (1983) Bolides and biostratigraphy. Geol Soc Am Bull 94:313–324
30a. Pojeta J Jr, Palmer TJ (1976) The origin of rock boring in mytilacean pelecypods. Alcheringa 1:167–179
31. Quinn JF (1983) Mass extinctions in the fossil record. Science 219:1239–1240
32. Radinsky LB (1982) Evolution of skull shape in carnivores. 3. The origin and early radiation of modern carnivore families. Paleobiology 8:177–195
33. Raup DM (1979) Biases in the fossil record of species and genera. Bull Carnegie Mus Nat Hist 13:85–91
34. Raup DM (1979) Size of the Permo-Triassic bottleneck and its evolutionary implications. Science 206:217–218
35. Raup DM (1981) Extinctions: Bad genes or bad luck? Acta Geol Hispanica 16:25–33

36. Raup DM (1984) Evolutionary radiations and extinctions. In: Patterns of change in earth evolution, eds Holland HD, Trendall AF. Dahlem Konferenzen. Berlin, Heidelberg, New York, Tokyo: Springer-Verlag, pp 5–14
37. Raup DM, Sepkoski JJ Jr (1982) Mass extinctions in the marine fossil record. Science 215:1501–1503
38. Russell D (1977) The biotic crisis at the end of the Cretaceous Period. Syllogeus, Natl Mus Nat Sci Can 12:11–23
39. Saunders WB, Swan ARH (1984) Morphology and morphologic diversity of mid-Carboniferous (Namurian) ammonoids in time and space. Paleobiology 10:195–228
40. Sepkoski JJ Jr (1984) A kinetic model of Phanerozoic taxonomic diversity. III. Post-Paleozoic families and mass extinctions. Paleobiology 10:246–267
41. Sheehan PM (1982) Brachiopod macroevolution at the Ordovician-Silurian boundary. Proc 3rd N Am Paleontol Conv 2:477–481
42. Sheehan PM (1985) Reefs are not so different – they follow the evolutionary pattern of level-bottom communities. Geology 13:46–49
43. Signor PW III, Brett CE (1984) The mid-Paleozoic precursor to the Mesozoic marine revolution. Paleobiology 10:229–245
44. Signor PW III, Lipps J (1982) Sampling bias, gradual extinction patterns, and catastrophes in the fossil record. Geol Soc Am Spec Paper 190:291–296
45. Stanley SM (1979) Macroevolution. San Francisco: WH Freeman
46. Stanley SM (1984) Marine mass extinctions: A dominant role for temperature. In: Extinctions, ed Nitecki MH. Chicago: University of Chicago Press, pp 69–117
47. Thayer CW (1983) Sediment-mediated biological disturbance and the evolution of marine benthos. In: Biotic interactions in recent and fossil benthic communities, eds Tevesz MJS, McCall PL. New York: Plenum, pp 479–625
48. Valentine JW (1974) Temporal bias in extinction among taxonomic categories. J. Paleontol 48:549–552
49. Valentine JW (1980) Determinants of diversity in higher taxonomic categories. Paleobiology 6:444–450
50. Valentine JW, Foin TC, Peart D (1978) A provincial model of Phanerozoic marine diversity. Paleobiology 4:55–66
51. Valentine JW, Jablonski D (1983) Larval adaptations and patterns of brachiopod diversity in space and time. Evolution 37:1052–1061
52. Vermeij GJ (1977) The Mesozoic marine faunal revolution: Evidence from snails, predators and grazers. Paleobiology 3:245–258
53. Vrba ES, Eldredge N (1984) Individuals, hierarchies and processes: Towards a more complete evolutionary theory. Paleobiology 10:146–171

Standing, left to right:
Karl Niklas, Doug Futuyma, Olli Järvinen, Dan Simberloff

Seated (center), left to right:
Franz Fürsich, Dick Bambach, Alec Panchen, Erik Flügel

Seated (front), left to right:
Tony Underwood, Konrad Weidich, Claude Babin

Patterns and Processes in the History of Life,
eds. D. M. Raup and D. Jablonski, pp. 331–350. Dahlem Konferenzen 1986
Springer-Verlag Berlin, Heidelberg

The Neontologico-Paleontological Interface of Community Evolution: How Do the Pieces in the Kaleidoscopic Biosphere Move?

Group Report

O. Järvinen, Rapporteur
C. Babin
R. K. Bambach
E. Flügel
F. T. Fürsich
D. J. Futuyma
K. J. Niklas
A. L. Panchen
D. Simberloff
A. J. Underwood
K. F. Weidich

Introduction

Expanding the Ecological Perspectives

The ecology of extant organisms can be studied in great detail, and the patterns emerging often show complex and unexpected interactions. Clearly the wealth of detail available for paleontologists is not equally impressive, but what paleontology has instead is the time perspective. This difference between ecological and paleontological studies may be illustrated with the following provocative calculation.

Let us think of a typical good ecological study that, giving a generous estimate, covers 3 years (out of tha 3½ billion possible) and 1 sq.km (out of the more than 10^8 possible on Earth). If the available area-time space is represented as a cube with sides of 1 m, the volume of a good ecological study constitutes a cube with sides of about 1 micrometer. Fiddling around with this invisible microcube (preferably with experiments, as the recent literature emphasizes) is the basis for statements like, "Ecologists have repeatedly found that X rather than Y is typical in ecological systems." These generalizations are often based on a few species out of the few million extant ones.

It seems that one way to proceed is through increased awareness of possible extensions of our ecological perspectives. Generalizations based on one species at one site in a brief period of time have three possible directions of test and extension: community studies, biogeography (an important and until recently a neglected field that will largely fall outside this report), and paleontology. Most of the report will attempt to illuminate the interface of ecology and paleontology, for the benefit of paleontologists and ecologists alike.

Perhaps two broad controversies in recent ecology should make a juxtaposition of ecology and paleontology especially fruitful for paleontologists. First, there has been extensive discussion about the importance of good data bases for generalization and testing. Conclusions cannot be better than the data upon which they are based. There has been an increased awareness in ecology that data should be collected and documented more carefully than has been customary in (American) ecology, and analyzed using more appropriate statistics than has been customary in (European) ecology. Second, considerable attention has been paid throughout the world to the distinction between scientific explanation and telling untested "just-so" stories. What we mean by this is that the *criteria* of establishing patterns have been debated. For example, how does one establish that competition (predation, parasitism, etc.) plays an important role in the system studied? Clearly, what is needed is something better than ad hoc stories that happen to fit the observations.

The following discussion starts with a clarification of basic conceptual issues, since sound analyses of "community patterns" require a good understanding of the nature of the evidence that different concepts imply. We conclude the conceptual discussion with critical remarks on the superorganism idea, which some readers may feel is like flogging a dead horse. Our group, however, found abundant evidence of superorganism-related thought patterns in recent papers on paleontology and on the community ecology of extant organisms. After the conceptual clarifications, we will proceed to a critical interpretation of real data on community evolution in the Phanerozoic.

Few of Us Study Communities

The standard practice both in paleoecology and in the ecology of recent organisms is to use the term *community,* but this implies organization, structure, and interactions. Most studies on "community structure" are not, in fact, on communities or on structure.

Neontologists will sometimes find difficulty in trying to understand the nature of paleoecological species assemblages. Useful distinctions include:

taphocoenosis – a sample of specimens found buried together, (an agnostic term that implies no evaluation whatsoever as to whether the remnants represent organisms that lived together at the site or were carried there from elsewhere);

symmigy – (see [1]) indicating a taphocoenosis where we suspect that the specimens found represent a mixture of organisms that actually lived in several places or times; and

thanatocoenosis – representing organisms that presumably did live together at the site where they were found.

If the data are filtered by excluding rare specimens of species that are abundant elsewhere and may have been carried to the site by later reworking and by taking into account the impact of diagenetic changes (which may remove families and even classes of ancient faunas and floras from a taphocoenosis), we have a *paleocommunity*. The term often refers to a recurring assemblage of species, but data on functional morphology are also available and are used in inferring features not directly observed. However, the criteria for establishing that there is indeed a community in the above sense should cause some worry, as pointed out by Underwood (this volume). He distinguished between a *community* and an *assemblage*, pointing out that a community is tightly structured and consists of many types of organisms at different trophic levels, whereas an assemblage is a neutral term and refers to a collection of organisms at a particular site at a particular time. We use this distinction in our discussion below. Our interpretation of observations may depend on what we think our observations constitute; interpreting assemblage data as if they were data on a true community may be a red herring.

A serious problem that may be even greater in neontological than in paleontological studies is the tendency to synonymize one taxocene (representatives of one taxon, such as birds, in a particular area) with the whole community. This tendency can partly be traced to the Darwinian article of faith that competitive interactions tend to be most intense among taxonomically closely related species. Insofar as recent ecology has focused on competitive relationships (the "consumer approach": With whom is this consumer likely to compete for food?), restricting the study to a single taxocene has been commonplace, in contrast to studies based on the "resource approachs" (Who eats this food?). However, an increasing literature suggests that taxonomic relatedness is not at all synonymous with ecological (competitive) similarity, one well-known example being the interkingdom competition between microorganisms and Daniel Janzen for a 95-cent avocado [13].

Of course, studies of single taxocenes continue to flourish in their own right, and they can *demonstrate* patterns even though *explaining* the patterns can be a problem; of course, taxocene studies are also often useful in studying environmental gradients. The fact that we then neglect intertaxocene interactions (of which those related to microbiology may be the most ignored in relation to their ecological importance) should nevertheless be a sign for caution.

Structure Is Deceptive

Some of the temporal heterogeneity of the neontological world is certainly due to vagaries of chance, but many short-term changes in species assemblages seem to have understandable and ecologically interesting causes. The paucity of long-term studies in ecology is nevertheless a problem as those studies that are available often show remarkable changes in the composition of assemblages or, in autecology, in the processes that are important in affecting the abundances and distributions of populations. This implies that the time-averaging that is inevitably present in paleontological samples can yield assemblages that actually never occurred at the site at any time. This is in part only a scale difference, for even ecological studies of recent assemblages are not able to report all comings and goings of the species. The time scale difference is, however, a double-edged sword: the neontological time perspective is so short that it is impossible to observe slow processes that may take apparently random twists over millenia but nevertheless lead to a definite result. For example, how long does a typical displacement of one taxon by another require? Moreover, could the time-averaging sometimes better represent what is generally going on over larger areas over certain periods of time? A central tendency (or an "equilibrium") may exist in the assemblage studied, but a high variance blurs such a tendency in neontological data sets usually spanning a period of only a few years.

Species assemblages are often said to possess some structure, which connotes organization. Often structure refers to a log-series or log-normal taxon-abundance distribution. These and other distributions were previously thought to reflect types of organizing processes but may be interpreted rather as epiphenomena (and therefore not represent structure at all). Of course, consistent rank order of taxa in such a distribution provides more, and more meaningful, information than the mere existence of a particular distribution.

In other contexts structure is ecologically more meaningful. For example, trophic structure (food webs or the presence of different *Lebensformen*) seems to show an increase in complexity over time, in the sense of increasing numbers of new types of feeding mechanisms (see below). In paleoecology, trophic structure is sometimes inferred indirectly, for example, from the existence of fractures or bore-holes in molluscan shells (indicating predation), from the appearance of puncture wounds on early land plant stems, or from the presence of bryozoans or other taxa that use seaweeds as their substrate.

Other types of structure include spatial structure and guild structure. Broadly defined guilds sometimes show interesting regularities (e.g., an approximately constant number of species) over time, but neontological data suggest that even experimentally demonstrated and clear interactions need not be repeatable in time and space. Also, it is clear that the same species belongs to markedly different assemblages in different places, as demonstrated by habitat selection studies over the geographical range of different species. Guild structure presupposes that guilds can be defined unambiguously, which for paleontological data raises difficulties: morphology does not correspond unequivocally to resource us, one pertinent example being Foraminifera (excepting some general tendencies, see [3, 4]). Paleobiological data therefore rarely meet the criteria of guild assignment sensu stricto, even though in some cases comparisons with extant organisms are of great value. For example, functional morphology can be examined and quantified on the basis of a good modern analogue (as, for example, the correspondence between extant and extinct plant grades) in order to define "guild structures" in the fossil record.

An obvious difference between neontological and paleontological studies of trophic structure is the differential preservation of species. For example, soft-bodied animals are seldom fossilized, so their contribution to paleoassemblages must be deduced indirectly or from unusual situations where soft-bodied organisms have been preserved well (*Lagerstätte*, Pompeii situations). *Lagerstätte* supply valuable information for positing criteria for determining the fidelity of preservation in other fossil assemblages. Another problem is that even the fossilizable part of the assemblages may not be represented in true porportions (varying rates of diagenetic destruction, varying probabilities of fossilization).

The Superorganism Is Dead

The study of ecological succession is an excellent example of a paradigmatic change in ecology from a holistic view (Clements) to an individualistic one (Gleason). In the individualistic view, succession is seen not as one superorganism replacing another, but as comings and goings of individual species whose responses to the environment are relatively independent of each other. It is true that some species have a particularly large effect on others: for example, the main tree species in forest succession alter the environment in several ways (light, temperature, soil structure and composition, wind, etc.) that are important for many other organisms. It is also clear that some species have undoubted interactions that coordinate their comings and goings in succession, epiphytes being an obvious example. The contrast between the individualistic and the superorganismic concepts is therefore not completely dictotomous; the main point is that succession is no longer seen as series of replacements of one superorganismic community by another.

Studies of keystone species such as *Pisaster*, kelp, or beaver should be particularly useful in evaluating the possible role of coordinated change of communities in evolutionary history. Cascading extinction effects will presumably be seen in tropical, moist forests where the elimination of a single tree species implies the extinction of its strictly monophagous herbivores. However, what is more frequently observed in ecological studies of keystone species is not a complete extinction of a number of species after the removal of the keystone species, but rather marked abundance changes in the species assemblages involved (e.g., [8, 9, 14]).

The paleontological evidence also offers no support for the superorganism concept. A perfect analogue to the study of ecological succession is provided by the history of the geoflora concept. The geoflora concept was originally proposed [5] to explain the repeated occurrence within the fossil record of a collection of fossil plant species. Biogeographically this collection or geoflora was seen to migrate in response to latitudinal climatic changes and continental migrations ("drift"). Careful analyses of the paleobiogeography of individual taxa, however, suggest that many species, genera, etc., have characteristic and different patterns of migration and climatic sensitivity (see [17]). While didactically useful, the geoflora concept has been replaced with the idea of taxonomic individuality.

An aspect that requires increased attention among paleoecologists is whether extinction events involving many species represent similar individualistic responses to an environmental change or cascading extinction waves caused by interspecific interactions. The decrease of diversity in Devonian reef assemblages may be a case in which it will be possible to make this distinction.

Diverse reef assemblages that had developed by addition of taxa from bryozoan-pelmatozoan assemblages in the Ordovician, to tabulate coral-stromatoporoid assemblages in the Silurian, to tabulate-rugose coral plus stromatoporoid frameworks by the Middle Devonian, disappeared in the Late Devonian (see [12]). This collapse of reef assemblages was part of a cascading decline in diversity of the major taxa that included the growth forms for framework-building in these reefs. Compound forms among the Rugosa and densely packed colonial forms among the Tabulata lost more diversity at the family level than did solitary Rugosa or loosely associated or encrusting Tabulata. Although organic structures that we would call reefs are common in the Frasnian and some examples are even known from the Famennian of Central Europe, Canada, and China, the diversity of families including frame-building morphologies had been declining since the Eifelian in the Middle Devonian. The decline of diversity seems to have been sequential. First the tabulate corals (the most abundant frame-builders) lost diversity in the Eifelian to Givetian, then the rugose corals started to decrease in diversity in the Givetian and Frasnian, and finally the stromatoporoids (the encrusting forms that had capped most reef structures) lost their importance in constructing reefs in the Frasnian and Famennian. Reef structures as topographic features and distinct assemblages persisted to the end of the Devonian as long as some frame-builders lasted, but the family diversity of these assemblages waned over an interval of some 10 million years. Because reef systems could persist even as the constituent taxa were declining in diversity, it seems clear that the reefs were not "superorganisms." The protracted decline of reef diversity over millions of years does not suggest the collapse of a highly integrated system, but rather the independent demise of its constituents.

As possible cases of cascading extinction waves suggesting a superorganismic unity seem to be rarities in the fossil record (if undisputed examples can be found at all), the burden of proof now seems to reside on the shoulders of those who wish to drag the superorganismic corpse along, despite abundant evidence for the individualistic view.

The presumed emergent properties of communities are frequently used to justify community studies but may rather be viewed as statements of ignorance. A useful simile is to regard emergent properties as something like interaction term in ANOVA. For a more detailed discussion, see Underwood (this volume).

The term "emergent property" has been used in so many ways that it is confusing. Certainly not all group properties would be defined as emergent. For example, some traits are defined only for a group, not for an individual: a group, not an individual, has a mean. Some term other than "emergent" that captures this distinction is needed; "collective property" seems unambiguous.

After this conceptual clarification, we now turn to a discussion of large-scale changes of species assemblages in the Phanerozoic, trying particularly to integrate recent ecological insights with paleontological data.

Coevolution Is Often Diffuse

If species assemblages are typically characterized by collective as opposed to emergent properties, we may expect that relationships between taxa do not constitute a tightly organized web of interaction. Pleistocene beetle assemblages in Britain seem a pertinent example: the species that constituted Pleistocene species assemblages are still present, but the assemblage has fragmented because the responses of the individual species to climatic and other environmental changes have been vastly different.

Of course, the idea of a web of interactions connotes the concept of *coevolution,* which, however, has been applied by one person or another to each of the following concepts (the list is probably not exhaustive):

1) Coincident evolution, i.e., evolutionary change of two or more lineages in the same region at the same time, as might be revealed in vicariance biogeography by congruence of several cladograms with a history of separation of land masses. No ecological interaction is implied.
2) Coincident evolution of ecologically associated taxa, but with no selective impact of either taxon on the other. Conceivably, birds and their mallophagan associates (feather lice) do not exert selection on each other yet might have congruent cladograms.
3) Unilateral evolution of adaptation of one taxon to the properties of another. Danaid butterflies (e.g., monarchs) have adaptations to properties of their host plants (Asclepiadaceae), but there is no evidence that milkweeds have evolved specifically in response to danaids.
4) Reciprocal adaptive response of specific species or lineages to each other. Examples include laboratory systems (e.g., bacteria and phage), rabbit and myxoma virus in Australia, *Acacia* and *Pseudomyrmex* ants. Such specific reciprocal responses are most likely for highly species-specific associations of symbionts.

5) Diffuse reciprocal influences, i.e., between sets of often taxonomically diverse species with qualitatively similar ecological influences. Examples might include reciprocal influences between groups of pollinating insects and plants, or the evolution of increased molluscan armor in response to increased predatory capabilities in several predatory taxa. "Diffuse coevolution" appears to embrace most cases of "coevolution" (see also Futuyma, this volume).

Empirical evidence on coevolution is not easy to obtain. For example, it is difficult to exclude the possibility that adaptations in a seemingly reciprocal system have actually arisen because of adaptation to a third party, or that coevolution exists but involves such asymmetrical influences that the weaker effects are difficult to observe.

Many paleontological examples (horse teeth – tougher grasses, floral structure – insect pollinators; locomotory speed in prey-predator systems; armor in snails) are well-known and belong under the concept of diffuse coevolution. What seems less well established is how much of the diffuse coevolutionary trends take place within species, and how much should be accounted for by a replacement of taxa, e.g., of less efficient competitors by more efficient ones, i.e., exactly what ecological process is involved.

An equally open problem seems to be how much of adaptive radiations can be accounted for by coevolution, for positive and negative examples can be easily found. For example, the adaptive radiations of angiosperms and some insects seem to be related (see Niklas, this volume), whereas the adaptive radiation of Hawaiian drosophilids seems to be independent of any corresponding radiation in the resources.

If species coexisting at a certain moment have typically relatively independent biogeographic histories and live in assemblages that tend to change more or less continuously, it is plausible to suggest that most coevolution is diffuse. Such biogeographic dynamism also suggests that the apparently coevolved features of the species in an assemblage have frequently evolved in other arenas, rather than as responses to the members of the local aggregation under study. Moreover, this view implies that the surrounding species assemblage, and therefore the biotic selection pressures, tend to change continuously. As all of the changes will hardly be in the same direction, this provides one (but not the only) explanation for evolutionary stasis (cf. also the evolution/radiation of many insular taxa to forms that are unexpected in view of the uniformity of the mainland counterparts).

Life Diversifies Locally and by Adding Life-Styles

The overall increase of diversity of living organisms is a well established broad trend on the global level (for examples see Bambach and Niklas, both this volume). Data are fewer on the trends in species richness in well-defined local assemblages, but such data are needed in order to understand how the global increase in diversity translates to diversity on the local scale. In neo-ecological terminology, in order to understand gamma diversity one needs data on alpha and beta diversities (which in the present context can be interpreted as local diversity and the degree of endemism, respectively); see, however, Underwood (this volume) for a critical discussion of the different concepts of diversity.

Data from marine invertebrate communities seem to indicate that local and global trends in diversity have been roughly parallel over long periods of time, although exact conclusions are premature because of many technical difficulties (unequal representativeness of sampling in different periods; different taxonomic levels in the local and global comparisons; possibly confusing effects of continental drift on local diversity patterns). There seems to be an interesting exception, however, to the parallel trends on the local and glocal scales. Diversity has apparently changed little in at least some high-stress environments, despite vast taxonomic turnover at levels from species to family. It seems safe to conclude that the global increase in diversity has been accompanied by an increase in many local assemblages of species, but at this stage it is not possible to evaluate the relative importance of changing patterns of endemism and changing local diversity as components of global diversity. Neither is it possible to establish what are the cause-and-effect relationships between local and global trends in diversity, nor is there clear evidence as to the extent to which diversity increases by the addition of ecologically specialized versus generalized species (if such can be defined objectively).

Another pattern that emerges from an examination of diversity changes is the increase in new life-styles (broadly defined "guild types") during the history of life. An example based on Bambach's work on marine invertebrate assemblages is shown in Fig. 1. There seems to be no doubt that one of the important components in the increasing diversity of marine invertebrates was the evolution of novel life-styles. It may be argued that some of the missing life-styles in the early assemblages were represented by the non-fossilizable part of the fauna, but in many cases this can be ruled out (e.g., the existence or absence of the deep infaunal mode of life in certain habitats can be deduced from bioturbation or its absence).

How much of the increase in diversity has been due to an increased number of species sharing a similar life-style ("increased species packing within guild types"), and how much is due to an increased number of major life-styles, such as those shown in Fig. 1? The pattern that seems to be emerging is that the increasing number of life-styles is particularly important in explaining the increase of marine invertebrate taxa. In contrast, plants and terrestrial vertebrates seem to show a clear increase in both number of life-styles and number of species associated with each life-style.

Plant diversity increases within a particular depositional setting, perhaps characteristic of a particular environment, both as a function of the total number of species and as a function of the total number of "guilds" (or the botanical term, "grades"). Since the total number of species and the number of "guilds" of fossil plant assemblages is seen to increase throughout much of the Phanerozoic, iterative patterns of thanatocoenoses are hard to observe. Nonetheless, some major overall trends in global diversity are clearly evident (e.g., a trend toward arborescence, vertical stratification; ecologically functional haplobiontic-haploid plants).

The history of vertebrates, and particularly terrestrial vertebrates, has always been presented in terms of a succession of grades – the "age of fishes," "age of amphibia," "age of reptiles," "age of mammals," (but not the "age of birds"!). Even at a descriptive level this is oversimplification to the point of falsehood. The age of mammals is also the age of teleost fish: three quarters of mammalian history had elapsed before the beginning of the age of mammals. Nevertheless, as in the history of plants, the increase of vertebrate diversity has been enhanced, at least on land, by adding life-styles in the form of major adaptive types, including entirely new habitats such as the air. However, the number of species with a particular life-style appears to have increased during the diversification of terrestrial vertebrates.

Different Actors, the Same Drama?

Taxonomic turnover is commonplace in evolutionary or biogeographical comparison. Nevertheless, it is often felt that the same drama is being played on different and independent stages. This is especially so for the dramatis personae in the evolutionary drama. Many examples from the

MIDDLE & UPPER PALEOZOIC FAUNA

PELAGIC	SUSPENSION	HERBIVORE	CARNIVORE
	CONODONTOPHORIDA GRAPTOLITHINA ? CRICOCONARIDA		CEPHALOPODA PLACODERMI MEROSTOMATA CHONDRICHTHYES

EPIFAUNA	SUSPENSION	DEPOSIT	HERBIVORE	CARNIVORE
MOBILE	BIVALVIA	AGNATHA MONOPLACOPHORA GASTROPODA OSTRACODA	ECHINOIDEA GASTROPODA OSTRACODA MALACOSTRACA MONOPLACOPHORA	CEPHALOPODA MALACOSTRACA STELLEROIDEA MEROSTOMATA
ATTACHED LOW	ARTICULATA EDRIOASTEROIDA BIVALVIA INARTICULATA ANTHOZOA STENOLAEMATA SCLEROSPONGIA			
ATTACHED ERECT	CRINOIDEA ANTHOZOA STENOLAEMATA DEMOSPONGIA BLASTOIDEA CYSTOIDEA HEXACTINELLIDA			
RECLINING	ARTICULATA HYOLITHA ANTHOZOA STELLEROIDEA CRICOCONARIDA			

INFAUNA	SUSPENSION	DEPOSIT	CARNIVORE
SHALLOW PASSIVE	BIVALVIA ROSTROCONCHIA		
SHALLOW ACTIVE	BIVALVIA INARTICULATA	TRILOBITA CONODONTOPHORIDA BIVALVIA POLYCHAETA	MEROSTOMATA POLYCHAETA
DEEP PASSIVE			
DEEP ACTIVE		BIVALVIA	

CAMBRIAN FAUNA

PELAGIC	SUSPENSION	HERBIVORE	CARNIVORE
	TRILOBITA (AGNOSTIDS)		

EPIFAUNA	SUSPENSION	DEPOSIT	HERBIVORE	CARNIVORE
MOBILE		TRILOBITA OSTRACODA MONOPLACOPHORA	MONOPLACOPHORA OSTRACODA	
ATTACHED LOW	INARTICULATA ARTICULATA			
ATTACHED ERECT	EOCRINOIDEA			
RECLINING	? HYOLITHA			

INFAUNA	SUSPENSION	DEPOSIT	CARNIVORE
SHALLOW PASSIVE			
SHALLOW ACTIVE	INARTICULATA	TRILOBITA "POLYCHAETA"	"POLYCHAETA"
DEEP PASSIVE			
DEEP ACTIVE			

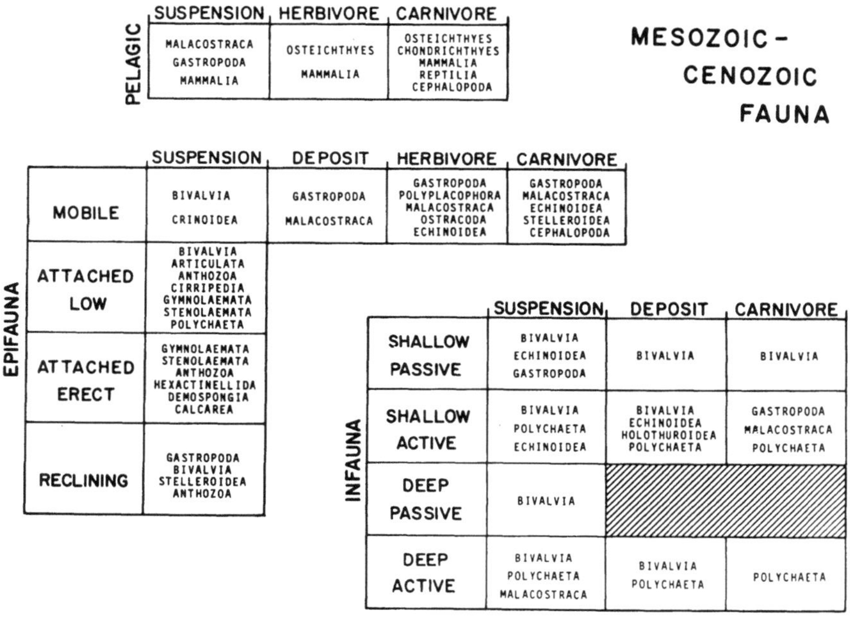

Fig. 1. Broad life-styles that can be distinguished in the fossilizable part of the three great evolutionary marine faunas of the Phanerozoic. The shaded boxes are non-viable life-styles. From Bambach [2]

paleontological record show that similar patterns occur repeatedly in similar environments (e.g., ammonites or marine benthic assemblages); a similar pattern refers here to the presence of the same assortment of morphotypes. Of course, the problems encountered in neontological studies of community convergence also arise here: How is one to judge equivalence of function, how to measure the degree of similarity?

Another major problem, exemplified by Connell's studies [6, 7] of intertidal barnacles, is that despite similar patterns, the major processes driving the systems may be strikingly different in different settings; congruent (or even convergent) patterns therefore may not indicate what is really going on. Another example from coral reefs emphasizes a similar point. It is known that in some reefs fish abundance and diversity are strikingly high and more or less dominate the ecological scene, while fish may be a relatively minor component in similar reefs elsewhere. Despite considerable similarity in the fossilizable parts of the reef assemblages, the major processes may thus be quite different.

Recent species assemblages have also been studied by analyzing morphometric measurements of numerous taxa using advanced multivariate statistics. It is then possible to construct multivariate morphospaces that depict the morphometric relationships within and among taxa. Attempts to construct such morphospaces and to find consistent patterns in them have proved a useful approach towards understanding community structure. Applying the methods of ecomorphology to paleontological data would be straightforward. For example, morphospace patterns in the different trilobite faunas could be a worthwhile object of study. Here the paleontological time perspective would possess the additional advantage of being able to show if there is indeed convergence towards congruent morphospace patterns over time.

Whether the plots in evolutionary dramas tend to recur is less clear. A classic generalization about the plots is the one given by Dobzhansky [10] and many later authors: in constant/stable/unperturbed environments, the main driving force is selection based on biotic factors, whereas in variable/unstable/perturbed situations abiotic selective forces dominate. While such a simplification can hardly be defended any more, no clear answers seem to have emerged in its stead. Many evolutionary changes can undoubtedly be traced to changes in the abiotic environment, but marked evolutionary changes are known to have taken place in the absence of major environmental changes. For example, one can recognize a gradual change in the qualitative and quantitative composition of reef (frame-building) assemblages from the Middle and lower Upper Triassic to the late Upper Triassic (calcareous sponges→calcareous sponges + some corals→corals). This change is paralleled by a continuous change in the epifauna. There seems to be no corresponding change in the environment (water depth). The same may apply to the transition from siliceous sponge to coral reefs in the Upper Jurassic (Kimmeridgian) in Franconia ([11] and E. Flügel, personal communication).

Assemblages experience major changes during mass extinctions, but relatively little is known of the patterns of survivorship. For example, during the Phanerozoic there was a general trend toward a decrease in the proportion of clonal plants. However, when lineages with both clonal and aclonal growth patterns pass through major episodes of global extinctions or major climatic disturbances, the clonal growth forms survive at disproportionately higher rates than to the aclonal growth forms. This suggests that conditions of relatively modest ecological disturbance and those of high ecological disturbance differ in their influence on the long-term course of evolution.

The relative role of biotic vs. abiotic factors as driving forces in assemblage change can also be examined from data on the effects of novel entities in species assemblages on the other species. An ecological review of the effects of introduced species shows relatively few cases of competitive displacement. However, it is difficult to observe resistance that succeeds in preventing introductions, and therefore reliable data on this point are hard to obtain (particularly because introductions may fail for reasons other than resistance).

The Great American Interchange that resulted in a striking "North Americanization" of the South American mammal fauna appears to be a case of displacement of one species set by an invading one, but actually the difference between the success of the two faunas seems to lie in a difference in the rate of diversification after the interchange and in the terminal Pleistocene extinction event [15]. Another example of possible displacement is the success of the angiosperms, although looking at the displacements at higher taxonomic levels than species is problematic and may be confusing. Also, what may look like a displacement may actually have been a response of the species to a secular change in the environment.

As discussed by Bambach (this volume), recent data suggest that there is a nearshore-offshore pattern in the history of different faunal elements of marine benthic invertebrates. First, Cambrian faunal elements and later, the faunal dominants of the Middle and Upper Paleozoic disappeared from nearshore assemblages that are now dominated by Mesozoic-Cenozoic dominants. Whether this is a large-scale displacement of one community with another seems questionable. One aspect that should receive more attention is the examination of the phylogenetic patterns in the taxa that show the changing nearshore-offshore distribution. For example, are the remaining offshore taxa descendants derived from most of the clade or merely from a branch of it? Interpretations of the changing nearshore-offshore distribution patterns, or other possible replacement patterns, evidently depend on the phylogenetic relationships of the surviving taxa to those that disappeared.

Another problem, already mentioned above, is that displacement above the species level are difficult to understand ecologically. Sepkoski [16] suggested that differences in origination rates and in "carrying capacities" (of the number of taxa) among the three different faunal elements may account for the pattern through which the Cambrian element is first replaced by the Middle and Upper Paleozoic elements and then by Mesozoic-Cenozoic elements. On the species level, the interpretation is that there are differences among species in the probability of extinction and of their splitting into two species. Perhaps the story is that the three major faunas are collections of species having different average rates of diversification and extinction. It is relevant and most interesting in this context that Sepkoski finds extinction rates to be taxon-specific so that even if generic extinction rates tend to increase by about 50% from nearshore to offshore environments, this is essentially a consequence of a different taxonomic composition of the nearshore and offshore assemblages of species.

It is quite possible that similar patterns can also be found in other taxa. For example, there seems to be a shift from wet to dry environments in the pteridophyte flora, but as discussed above, paleobotanists seem to view the concept of the geoflora in the individualistic vein. We should thus be careful in our interpretations of habitat shifts of "archaic" elements in the fauna and flora: Is the story really abount *community* displacements? As shown by many structuralist analyses, in traditional fairy tales the plots follow a limited number of schemes, as most of the stories that attain the best-seller level also do; but what is really the evidence that Mother Nature's bedside stories would be a collection of repetitive plots? Here we see the neontologico-paleontological interface: a search for more rigorous criteria in our attempts to infer processes that shaped and continue to shape the patterns in the kaleidoscopic biosphere.

Future Research Questions

Heterogeneity in Modern and Fossil Assemblages

How much of the perceived structure of a modern assemblage (in any environment) is consistent from site to site (or time to time)? Thus, how much spatial and temporal variance in modern assemblages is there relative to fossil assemblages? These questions ought to be addressed for whole assemblages as well as the fossilizable parts of an assemblage. As an example, the average structure of two modern assemblages, one dominated by mussels and one dominated by limpets, may not have any real existence – but might be conflated from both in a "time-averaged" fossil assemblage. Is it possible to quantify "time-averaging" patterns in different habitats (soft substrates, hard substrates, reefs, etc.)?

Representativeness of Fossilizable Material

How much of the "web of complex interactions" of an assemblage can be inferred from the fossilizable components of an assemblage? Which interactions among subsets of species are repeatable across different assemblages in similar environments? Might any of these be represented by the subset of dominant, fossilizable components of an assemblage? How much are the different constituents of ancient assemblages affected by diagenetic destruction in different environments?

Species-Abundance Curves from Fossilizable Parts of Modern Plant Assemblages

The quantification of the abundance of species in an environment is essential to ecological studies. Yet this quantification is difficult for fossil plants that are represented by isolated organs (fruits, seeds, leaves, twigs). Neotaphonomy, based on extant systems, ought to be explored so that the number of individuals within an assemblage and the number and type of organisms found in particular depositional settings are tracked through ecological time and along distance gradients.

New Criteria for Studying Fossil Terrestrial Biotas

There is very little work on assemblages of terrestrial organisms in the fossil record, except perhaps in the Pleistocene. Yet there are many scenarios in the literature about the coevolution of, e.g., plants, insects, and vertebrates, and many assertions arising from functional morphology about the mode of life and especially the feeding habits of fossil vertebrates. Two things are needed to put this type of speculation on a scientific basis:

a) detailed study of selected taphocoenoses to find evidence of, e.g., plant feeding and/or predation or other interactions between members of different taxa (several known *Lagerstätte* are promising for this work) and

b) the formulation of criteria, testable if possible, for stating that trophic or other interactions within the assemblage actually occurred.

Assemblage Changes in Clearly Defined Environments

Do assemblages within particular environments change through time simply by replacement of constituent taxa? How do diversity and morphotype composition change within particular environments through time? What happens when new ecological types or morphotypes enter such systems? Do major changes in assemblages coincide with major disturbances of the environments? Are there apparent consequences of the latter, such as extinctions or changes in species-abundance curves? Could such patterns not be perceived in a local site if one examined many strata very close to one another? It should be emphasized here that the focus is on patterns and processes within clearly defined environments and that the level of analysis should be that of the species or, if this is not possible, the genus.

Morphospaces of Fossil Assemblages

Could one examine whether species in a local assemblage occupy morphospace in predictable ways? Are species' morphologies independent of one another, or are such patterns as limiting similarities apparent? There exists a body of ecological literature on statistical methods for dealing with such problems.

Range Changes of the Constituent Species of Assemblages

Several ideas about the integrity of communities and coevolution are affected by the degree to which the geographic distributions of formerly associated species change independently over time. Several cases of independent movement have been cited, but more detailed studies such as these are desirable.

Detailed Studies of "Specialized" vs. "Generalized" Species

Both functional morphology and the observable association of a species with one or more habitats can be used to describe species as ecologically generalized or specialized. Some evidence has been advanced on the greater extinction rate of specialists, but more extensive and detailed data are needed.

Survivors of Mass Extinctions

How and why do some communities, assemblages, taxocenes, or parts of them survive mass extinctions? Several aspects need attention, such as the data base (new field sampling!) and the taxonomic level (preferably the species or perhaps the genus level; patterns based on higher taxa may be difficult to interpret). What "structures" survive? Do these depend on morphology, "community structure," etc.? Why are the survivors (species, assemblages) not fatally affected by the mass extinction event? Of particular importance is that increased attention be paid to the (short?) time interval between the mass extinction event and the beginning of a new radiation: data immediately after the mass extinction are needed.

Suggested General Reading

Dodd, J.R., and Stanton, R.J., Jr. 1981. Paleoecology. Concepts and Applications. New York: John Wiley.
Futuyma, D.J., and Slatkin, M., eds 1983. Coevolution. Sunderland: Sinauer.
Panchen, A.L., ed 1980. The Terrestrial Environment and the Origin of Land Vertebrates. Systematics Association Special vol 15. New York: Academic Press.
Saarinen, E., ed 1982. Conceptual Issues in Ecology. Dordrecht: Reidel Publishing.
Strong, D.R., Simberloff, D., Abele, L.G., and Thistle, A., eds 1984. Ecological Communities: Conceptual Issues and the Evidence. Princeton: Princeton University Press.
Tevesz, M.J.S., and McCall, P.L., eds 1983. Biotic Interactions in Recent and Fossil Benthic Communities. New York: Plenum.
Valentine, J.W., ed 1986. Phanerozoic Diversity Patterns: Profiles in Macroevolution. Princeton: Princeton University Press, in press.

References

1. Babin C (1971) Eléments de Paléontologie. Paris: Armand Colin éd
2. Bambach RK (1983) Ecospace utilization and guilds in marine communities through the Phanerozoic. In: Biotic interactions in recent and fossil benthic communities, eds Tevesz MJS, McCall PL, pp 719–746. New York: Plenum
3. Bandy OL (1960) General correlation of foraminiferal structure with environment. Int Geol Congr, Rept 21st Session, Copenhagen, Proc Int Paleontological Union, p 22
4. Boltovskoy E, Wright R (1976) Recent foraminifera. The Hague: Junk W Publishers
5. Chaney RW (1959) Miocene floras of the Columbian Plateau. Part 1: Composition and interpretation. Contr Paleontol Carnegie Inst Washington 617:1–134
6. Connell JH (1961) The influence of interspecific competition and other factors on the distribution of the barnacle Chthamalus stellatus. Ecology 42:710–723
7. Connell JH (1970) A predator-prey system in the marine intertidal region. I. Balanus glandula and several predatory species of Thias. Ecol Monogr 40:49–78
8. Dayton PK (1971) Competition, disturbance, and community organization: the provision and subsequent utilization of space in a rocky intertidal community. Ecol Monogr 41:351–389
9. Dayton PK (1984) Processes structuring some marine communities: are they general? In: Ecological communities: Conceptual issues and the evidence, eds Strong DR, Simberloff D, Abele LG, Thistle AB, pp 181–197. Princeton: Princeton University Press

10. Dobzhansky T (1950) Evolution in the tropics. Am Sci 38:208–221
11. Flügel E (1982) Evolution of Triassic reefs: current concepts and problems. Facies 6:297–327
12. James NP (1984) Reefs. In: Facies models, ed Walker RG, pp 213–245. Geoscience Canada Reprint Ser 1. Toronto: Geological Association of Canada
13. Janzen DH (1977) Why fruits rot, seeds mold, and meat spoils. Am Nat 111:691–713
14. Knudsen GJ (1962) Relationship of beaver to forest, trout and wildlife in Wisconsin. Wisconsin Conservation Dept Tech Bull No 25
15. Marshall LG, Webb SD, Sepkoski JJ, Raup DM (1982) Mammalian evolution and the great american interchange. Science 215:1351–1357
16. Sepkoski JJ Jr (1984) A kinetic model of Phanerozoic taxonomic diversity. III. Post-Paleozoic families and mass extinctions. Paleobiology 10:246–267
17. Wolfe JA (1981) Vicariance biogeography of angiosperms in relation to paleobotanical data. In: Vicariance biogeography: A critique, eds Nelson G, Rosen DE, pp 413–445. New York: Columbia University Press

Patterns and Processes in the History of Life,
eds. D. M. Raup and D. Jablonski, pp. 351–367. Dahlem Konferenzen 1986
Springer-Verlag Berlin, Heidelberg

What Is a Community?

A. J. Underwood

Dept. of Zoology, School of Biological Sciences
University of Sydney, Sydney, NSW 2006, Australia

Abstract. Attempts are often made to distinguish a community of organisms from a haphazard assemblage of populations of various species that happen to be together at any place and time. These depend on definitions of those organisms to be included in the community and of the appropriate temporal and spatial scales that might demonstrate consistency in coincidence and interdependence of the set of organisms. Some of the problems with such definitions are discussed, and it is concluded that much of the definition of a community is somewhat arbitrary. The historical debate between proponents of integrated communities and those who favor individual populations as the appropriate units for study continues. The argument that communities might have emergent properties not revealed by the more reductionist studies of populations is considered to be without firm foundation and of no relevance to the design of ecological studies. Patterns and processes should be investigated at several spatial and temporal scales, and community-level attributes of assemblages of species only become relevant where less holistic research programs fail.

Introduction

Ecologists collectively display a form of temporal schizophrenia about the nature of their science. This is particularly manifested in debate about the actual units of study, with considerable energy expended on disagreement about populations versus communities as the appropriate levels of organization of nature to study. One view holds that organisms are assembled into integrated communities that consistently recur in time, are repeated in space, and show complex but interdependent sets of interactions among the species. The alternative viewpoint is that communities are simply convenient human descriptions of sets of organisms that tend to be found in the same place at the same time (because of similarities in physiology and requirements for habitat and resources such as food) but are not interdependent in any sense, and the populations are not integrated in any way. There have been historical shifts of emphasis and dominance of prevailing views, and there is never, at any given time, any resolution that satisfies all participants. Against this historical disagreement, this essay cannot succeed in producing unarguable definitions for communities. Many ecologists opt for a middle ground between the two extremes (i.e., between completely integrated, superorganismic communities and completely random assemblages of co-occurring species). Such communities are, however, arbitrarily defined, if at all, by many ecologists. In many studies, the "community" investigated appears to consist simply of the particular set of coexisting species that are being studied. Among these species there may, of course, be important and complex interactions that are worthy of study insofar as they shed light on patterns of distribution and abundance of the species – regardless of whether or not these are part of an integrated larger community.

For this discussion I shall distinguish between a tightly structured, integrated *community* and a simple haphazard *assemblage* of organisms in any particular habitat at a particular time. For the former, the botanical term "association" would be useful, but this term generally means an assemblage with a particular composition of plant species. This usefully subsumes the notion that it is a repeatable and consistent entity in time and space, but associations of plants are traditionally dealt with by plant ecologists as though animals and microorganisms were irrelevant or absent. This reflects my personal bias that assemblages or communities to be studied consist of many different types of organisms – not solely of taxonomic or trophically related types of organisms. Thus, a community may be of any size and will inevitably consist of different types of organisms involving different levels of trophic organization. Assemblages of a particular taxonomic grouping ("bird communities," "fish communities," etc.) are probably best referred to as guilds (sensu [16]) if they share common resources; otherwise they can be defined without loss of meaning simply as taxonomic and habitat groupings ("birds of the tundra," "fish on the reef"). These groupings are sometimes referred to as "taxocenes," i.e., members of any taxonomic grouping greater than a species (see [15], p 289, where it was suggested, for no compelling reason, that these are the proper units for study).

Early Superorganisms

Several reviews [8, 10, 19] have covered the botanical ground from which some early concepts of communities sprouted. Plant ecologists recognized many different associations, each consisting of several different species of plants considered as an integrated unit – a kind of "superorganism." The various species were thought to coexist because of similar requirements for habitat and resources and to develop together along deterministic paths of succession, leading eventually to stable, climax communities [19]. These processes were sometimes considered akin to the ontogenetic development and growth of a single organism and thus equally predetermined from conception. Another property of such "superorganismic" associations (i.e., plant communities) was that they contacted other associations at narrow boundaries (ecotones). Any interactions among the organisms within an association are evolved responses by the component populations (or stands of plants) leading to homeostasis (and stability and predictability of structure) of the community [10].

This view prevailed for many years, despite early and persistent criticism (as summarized by McIntosh [10]). It was, however, a mode of thought that had serious influences on the methods of sampling of natural assemblages of plant species. Obviously, under an accepted prevailing paradigm that associations of plants (or animals, for that matter) were repeatedly recognizable in space and time, any quantitative descriptions of the assemblages or any attempts to investigate successional or other temporal changes should most efficiently be done by sampling within areas containing the recognized associations. Where the numbers and types of species recorded in samples within an apparent association are compared with similar samples taken within a supposedly different association, the apparent differences and demarcation between the two associations will be confirmed. Similarly, if two areas are designated as being occupied by the same association, the similarity of the populations of plants within the two areas can readily be "confirmed" by any form of sampling within the two plots (confirmation is likely because the observer has already identified the two areas as containing similar compositions of species). This circularity of reasoning is inevitable wherever the samples taken are those dictated by preconceived notions about similarity or dissimilarity of structure of assemblages in different places [25].

Fragmentation of the Superorganism

Early critics of the superorganism concept of natural communities included Gleason in the United States and Ramensky in Russia [8, 10], who were largely ignored. Without superorganismic communities and deterministic patterns of succession in natural communities (as discussed in [19]), ecologists would not have had much structure to work with. Eventually, however, the inevitable appearance of objectively sampled critical data had to cause some change in thinking. First, data could easily be accumulated that were sampled randomly or at intervals along environmental gradients regardless of the presence or absence of particular associations. Many such data were examined, and many were designated as "atypical," "mixed stands," "transitional samples," etc. ([8], pp 386–406). The ecologists of the day resorted to the familiar grandiose glossary of post-hoc modifications of descriptions and subcategorizations of definitions of associations and ad hoc reasons for dismissing evidence that categorizes much of debate in science.

Finally, sufficient change of thinking occurred to allow the evidence to stand. McIntosh [10] has summarized the major attacks on the superorganism, and Whittaker's comprehensive analysis of vegetation along environmental gradients is still one of the best accounts of the resurgence of individual populations as units of study [25].

Essentially, the analyses by supporters of the "individualistic hypothesis" suggest that each species is distributed individually in space in accordance with its physiological requirements and responses to the presence of other species, according to its own genetics and population dynamics. As a result, assemblages of species tend to merge and intergrade together (in contrast to the abrupt discontinuities identifed as ecotones). Similar analyses in other habitats, and involving animals in addition to plants, have also rejected the concept of clear demarcation between successive communities in favor of relatively independent distributions of populations along environmental continua (e.g., [22]).

Nevertheless, the debate between adherents of the individualistic hypothesis (who advocate that because species are distributed according to their individual evolutionary histories, they are the natural units of study) and the supporters of some form of tightly coevolving, superorganismic community continues. As suggested at the beginning, the inability of ecologists to agree has led to a resurgence of the community being considered as a superorganism, particularly by those who advocate that the proper unit for study by ecologists is the ecosystem (see particularly the examples and discussion in [19]). Considerable effort has been expended in the development of theory, some of it very sophisticated theory, about diversity of species in communities, the degree of coevolution and limiting similarity of functional morphologies, and many aspects of community relationships (often discussed with no suggestions of doubt about whether integrated communities actually exist; see [2]).

There is a certain irony that the work tending to cast such doubt on the plant ecologists' superorganisms was being published at about the same time as much of the work that has led to zoologists developing new definitions and conceptualizations of niche and community relationships and concurrently with a resurgence of interest in coevolved patterns of resource usage [7]. Whilst much early plant ecology ignored animals, much of the later animal ecology appears to have ignored plant ecologists [7].

When Might an Assemblage be a Structured Community?

There is no simple method to determine whether an assemblage of organisms in a particular habitat or at a given time is anything other than that set of individuals that happen to be present. Most authors apparently accept the notion that the community being studied is simply that assemblage of organisms present, with perhaps some notion that they share some component of the environment (e.g., space on a rocky shore) as a resource, so that they are not just transient individuals wandering through a study site. For this majority (which includes most field experimental ecologists), the term "community" is essentially a useful description that carries no connotations of coevolution or tight integration. The organisms in such assemblages may interact with one another, but the assemblage would continue to thrive without some of the members, or is known to vary in composition from time to time and place to place. For many workers, the assemblage can be defined by providing detailed sampling data on its "structure": the patterns of distribution and abundance, sizes of organisms, trophic relationships, and species diversity [11]. These attributes can all be quantified by one form or another of sampling of the organisms. Descriptions of structure of an assemblage must then include definitions of the spatial and temporal scales chosen for study (see below). Thus, the dimensions of the assemblage are defined in such a study.

The patterns identified as structure can then be studied to determine their causes. This is the study of "organization" of the assemblage and invovles determining any dynamic effects of interactions among the species and the effects of physical environmental variables [11]. Note that the latter investigations of the organization of an assemblage of species are often desirable, or can be successful, regardless of whether the assemblage has the sort of structure and boundaries that might reveal it to be a tightly integrated community or association.

For those who accept this definition of community as simply the assemblage of things present, the problem of defining the unit of study is one of definition of which organisms to include and which to omit from study. This is not a trivial task and must be dependent on the nature of the problem being addressed. Coherent definitions of the scale and type of study being undertaken will help others understand the temporal and spatial scales chosen for study. Note that ecologists concerned with the study of a single species face the same problems as does the "community ecologist." No study of patterns and processes affecting any single species can proceed without due regard to the fact that the organisms are entangled in Darwin's "web of complex interactions." Other species consume the target species, provide or use its food or otherwise affect its survival, growth, and reproductive output, etc. Hence the role of species in any assemblage must be considered in the context of that assemblage.

As two examples of why care must be exercised in the choice of organisms to include in a study, consider rare species and organisms that are difficult to see because of small size. The experimental work of Paine [12] has led to the notion of "keystone species," species that, regardless of their densities or biomass, can exert considerable influence on the structure and/or organization of an assemblage. The predatory starfish *Pisaster ochraceus* has been identified as a keystone species on rocky shores of the northwestern United States. Experimental removal of starfish led to increased cover of rock by mussels, leading to elimination of many other species because the mussels were superior competitors for space and excluded most other species from the shore [12]. Thus, much of the structure of the assemblage on the shore was determined by the predatory behavior of *Pisaster*. Such keystone species may be important in a wide variety of habitats, but might also be rare, or may have long-term effects that persist for long periods after their own abundance has decreased. Considerable care must therefore be exercised in decisions to include or exclude species from study. It is common to ignore rare species (they do not generate data as quickly as is possible with abundant species and thus are often excluded by the methods of funding and by the temporal constraints on research programs of graduate students).

By choosing to omit from study organisms that are technically difficult to quantify, analyses of the organization of assemblages will be incomplete and potentially incorrect. I have found that variations in the intensity and outcome of competitive interactions between grazing gastropods at different heights on a seashore and in different seasons of the year can be interpreted if enough is known about their microalgal food resources [23]. In many studies of competition, the resources are microscopic or otherwise technically difficult to observe, or are tedious to examine and therefore remain unquantified. Thus, the observer's decisions about which types of organism to include in any study are crucial to the success of the study, even when the unit of study is considered to be a simple assemblage.

These cautionary examples indicate the extreme difficulties faced by a paleoecologist confronted with incomplete fossil assemblages, with underrepresentation of soft-bodied taxa and unclear temporal and spatial mixing of the fossilizable forms.

If the assemblage is to be considered as more than the simple juxtaposition and temporal coincidence of various organisms, several types of information should be sought. To identify a community, it must be possible to demonstrate some spatial and/or temporal consistency in the presence of a set of organisms arranged together with some particular structure.

Consistency Through Time

Any attempt to identify a consistent structure of a community through time will first have to solve the problem of an appropriate time scale. Unless the structure of the assemblage recurs through successive generations, responses by members of the assemblage to the presence and activities of other members are unlikely to have evolved in response to those particular species. Such interactions are still of great interest for explaining existing patterns in the assemblage and of value for predicting events whenever a similar assemblage happens to be together. In contrast, where sets of species become interdependent, recognition of their organization into communities might allow more useful predictions to be made, because there can be reasonable expectation that the patterns and processes in the community will continue to recur.

A persistent assemblage must have a structure in some equilibrial state in which it is observed to persist. This requires mechanisms enabling the assemblage to withstand perturbations and/or to return to the equilibrial state after a perturbation [3]. The problem inherent in determining whether a particular assemblage is persistent is that of knowing the appropriate temporal scale for the study. Some communities will appear to have consistent structure (i.e., the assemblage of species in a given habitat persists through considerable periods of time) simply because the organisms are long-lived relative to the length of study. Thus, assemblages of long-lived organisms will appear highly structured unless studied for appropriately long times, and the identification of assemblages of long-lived organisms as a community is tautologous [5]. One way of determining the appropriate time scale is to sample the assemblage for long enough for there to have been a turnover or replacement of all the adult organisms in the area [3]. Another method consists a calculations of transition probabilities of replacements of adult organisms in the future from their reproductive outputs, from the relative abundances of their recruiting juveniles, and from age-specific mortality schedules – but such calculations would depend on several, probably unlikely, assumptions [3].

Using this criterion of turnover time, a review of extant studies suggested that they were mostly too short to allow any determination of persistent structure of a community [3]. The conclusion using this very rigorous definition of appropriate time scales was that little or no evidence had been published to indicate which assemblages might exist in persistent patterns (i.e., have structured integrated communities) and which do not. Objections to the rigor of definition of turnover times must be accompanied by an alternative identification of an appropriate time scale, otherwise evidence gathered in support of persistent temporal structure of communities will be subject to the potential tautology mentioned earlier.

Two further approaches might be alternatives to the rigorous definition just described. First, experimental perturbations of communities should lead to convergence of experimental plots with the structure found in control (unperturbed) plots. If no convergence occurs, it is not possible to claim a consistent structure for the assemblage of species originally in the experimental (and control) plots. Such convergence experiments are often done for single populations in order to determine processes of density-dependent regulation of numbers (e.g., [15]). There is, however, still the inherent problem of defining in advance the appropriate time scale for the experiment. If the experiment is not done for long enough, no convergence can occur (thus biasing results in favor of the individualistic hypothesis). In contrast, the experiment should not stop as soon as the assemblages in control and experimental plots are considered similar, in case the processes of change in the experimental plots have not yet finished and the ultimate structure would stabilize in some different manner. Such timing would bias results in favor of the community hypothesis. A rigorous definition of an appropriate time scale is also necessary for such experiments.

One further piece of evidence that might bear on the question of temporal persistence is the experimental introduction of new species into the assemblage. If the species already present are well integrated into a "biologically accommodated" community, introduction of new consumers or producers ought to change patterns of utilization of resources, competitive hierarchies, or food webs to some significant degree. Experimental introductions are often unethical or impracticable. There have, however, been numerous accidental or deliberate introductions, such as in biological control programs. Of cases reviewed recently, 678 of 854 introductions of species had no measurable effects on the structure of the preexisting assemblage [20]. This suggests that many apparent communities are not tightly structured and can tolerate the arrival of completely new species without any change. The remaining cases are examples of some structure amongst sets of coexisting species that is upset when new species appear. In some instances, however, several variables were altering when the introductions occurred (e.g., many species have been introduced by man, but man also changes the habitat in various ways in addition to bringing in new species), and very detailed information is needed before such confounding variables can be eliminated. The available evidence on introductions is, however, biased against situations where communities are tightly structured. Numerous introductions fail and are therefore not noticed (only the appearance of a new species alerts observers to the fact that an introduction has occurred). Many introductions probably fail because the habitat and resource requirements of the species are lacking in the new environment. Others undoubtedly fail because of the small number of individuals introduced (see Simberloff, this volume). Yet other accidental introductions are probably resisted by the structure and organization of existing communities. There is insufficient evidence available to determine how often introduced species might fail to become established because of these various causes.

Consistency in Space

As with temporal scales, sensible identification of a community must depend on defining an appropriate spatial scale. At one end of possible scales is the Earth as a single entity (for example, dependence on oxygen by terrestrial animals is, at least partially, dependence on production of oxygen by marine phytoplankton). Community ecologists are not usually concerned with the study of all life on Earth. There are smaller biogeographic scales. Outside the biogeographic range of a particular species, other species exist in assemblages that do not contain it. The degree of mutual history and integration of assemblages of these species must therefore be less than for a group of similar species that all have coincident geographic boundaries. Within any geographic area or any size of habitat, chance vagaries of dispersal and recruitment of juveniles or patchy local extinctions make the structure of assemblages variable at several spatial scales. This is inherent in discussions of different types of species diversity (one measure of structure of assemblages), which are sometimes described as alpha, beta, and gamma diversity [15]. Alpha diversity is defined as that within an identified natural community (in whatever way that is to be identified!); gamma diversity refers to diversity in a large heterogenous region (presumably potentially containing several different communities each with its own alpha diversity). This would not be a problem if the former could be defined unambiguously. What sets the angels tap dancing on the head of their pin is that beta diversity refers to spatial variability in composition of species within a single community. Beta diversity is nonzero when differences occur from place to place *within* a particular area supposedly containing a recognizable community. There seems no way to define objectively the appropriate spatial scale for measuring alpha diversity; there is obviously no method for distinguishing between gamma diversity (at some large spatial scale) and alpha diversity of a community with great beta diversity.

Measures of diversity, or any other structural attribute of an assemblage of species, in order to compare it with other assemblages or to determine its uniqueness or where its boundaries might occur, must depend on the spatial scale over which it is sampled. As a very simple example, consider two species of plants arranged in monospecific, small, discrete patches in a field; the patches containing plants are mostly located in a single corner of the field. In any sampling with quadrats (or other sampling units) of the same or smaller size than that of the natural patches of plants, there will be a negative association between the two species (i.e., they are not coexisting in a two-species community). In large samples (of the size, say, of one quarter of the area of the field) the two species will probably be positively associated ([8], pp 375–381) (they are found together in any large sampling unit in the corner where patches of each species occur) and are then potentially present in a definable community. Plotless sampling strategies [8, 15] solve some of these problems, but problems of suitable spatial scale confuse most discussions of ecological phenomena (e.g., [3, 8, 15] and especially

[1]). Clearly, one sensible approach is to analyze patterns at several spatial scales (ideally from very small to large, biogeographic scales) simultaneously – although this must inevitably blur any possible boundaries or distinctions between potential communities.

Alternatively, wherever assemblages of species do differ from place to place (or from time to time) there remains the possibility of identifying each assemblage in some defined area at a particular time as a recognizably different community, and thereby ignoring temporal changes and spatial differences in structure. This would provide rigorously definable community units for study – but they would have no temporal or spatial organization, and therefore must ignore any migration or dispersal across the boundaries. Such communities would probably satisfy no one as appropriate topics for investigation.

Do Communities Have Emergent Properties?

Several authors have claimed that communities can be demonstrated to be more than simple assemblages because they have emergent properties, i.e., attributes or properties that are not identifiable by study of the individual populations comprising the community. Considerable confusion can exist about the nature or identification of emergent properties of communities. As with everything else concerning community ecology, the phenomenon of emergence has come to mean several things, as indicated by the several definitions from Allen and Starr [1]: a) properties that become apparent when the observer changes the scale of observation, b) properties that are unexpected because the observer had incomplete information about the components of a system, or c) properties that are strictly not derivable *a priori* from the known properties of the parts of a community.

The first of these is obviously a function of the scale of observation of the unit studied, so that once this has been defined the property is no longer emergent. For example, competition for food between two species obviously cannot be observed by studies on small areas, each with only one species present. Its discovery when examining a community comprising larger patches of mixtures of the two species is not a new property of the community.

The second definition is more problematic, because it is potentially the explanation for some of the claimed emergent properties of natural communities. For example, some of the examples discussed by Salt [17] could be emergent only in the sense that not enough was known about the component populations in the communities. Salt commented that exploitative competition in a two-species community could be inferred from knowledge of each species alone, but that interference competition and character displacement were emergent properties because they could not be so deduced. This is an unclear distinction. Are the processes of interference competition present in intraspecific encounters amongst members of each species, regardless of the presence of other users of the same resources? If so, interference competition between species is easily predicted from knowledge of each species alone. Greater information about the morphology of each species under different conditions of utilization of resources (in the absence of a putative competitor) should surely allow predictions to be made about the circumstances under which morphology would vary in response to different availability of resources. Then, studies of the species in assemblages with other species that might have effects on the availability of resources should not lead to surprises about measurable characters. In addition, knowledge that character displacement has been observed in some assemblage or community, when coupled with appropriate knowledge about the resources and biology of another set of species, allows prediction of its presence in another assemblage or community. Thus, it was emergent only once!

The same argument applies to the notion that emergent properties are similar to unexpected interaction terms in an analysis of variance (Futuyma and Sousa, personal communication) – i.e., combinations of factors produce synergisms and antagonisms that had not been predicted from knowledge of the factors in isolation. Once such an interaction has been identified in any study, however, subsequent studies should always be planned with the possibility of such interactions in mind. Thus, interactions among factors can only "emerge" in the first study because of previous ignorance of their existence.

This leaves the final definition – an important one because if emergent properties really exist that are not derivable *a priori* from the component populations, then the community is not only a desirable, but also a necessary, unit for study. Natural selection may indeed be operating at the level of the community to produce the emergence of attributes that are not those of the populations. Some authors have suggested that various statistical measures emerge only by the study of communities (e.g., measures of species diversity), but these are collective properties determined by summing the properties of component populations [17]. Some more realistic candidates have been proposed [1, 18, 20], e.g., the notion that communities in some habitats may have constant trophic structures even though the composition of the species changes [6], although this has been denied by other workers [18]. It is not clear what this "emergent" attribute is a property of (because the community structure must itself be independent of the identity of the species that make up the community – see [18]).

Of considerable interest is the counter-notion from more logical grounds that classifying certain features of communities as emergent properties is unlikely to add anything to our understanding of an assemblage. Consider the following statement from Edson et al. [4]: "It is because claims of emergence can tell us nothing about the objects of our inquiry that such claims are irrelevant to the practising ecologist, in respect to both research methodology and ultimate explanation" (p 594). The only obvious difference between an emergent property of a community and some attribute that is readily explained by knowledge of the component populations is that the latter can be explained and the former cannot (yet!); calling the former property by another name and implying that it has special status does not advance our understanding of it. I am persuaded that "continued dispute over the emergent status of various ecological phenomena can serve only to divert limited time, attention, and resources from the real focus of ecological inquiry" ([4], p. 595).

Some Other Considerations

There are many ecologists whose holistic views about the nature of integrated communities stem from consideration of energy flow (tropho-dynamics) and systems analysis. I have not ventured into this topic here for three reasons. First, patterns of energy flow leave no trace in the geological record (unless links between energy pathways, food webs, and evolutionary patterns of the organisms can be identified unambiguously). Thus, these properties of ecosystem analysis are of limited value in the interpretation of patterns in assemblages of species (extant or extinct).

Secondly, a cogent argument has recently been developed that there are three distinct methods for analyzing food webs and trophic relationships. Of these, webs based on energy flow were considered of little value because this approach is "remote from intimate biological details, pays little heed to competitive cross-links and has generated few or no insights into ecological processes" ([13], p 682). Paine argued that an analysis of functional relationships among members of an assemblage, by experimental manipulations to determine magnitudes and directions of trophic and other links among species, is the most profitable method for determining processes organizing assemblages [13]. Such analyses can take account of the important organizational roles of keystone species (see above) that would never be revealed by other modes of analysis.

Finally, authors who fundamentally disagree about the appropriate units for study of natural assemblages [9, 19] and disagree philosophically about the very nature of systems ecology agree that it has been unproductive ([9], p 111; [19], p 88). With these views I concur (indicating personal bias, but falling amongst august company!).

Conclusions

Clearly, much of this discussion is directed towards identifying problems with the concept of integrated communities. My bias in favor of species being investigated individually does not, however, make it necessary to rush in the direction of complete reductionism. Populations and species themselves have ambiguous definitions and blurred boundaries and are not necessarily wholly integrated entities. As discussed above, no species exists in the absence of any biological interactions with surrounding organisms; studies of single species will be incomplete. Sometimes it will be necessary to study an assemblage in the contects of larger units of study such as biogeographic provinces [9]. Therefore, I urge that patterns of distribution and abundance of organisms in natural assemblages must be dealt with at several different scales of investigation according to the type and nature of question being asked. It is, however, crucial that spatial and temporal scales and the subsets of the assemblage of species are chosen to be appropriate for the problem being addressed. It is also important that results obtained on selected components of an assemblage at certain spatial and temporal scales should not be extrapolated to other scales and other sets of species without some proper justification. In this, I agree with the conclusions arrived at from different arguments by several thoughtful reviewers (e.g., [1, 9]).

Do we need communities as units of study? Note that despite the difficulties associated with identification of clearly definable, consistent, and repeatable patterns of structure of communities, organisms in many assemblages clearly evolve in response to the presence of other species. Coevolution within communities has been dealt with elsewhere (Futuyma, this volume) but serves as an example of a large-scale phenomenon that is not contradicted by a lack of coherent communities to study. Species in some haphazard natural assemblage might have evolved behavioral or morphological attributes in response to competitors (or predators) even though they do not consistently co-occur in tightly bonded communities. For example, selection for traits in response to shortages of resources will still occur even though the competition for resources (if it occurs at all) is among varying numbers of many different species, in varying mixtures from generation to generation over many habitats (i.e., in different and varying assemblages). Such selective agents will not be identified if a community is considered to have a constant structure (by definition) at some particular spatial scale. Other areas and other spatial scales will not be examined.

Rather than persisting with the development of theory and concepts for many types of communities that may not exist (or cannot easily be defined or recognized), it is probably more profitable for ecologists to pay more attention to how often, and how consistently, various combinations of species occur together. The importance of their interactions and the degree of their interdependence could then be evaluated properly in biogeographic and evolutionary contexts. This will require studies on populations at various spatial and temporal scales, interacting in diverse and varying combinations with other species. Each of these scales and assemblages can be described fully without recourse to invention of organized communities or superorganisms. Only where patterns in natural assemblages, and the processes organizing them, cannot be understood at the level of populations will it be necessary to invoke community-level patterns and community-wide processes as appropriate higher levels of study. The experimental analysis of patterns and processes determining structure and organization of natural assemblages has been proceeding in various contexts and with varying degrees of relationship to theories of different kinds (see, e.g., [14, 21, 24]), despite the fact that rationally definable, integrated communities are not readily identifiable in nature.

Acknowledgements. The preparation of this paper was aided by funds from the Australian Research Grants Committee and the University of Sydney Research Grant. D. T. Anderson and P. A. Underwood helped by discussing parts of the morass. D. Simberloff, D. Futuyma, and W. P. Sousa suggested useful revisions to the manuscript. I thank G. Chapman, P. Fairweather, and K. McGuinness for their help, and my students and colleagues in Ross Street for their support.

References

1. Allen TFH, Starr TB (1982) Hierarchy: perspectives for ecological complexity. Chicago: University of Chicago Press
2. Cody ML, Diamond JM (eds) (1975) Ecology and evolution of communities. Cambridge: Harvard University Press
3. Connell JH, Sousa WP (1983) On the evidence needed to judge ecological stability or persistence. Am Nat 121:789–824
4. Edson MM, Foin TC, Knapp CM (1981) "Emergent properties" and ecological research. Am Nat 118:593–596
5. Frank PW (1968) Life histories and community stability. Ecology 49:355–357
6. Heatwole H, Levins R (1972) Trophic structure stability and faunal change during recolonization. Ecology 53:531–534
7. Jackson JBC (1981) Interspecific competition and species' distributions: the ghosts of theories and data past. Am Zool 21:889–901
8. Krebs CJ (1978) Ecology: The experimental analysis of distribution and abundance. New York: Harper and Row
9. Levins R, Lewontin R (1980) Dialectics and reductionism in ecology. In: Conceptual issues in ecology, ed Saarinen E. Dordrecht: Reidel Publishing, pp 107–138
10. McIntosh RP (1980) The background and some current problems of theoretical ecology. In: Conceptual issues in ecology, ed Saarinen E. Dordrecht: Reidel Publishing, pp 1–62
11. Menge BA (1976) Organization of the New England rocky intertidal community: role of predation, competition and environmental heterogeneity. Ecol Monogr 46:355–393
12. Paine RT (1974) Intertidal community structure: experimental studies on the relationship between a dominant competitor and its principal predator. Oecologia 15:93–120
13. Paine RT (1980) Food webs: linkage, interaction strength and community infrastructure. J Anim Ecol 49:667–685
14. Paine RT (1984) Ecological determinism in the competition for space. Ecology 65:1339–1348
15. Pielou EC (1974) Population and community ecology: principles and methods. New York: Gordon and Breach
16. Root RB (1967) The niche exploitation pattern of the blue-gray gnatcatcher. Ecol Monogr 37:317–350
17. Salt GW (1979) A comment on the use of the term emergent properties. Am Nat 113:145–148
18. Simberloff DS (1976) Trophic structure determination and equilibrium in an arthropod community. Ecology 57:395–398
19. Simberloff DS (1980) A succession of paradigms in ecology: essentialism to materialism and probabilism. In: Conceptual issues in ecology, ed Saarinen E. Dordrecht: Reidel Publishing, pp 63–99
20. Simberloff DS (1981) Community effects of introduced species. In: Biotic crises in ecological and evolutionary time, ed Nitecki M. New York: Academic Press, pp 53–81
21. Sousa WP (1979) Disturbance in marine intertidal boulder fields: the nonequilibrium maintenance of species diversity. Ecology 60:1225–1239

22. Underwood AJ (1978) A refutation of critical tidal levels as determinants of the structure of intertidal communities on British shores. J Exp Mar Biol Ecol 33:261–276
23. Underwood AJ (1984) Vertical and seasonal patterns in competition for microalgae between intertidal gastropods. Oecologia 64:211–222
24. Underwood AJ, Denley EJ, Moran MJ (1983) Experimental analyses of the structure and dynamics of mid-shore rocky intertidal communities in New South Wales. Oecologia 56:202–219
25. Whittaker RH (1956) Vegetation of the Great Smoky mountains. Ecol Monogr 26:1–80

Patterns and Processes in the History of Life,
eds. D. M. Raup and D. Jablonski, pp. 369–381. Dahlem Konferenzen 1986
Springer-Verlag Berlin, Heidelberg

Evolution and Coevolution in Communities

D. J. Futuyma

Dept. of Ecology and Evolution
State University of New York
Stony Brook, NY 11794, USA

Abstract. Communities, broadly defined, are evident virtually from the beginning of the fossil record; the most general features of their structure may be said to arise automatically, but this cannot be said for more detailed aspects of structure, such as resource partitioning, even for contemporary communities. Diversity, although interrupted by major extinction events, tends to increase globally and probably locally. Although some guilds may approach the limits of species packing, new resources (species) emerge during evolution: these species and those that evolve to use them contribute to long-term increases in diversity. Trends toward greater ecological specialization may exist but are uncertain. Analogies between these trends and those that characterize ecological succession are probably just that. The taxonomic composition of a community is determined largely by ecological processes of invasion and extinction of species that owe their attributes more to phylogenetic history, to a long, extremely diffuse history of spatially, temporally, and phylogenetically discontinuous ecological interactions, than to immediate coevolutionary responses in situ or to coevolutionary adjustment of particular lineages over extended periods of ecological time.

Introduction

The definition of "community," the spatial delimitation of communities, the existence of predictable community structure, and the role of interspecific interactions and coevolution among species in determining such structure as may exist are all subjects of debate among ecologists who study contemporary communities. These problems are enormously more difficult in a paleontological context. The limitations of paleontological data (to say nothing of my limited familiarity with the paleontological literature) will make whatever generalizations I offer extremely tentative.

Properties of Communities

Because the spatial limits of a community are very difficult to define, the community under study will generally be a set of species that coexist over some rather arbitrarily delimited region and time span. It is important to recognize in analyzing contemporary assemblages that many of the associations among species are geologically recent (often post-Pleistocene; e.g., [9, 10]) and that the properties of many of the species are anachronistic, having been shaped by interactions that no longer exist because of extinctions (e.g., [19]).

Most communities include one or more species of primary producers and one or more trophic levels of consumers (grazing food chain) and decomposers (detritus food chain), although some species may occupy several trophic levels and even both food chains, and a few communities do not fit even this minimal description (e.g., those based on allochthonous primary production). Primary producers and decomposers are found in very early Precambrian deposits and consumers are found in the Vendian, so communities, broadly defined, are recognizable very early in evolutionary history. Minimal community organization of this kind may be presumed to arise automatically in any locality or habitat, unless suitable species for these roles fail to invade because of limitations of dispersal or of physiological ability to withstand environmental conditions (e.g., the virtual absence of a grazing food chain in hot springs). Even minimal community structure may therefore await the evolution and immigration of suitable organisms.

The more debatable properties of community organization include the theoretical propositions that predation stabilizes coexistence among species at lower trophic levels, that competition for resources exists among species at a trophic level and determines the number and variety of species that can coexist, and that community stability (loosely defined here as persistence of all the included species) increases as connectance (the proportion of other species with which an average species strongly interacts) decreases [23]. These propositions, which require more empirical work before they can be said to apply generally, predict that the abundance, diversity, and often the trophic specificity [18] of predators enhances the diversity of their prey, that species at a given trophic level will display predictable patterns of resource specialization and of dispersion over the resource spectrum, that a greater variety of resources (e.g., prey species) will support a greater variety of consumers, and that communities may be organized into "component communities" within which species interact strongly and between which, weakly. While empirical instances of each of these predictions can be cited, the universality of such structure, especially of the prevalence and importance of interspecific competition, is the subject of considerable debate among ecologists. Nevertheless, numerous instances of regularity in the structure of communities exist: e.g., it has been suggested that the ratio of the number of predator and prey species is independent of total diversity [6], and independently evolved guilds of consumers sometimes show convergent niche-packing structure (e.g., [12]), although sometimes they do not [26]. Interspecific competition may contribute, but specialized habitat requirements and demographic stochasticity may suffice, to explain some ecological regularities such as the relation between species diversity and area, which is evident both in contemporary and in some fossil [22, 36] biotas.

Coexistence of numerous specialized species need not constitute evidence that their coexistence depends on reduced niche overlap nor that they evolved specialization in response to competition; for example, many phytophagous insects appear to experience little competition or resource scarcity [43], and their host specificity probably has other evolutionary causes (see below). In some groups, though, diversification in regions that have few potential competitors (e.g., on archipelagoes) strongly suggests that diversification elsewhere may be constrained by competition from species that have preempted the resources; however, this does not necessarily imply that competition among the members of an adaptive radiation is responsible for their divergence. The existence of "empty niches" does not necessarily guarantee the emergence of species to fill them, for many adaptive zones remained vacant for long periods of evolutionary time (e.g., volant insectivores) and many are still vacant (cf. the distribution of sanguivorous bats). Thus the detailed aspects of community organization need not automatically arise: some resources are not intercepted before they become available to the detritus food chain. Thus the particular organization of a community depends on phylogenetic history. Nevertheless, given a set of potential members of a community, ecological processes (predator/prey dynamics at the very least, and in many instances competition as well) determine which can invade and persist, so some degree of organization is guaranteed.

Do Communities Evolve?

Some marine benthic communities have retained a remarkable stability of composition for long periods of geological time and may shift in space as environmental conditions change, yet retain their character [4]. Nevertheless, the composition of most communities changes over time, as do the features of their members. On a global level, species diversity has declined precipitously during periods of mass extinction, yet has increased rather steadily for a considerable time after these events. Species diversity in the middle to late Cenozoic appears higher than ever before [15, 25, 35, 45], both because of the diversification of major new taxa (e.g., angiosperms and associated insects) and because of continuing diversification of long established taxa such as various groups of benthic invertebrates. Global diversity has increased in part because of increasing provincialization associated with the separation of land masses [45], although diversity has fluctuated in concert with the area of shallow seas [33, 36]. There is some evidence of a long-term increase in diversity within communities of benthic invertebrates [1] and possibly of Cenozoic insects [50]. The global increase in diversity may be associated in part with a long-term decline in the rate of "background" extinctions [28], although the relation of this discovery to Van Valen's [46] evidence of constant taxon-specific extinction rates is unclear. Diversity may be loosely regulated by diversity-dependent limiting factors [42], as the correlation between origination and extinction rates [38, 49] and the diversity-dependence of origination rates [34] suggest. Although the history of evolution has been marked by a progressive loss of major "adaptive types" (41), it is equally clear that major new "types" have appeared throughout the Phanerozoic, at least in the terrestrial biota (plants, insects, vertebrates).

The effect of morphological and ecological properties of a lineage on origination and extinction rates appears to be complex. There is evidence that taxa whose morphology suggests ecological specialization are more susceptible to extinction than more generalized forms (e.g., [21]), although this is not invariably true (e.g., [48]); that offshore communities have more stenotopic species, with higher extinction rates, than onshore communities [5]; and that both extinction and origination rates of endemic, stenotopic, nonplanktotrophic molluscs are higher than those of widespread, eurytopic forms with planktotrophic larvae [17]. This evidence suggests that "generally adapted" [7] lineages persist for long periods, generating "specially adapted" lineages that increase in diversity more rapidly than generally adapted forms (because of higher speciation rates consequent on their more ready geographic isolation and habitat specialization, and perhaps because of lower susceptibility to competitive exclusion), but which have shorter tenure times. The diversification of higher taxa (e.g., orders of mammals) appears sometimes to follow from the evolution of novel general adaptations (e.g., Rodentia) but sometimes not (e.g., Carnivora) [27]. Nevertheless, it is clear that unused resources invite the evolution of organisms able to exploit them (e.g., insect diversification associated with the rise of angiosperms), that interactions among species in themselves create new niches (e.g., for parasites), and that the evolution of new species and ways of life (hence, of the species composition and structure of communities) does not require change in the physical environment.

In the absence of major extinction events that reset the stage and leave ecologically generalized forms as the major players, it appears likely that species diversity tends toward consistent increase (both globally and locally), and that for reasons cited above the increase is predominantly in specialized species. (Both of these trends may be analogous to trends in ecological succession, although it is not entirely clear that diversity and average degree of stenotopy universally increase in succession; even if they do, the mechanisms are probably entirely different and the similarity is to be viewed as strictly analogical.) Increase in diversity arises from finer partitioning of resources that are already utilized (in groups in which competition is important), but probably more importantly, as documented for marine benthic paleocommunities [2], from expansion of the resource spectrum as new resources (species and their interactions) come into existence and as taxa diversify to utilize them. Although paleontological and neoecological evidence both suggest that the diversity of some guilds approaches an upper limit (at least within the relatively short intervals between major extinctions), it is not evident that all guilds are so limited; it is not unusual for more than 100 species of insects to feed on a single species of tree in a local community.

The Role of Coevolution

"Coevolution" is very difficult to define ([14], also see Järvinen et al., this volume), but the role it plays in community evolution depends on the strictness of definition. Broadly defined, it includes sequential evolution [20], i.e., any evolutionary response of a species or taxon to prior evolutionary changes in some other taxon, no matter how far removed in space, time, or phylogenetic continuity. For example, coevolution thus envisioned can embrace the evolution of toxic secondary compounds by plants in response to fungal attack, followed millions of years later by the diversification onto this chemically diverse array of resources of a group of insects that did not exist when the plants underwent chemical diversification. Thus broadly defined, coevolution has been preeminent in the history of community change, because much of the diversification of every taxon has occurred in response to biotic factors (diversity of prey, predators, mutualists), and each such taxon, once diversified, is the foundation of new predatory or mutualistic relationships at some later (perhaps much later) time.

Defined in the narrowest sense, coevolution is a more or less synchronous, spatially localized, series of reciprocal evolutionary responses between particular evolutionary lineages, leading to ever more refined mutual adaptation (whether mutualistic or antagonistic, as in the "arms race" scenario of predator-prey interactions). If this kind of coevolution strongly influences the structure of local communities, the interaction between particular lineages must be persistent, either because they remain *in situ* for long periods or because their distribution changes in concert as environmental conditions change. Moreover, coevolutionary responses will proceed on course only to the extent that conflicting selection pressures imposed by other species or by the physical environment do not vary too greatly. None of these conditions is likely to obtain for most pairs (or larger sets) of individual species. If we discount these complications by considering coevolutionary responses between two groups of ecologically substitutable species (e.g., a genus of plant and genus of pollinating bee), new complications arise from the historical contingency inherent in genetic change, which assures that two species will respond to the same selection pressure by different genetic changes, moving toward different peaks on the genetic "adaptive landscape." These considerations lead to the conclusion that coevolution *sensu stricto* will seldom be prolonged in time, have much long-lasting impact on community structure, or generate highly refined reciprocal adaptive responses except among small groups of species that have strong, persistent, virtually exclusive interactions, as among certain species-specific mutualistic and parasitic associations. Coevolution *sensu lato*, diffuse in nature, sporadically distributed in space and time, largely lacking in phylogenetic continuity, and hardly deserving of the name, is in contrast a major feature of evolutionary history and consequently affects community structure in the long run.

Evidence on Coevolution

Long-persistent coevolution between prey and predators or hosts and parasites (or mutualists) should result in a certain degree of congruence between their respective phylogenies, if divergence among species in one group is followed by adaptive divergence in the other ("resource tracking"); but except in a few groups of highly host-specific parasites, there is little evidence for such phylogenetic congruence, and considerable evidence against it [24]. Within any group of phytophagous insects, for example, related species often feed on closely related hosts, but the phylogenetic relationships of those that differ in host affiliation bear little relation to the phylogeny of their host plants; rather, most such groups of insect appear to have radiated to use plants that had diverged, chemically and otherwise, long before. The evolution of a new host affiliation in insects appears to be guided largely by the habitat association and abundance of the host relative to that of the host used by this insect's ancestors, by the insect's preadaptation to chemical stimuli and toxins (adaptation that may be refined once specificity on the new host evolves), and by numerous other ecological factors [13].

While there is good reason to believe that many chemical and other features of plants are adaptive responses to herbivory (or to pathogens), there are few cases in which a particular plant's features or the chemical diversification of a plant taxon can be ascribed to the impact of a particular group of insects. A plant's defense system appears to evolve in response to herbivores that are ecologically similar, but phylogenetically diverse, from place to place and time to time. Certain plant adaptations that are effective against one or more herbivores in one locality will be effective against others elsewhere or subsequently, and such traits are likely to become persistent features of the plant lineage and its descendants. Hence many plant compounds effectively deter a broad spectrum of insects. Other features are selected locally in response to a spatiotemporally restricted enemy (as illustrated by gene-for-gene systems in some plants and their pathogens) but do not provide cross-resistance to other species, and are evolutionarily evanescent. These considerations apply with even greater force to prey and predators that have less specialized interactions than many parasites and phytophagous insects. They apply also to mutualistic associations, which are usually many-on-many interactions (as in most pollination or seed dispersal systems) rather than one-on-one [16, 31, 44], and in which the evolution of species-specificity and refined mutual adaptation is similarly constrained by variation in the number and properties of interacting species in time and space.

Likewise, although examples can be provided of evolutionary responses to competing species (resulting in divergent character displacement, competitive exclusion, or change in niche breadth), the examples are few indeed, despite considerable search. Some groups of lizards and birds on archipelagoes [29, 32] provide evidence that the resource partitioning among potential competitors in a community is the consequence of coevolutionary adjustments within that community. However, such evidence is difficult to obtain, because a pattern of resource partitioning among putative competitors can arise either by coevolution or by processes of colonization and extinction.

Nevertheless, resource specialization, resource partitioning (in at least some cases), and adaptations to predators, pathogens, prey, and mutualists are evident in all communities. They come about through a complex interplay of ecological and evolutionary dynamics.

The Evolutionary Origin of Ecological Structure

The species composition of a local community is the outcome of recent ecological processes of invasion and local extinction, the latter resulting in an assemblage that presumably approaches some degree of stability. The features of the individual species in the assemblage have been shaped by their past evolutionary history (which determines the properties they have upon arrival at the locality) and by their recent evolution *in situ*. Their *in situ* evolution (including genetic responses to the local species assemblage) accounts, I suggest, for very little: generalizing grandly, we may assume that the species have not occupied the study site, nor been associated with each other, for very long. Moreover, a species will not be able to invade and form part of a stable assemblage if it has not already evolved the features required for successful invasion before it arrives: genetic changes in the colonizing propagule will not be rapid enough to counter the ecological dynamics of competitive exclusion or effective predation. Thus the structure of a local community (e.g., distribution of species over a resource spectrum) is largely a consequence of ecological sorting among species that evolved their properties in a variety of other arenas and is only slightly attributable to local coevolution. Unfortunately, there exist few tests of this proposition, as the theory of how to distinguish the relative impact of ecological sorting and coevolution is only in its early stages [8, 30].

As an example, let us consider in a hypothetical way the evolution of interactions between a group of moderately host-specific insects such as papilionid butterflies [11] and their hosts (chiefly Aristolochiaceae, Rutaceae, and Apiaceae). Although closely related papilionids tend to feed on members of the same plant family, some related genera or even congeneric species feed on different plant families, which surely diversified before the papilionids did. The plants have defensive compounds such as alkaloids, but preadaptation to alkaloids of the Aristolochiaceae appears to be a plesiomorphic trait that has permitted at least two independent shifts of papilionids onto these hosts (J. S. Miller, personal communication). Host choice is governed by responses to host recognition stimuli that may or may not include plant toxins. Within some species, there is some geographic variation in behavioral preference for and physiological adaptation to different locally abundant hosts [3].

If such a species follows the rules of optimal foraging theory and becomes moderately host-specific on one or more abundant plants in the southern region of a continent, it brings its host specificity with it to a northern community. There it persists or not according to the availability of its host (or a closely similar one) and the impact of the species it encounters, and if successful, contributes to the list of specialized species – specialized not because of competitive interactions or plant/insect coevolution *in situ,* but because of its autoecological evolutionary history. Once incorporated into the community, it may evolve in response to the peculiarities of the local population of the host and may well become more highly host-specific if only one of its hosts is available. If its impact on the host is severe, the plant may evolve some resistance to this butterfly if it does not thereby become more susceptible to any of the numerous other phytophagous insects that attack it. But the magnitude and long-term effect of these genetic changes may be slight or considerable. In all likelihood, genetic change will be slight and temporary in each species because other hosts or phytophagous insects invade and impose conflicting selection pressures, and because habitats become unsuitable so that local extinction and recolonization maintain gene flow [37] among populations that have been adapting to different local hosts or insects. Advances in adaptation of either plant or insect are more likely to be permanent if the population achieves reproductive isolation and becomes an independent player on the ecological stage, a potential invader of other communities (cf. arguments for punctuated equilibrium). Thus the precision and long-term permanence of adaptation to one or more other species depends in part on the degree of permanence of regional associations relative to the time required for speciation (which will seldom be the same for the various participants in the interaction). The specificity of adaptation to other species will be lower, the greater the number of other species (e.g., of hosts) with which a given species interacts (either locally or regionally).

Long-Term Trends

What little we know of coevolution over the long term from the fossil record accords with the very diffuse kind of coevolution described above. Many mollusc lineages increased in armature in concert with improved predatory capacities of various fishes and crustaceans during the Mesozoic [47], but no arms races between particular pairs of lineages have been described. Solenoporacean algae were largely replaced by calcareous reds when more efficient grazers evolved, but the traits that favored the calcareous algae had evolved previously [40]. Increased hypsodonty in equids evolved as grasslands became widespread, and there is reason to believe that grazing favored a replacement of caespitose by rhizomatous grasses [39], but there appears not to be any evidence that growth form or opalith structure evolved within any grass lineages in response to the evolution of horses or other grazers.

Although coevolution in the strictest sense appears to be rather rare, and restricted primarily to highly exclusive interactions among small numbers of species, adaptation to biotic features of the environment is pervasive throughout evolutionary history and is the direct or indirect source of much evolutionary novelty. Many adaptive radiations can be ascribed either to release from competition (e.g., many taxa of plants and animals in the Hawaiian archipelago; perhaps the Cenozoic radiation of mammals) or to the utilization of another adaptive radiation as a resource (e.g., diversification of almost any group of parasites or phytophagous or pollinating insects). Organisms that constitute or provide abundant resources (e.g., corals, ants, termites) are almost always used by a variety of other, often highly specialized, taxa. Predation, parasitism, and mutualism are clearly responsible for many dramatic patterns of adaptation (e.g., aposematism and mimicry, the vertebrate immune system, zoophily and zoochory in the angiosperms). In many cases the diversification of one or more clades (e.g., pollinating insects and birds) in response to the diversification of a "resource clade" (e.g., angiosperms) has clearly contributed to the further diversification of the latter. In most such cases, however, it appears likely that the long-term, macroevolutionary pattern of coevolution arises from evolutionary events that are widely scattered through time, space, and phylogeny. This macroevolutionary history may require a more hierarchical theory of coevolution than current coevolutionary theory, focussed at the population level, provides.

Despite interruptions by mass extinctions, diversity tends to increase over geological time, so one might expect associated trends toward increasing specialization of resource use and consequently decreased connectance (implying higher stability), finer resource partitioning, and increased specificity of parasitic and mutualistic associations. The data are hardly sufficient to make comparative statements about contemporary communities, and inferences about long-term trends in these community properties are necessarily even weaker. Within higher taxa, morphologically specialized and generalized forms have appeared throughout the history of diversification. The Orchidaceae, the most species-rich family of plants, have more specialized associations with their pollinators than the Cretaceous (or more recent) Magnolioideae; but the second most diverse plant family, the recently evolved Asteraceae, is generally open-pollinated. Presently available data appear to be insufficient to tell whether or not there are long-term directional trends in connectance, stability, or resource partitioning in communities.

Acknowledgements. I am grateful to P. Feeny and K. Niklas for comments on this manuscript and for stimulating, informative discussion. Many of the ideas expressed here have been derived from my research on phytophagous insects, supported during this period by grants from the National Science Foundation (BSR 8306000) and the Whitehall Foundation.

References

1. Bambach RK (1977) Species richness in marine benthic habitats throughout the Phanerozoic. Paleobiology 3:152–167
2. Bambach RK (1983) Ecospace utilization and guilds in marine communities through the Phanerozoic. In: Biotic interactions in recent and fossil benthic communities, eds Tevesz MJS, McCall PL. New York: Plenum, pp 719–446
3. Blau WS, Feeny P (1983) Divergence in larval responses to food plants between temperate and tropical populations of the black swallowtail butterfly. Ecol Entomol 8:249–257
4. Boucot AJ (1978) Community evolution and rates of cladogenesis. Evol Biol 11:545–655
5. Bretsky P (1969) Evolution of Palaeozoic marine benthic communities. Palaeogeog Palaeoclimatol Palaeoecol 6:45–59
6. Briand F, Cohen JE (1984) Community food webs have scale-invariant structure. Nature 307:264–267
7. Brown WJ Jr (1959) General adaptation and evolution. Syst Zool 5:49–64

8. Case TJ (1982) Coevolution in resource-limited competition communities. Theoret Pop Biol 21:69–91
9. Coope GR (1979) Late Cenozoic fossil Coleoptera: evolution, biogeography, and ecology. Ann Rev Ecol Syst 10:247–267
10. Davis MB (1976) Pleistocene biogeography of temperate deciduous forests. Geosci Man 13:13–26
11. Feeny P, Rosenberry L, Carter M (1983) Chemical aspects of oviposition behavior in butterflies. In: Herbivorous insects: Host-seeking behavior and mechanisms, ed Ahmad S. New York: Academic Press, pp 27–76
12. Fuentes ER (1976) Ecological convergence of lizard communities in Chile and California. Ecology 57:1–17
13. Futuyma DJ (1983) Evolutionary interactions among herbivorous insects and plants. In: Coevolution, eds Futuyma DJ, Slatkin M. Sunderland: Sinauer, pp 207–231
14. Futuyma DJ, Slatkin M (1983) Introduction. In: Coevolution, eds Futuyma DJ, Slatkin M. Sunderland: Sinauer, pp 1–13
15. Gingerich PD (1977) Patterns of evolution in the mammalian fossil record. In: Patterns of evolution as illustrated by the fossil record, ed Hallam A. Amsterdam: Elsevier, pp 469–500
16. Howe HF (1984) Constraints on the evolution of mutualisms. Am Nat 123:764–777
17. Jablonski D, Lutz RA (1983) Larval ecology of marine benthic invertebrates: paleobiological implications. Biol Rev 58:21–89
18. Janzen DH (1970) Herbivores and the number of tree species in tropical forests. Am Nat 104:501–528
19. Janzen DH, Martin PS (1982) Neotropical anachronisms: the fruits the gomphotheres ate. Science 215:19–27
20. Jermy T (1976) Insect-host-plant relationship – co-evolution or sequential evolution? Symp Biol Hung 16:109–113
21. Lipps JH (1970) Plankton evolution. Evolution 24:1–21
22. Marshall LG, Webb SD, Sepkoski JJ Jr, Raup DM (1982) Mammalian evolution and the great American interchange. Science 215:1351–1357
23. May RM (1981) Patterns in multi-species communities. In: Theoretical ecology: Principles and applications, ed May RM. Sunderland: Sinauer, pp 197-227
24. Mitter C, Brooks DR (1983), Phylogenetic aspects of coevolution. In: Coevolution, eds Futuyma DJ, Slatkin M. Sunderland: Sinauer, pp 65–98
25. Niklas KJ, Tiffney BH, Knoll AH (1980) Apparent changes in the diversity of fossil plants. Evol Biol 12:1–89
26. Orians GH, Paine RT (1983) Convergent evolution at the community level. In: Coevolution, eds Futuyma DJ, Slatkin M. Sunderland: Sinauer, pp 431–458
27. Radinsky LB (1982) Evolution of skull shape in carnivores. 3. The origin and early evolution of the modern carnivore families. Paleobiology 8:177–195
28. Raup DM, Sepkoski JJ Jr (1982) Mass extinctions in the fossil record. Science 215:1501–1503
29. Roughgarden J (1983) Coevolution between competitors. In: Coevolution, eds Futuyma DJ, Slatkin M. Sunderland: Sinauer, pp 383–403
30. Rummel JD, Roughgarden J (1983) Some differences between invasion-structured and coevolution-structured competitive communities – a preliminary theoretical analysis. Oikos 41:477–486
31. Schemske DW (1983) Limits to specialization and coevolution in plant-animal mutualisms. In: Coevolution, ed Nitecki MH. Chicago: University of Chicago Press, pp 67–111

32. Schluter D, Grant PR (1984) Determinants of morphological patterns in communities of Darwin's finches. Am Nat 123:175–196
33. Sepkoski JJ Jr (1976) Species diversity in the Phanerozoic: species-area effects. Paleobiology 2:298–303
34. Sepkoski JJ Jr (1978) A kinetic model of Phanerozoic taxonomic diversity. I. Analysis of marine orders. Paleobiology 4:223–251
35. Sepkoski JJ Jr, Bambach RK, Raup DM, Valentine JW (1981) Phanerozoic marine diversity and the fossil record. Nature 293:435–437
36. Simberloff D (1974) Permo-Triassic extinctions: effects of area on biotic equilibrium. J Geol 82:267–274
37. Slatkin M (1977) Gene flow and genetic drift in a species subject to frequent local extinctions. Theoret Pop Biol 12:253–262
38. Stanley SM (1979) Macroevolution: pattern and process. San Francisco: Freeman
39. Stebbins GL (1981) Coevolution of grasses and herbivores. Ann MO Bot Gard 68:75–86
40. Steneck RS (1983) Escalating herbivory and resulting adaptive trends in calcareous algal crusts. Paleobiology 9:44–61
41. Strathmann RR (1978) Progressive vacating of adaptive types during the Phanerozoic. Evolution 32:907–914
42. Strathmann RR, Slatkin M (1983) The improbability of animal phyla with few species. Paleobiology 9:97–106
43. Strong DR, Lawton JH, Southwood R (1984) Insects and plants: community patterns and mechanisms. Cambridge: Harvard University Press
44. Thompson JN (1982) Interaction and coevolution. New York: Wiley
45. Valentine JW, Foin TC, Peart D (1978) A provincial model of Phanerozoic marine diversity. Paleobiology 4:55–66
46. Van Valen L (1973) A new evolutionary law. Evol Theory 1:1–30
47. Vermeij GJ (1979) The Mesozoic marine revolution: evidence from snails, predators, and grazers. Paleobiology 3:245–258
48. Ward PD, Signor PW III (1983) Evolutionary tempo in Jurassic and Cretaceous ammonites. Paleobiology 9:183–198
49. Webb SD (1969) Extinction-origination equilibria in late Cenozoic land mammals of North America. Evolution 23:688–702
50. Wilson MHV (1978) Evolutionary significance of North American Paleogene insect faunas. Quaest Entomol 14:35–42

Patterns and Processes in the History of Life,
eds. D. M. Raup and D. Jablonski, pp. 383–405. Dahlem Konferenzen 1986
Springer-Verlag Berlin, Heidelberg

Large-Scale Changes in Animal and Plant Terrestrial Communities

K. J. Niklas

Div. of Biological Sciences
Cornell University, Ithaca, NY 14853, USA

Abstract. The fossil record of terrestrial plants and animals is reviewed and discussed within the context of large-scale patterns in the composition and taxonomic richness of assemblages through time. The evolution of vertebrates is seen to be closely tied to the use of terrestrial plants as food. However, until the Permian it appears that the major herbivores were terrestrial arthropods. From the Middle to Upper Devonian, the tetrapods may have been limited to the primary productivity of seas and lakes, since terrestrial plant communities were limited in biomass and geographic range. Simultaneous with the appearance of arborescent pteridophyte communities and the diversification of insects, tetrapods diversified and radiated into numerous niches. By the Permian, some tetrapods appear to have become specialized as herbivores. The earliest reptiles appear to have been insectivores. Tetrapod herbivory coincides with a decline in the pteridophyte-dominated flora and the Paleoptera-dominated insect fauna. These latter two changes are explicable in terms of a general shift in global climate to mesic and xeric conditions. The radiation of reptiles shows the repeated occurrence of herbivorous lineages. By the Triassic-Jurassic, synapsid amniotes achieved mammal-like dentition and possibly endothermy. By the end of the Mesozoic, tetrapods appear to have diversified into most if not all of the ecologic roles currently occupied by mammals and cursorial birds. The physiologic advantages of endothermy appear not to have become significant until the end of the Mesozoic and the beginning of the Cenozoic. With a global change to cooler conditions, nocturnal activity capable of capitalizing on the protein-rich and diverse insect fauna appears to have become important.

Introduction

This paper discusses the fossil record of plants and animals within the context of evolutionary changes in terrestrial communities. "Community evolution" is defined as large-scale changes in the composition (relative abundance and diversity of grades) and taxonomic richness (number and types of clades or species) of assemblages through geologic time. This definition is used because at least three limitations are imposed on the interpretation and even the recognition of a community in a paleontologic context. First, the fossil record preserves morphologies. Interpretations of ecologic relationships among fossil organisms are, therefore, limited and based primarily on inferences from extant, presumedly functional analogues. Given this limitation, fossil assemblages are collections of morphologies, and not necessarily complete collections. Second, terrestrial organisms are preserved as a result of deposition, which may be sporadic both temporally and spatially [21, 29]. Thus, many fossil assemblages contain morphologies that are "time- and space-averaged" and do not necessarily represent organisms from a single community. Third, transport and deposition remove organisms from their native environments, thereby obscuring or removing the abiotic attributes of a community's structure. These limitations are countered by at least three advantages. First, the fossil record documents evolutionary changes that cannot be inferred form the temporal scale of ecologic studies. As such, the fossil record can reveal which of the many processes that regulate community structure in ecologic time (10^1–10^3 yr) are significant in the development of large-scale patterns through geologic time (10^4–10^6 yr). Second, the fossil record provides insight into unique combinations of characters that were successful temporarily but are presently unrepresented. Although some facets of present and past community structures may be consistent, it is reasonable to argue that the structure and composition of past communities differed in significant ways from those of today. In this regard, a third advantage to studying the fossil record comes from paleontologic evidence for changes in morphology, the assessment of the relative evolutionary success of these changes, and the construction and testing of adaptationist arguments for these changes. In summary, the concept of "community" is flexible enough to be described as "a collection of coexisting structural and reproductive grades distributed in a matrix of taxonomic diversity." Changes in both of these facets of community structure can be observed in the fossil record.

The Fossil Record of Terrestrial Plants

Based on the analysis of terrestrial plants, a number of trends during the Phanerozoic are discernable [15, 17–19]. These trends can be categorized as a) vegetational, in which species turnover occurs without producing significant changes in the number or type of grade levels, or b) floristic, in which species or suprageneric group turnover changes the dominance or distribution of clades. However, for plants, the distinction between grade- and clade-level changes are closely related, particularly with regard to reproductive modes, which have figured prominently in the systematic treatments of plants. Two major trends in the taxonomic richness of terrestrial plant communities are evident. First, the total species diversity of plants on a worldwide basis has increased through the Phanerozoic. This resulted from the successive evolutionary appearance and proliferation of four major floristic components: a) a Silurian to mid-Devonian proliferation of early vascular land plants that were characterized by simple and presumably primitive morphologies, b) a subsequent late Devonian to Carboniferous radiation of derived seedless lineages ("pteridophytes"), c) the appearance of seed plants in the Late Devonian and their diversification in the Late Paleozoic, culminating in a gymnosperm-dominated Mesozoic flora, and d)

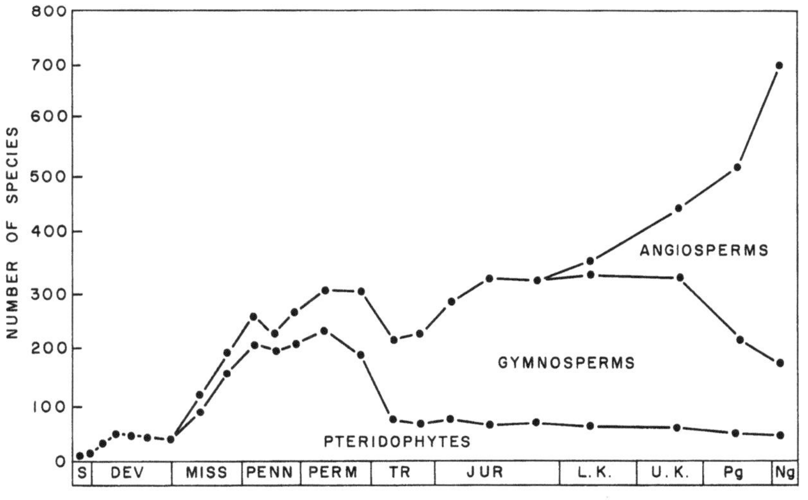

Fig. 1. Global species-diversity changes in vascular plants during the Phanerozoic. Data for three of the four major plant group radiations are taken from [17–19]

the appearance and rise of flowering plants in the Cretaceous and Tertiary (Fig. 1). Taxonomic richness in the world's paleofloras increased by the superimposition of species from all but the first group of primitive plants, which became extinct by the end of the Devonian. The second trend in the data is an increase in the mean number of species in individual fossil assemblages (=paleocommunities) [15]. The number of species per assemblage increases from the Late Silurian to the early Permian, plateaus through most of the Mesozoic, and rises once again from the Lower Cretaceous to the Paleocene (Fig. 2).

Five major trends with regard to compositional (grade-level) changes are also evident through the Phanerozoic. These trends are sometimes closely tied to floristic or taxonomic trends: a) the number of developmental models or Baupläne upon which plants are constructed increases through the Phanerozoic, with flowering plants (the last evolutionary plant group to appear) possessing collectively all Baupläne seen in other lineages [8], as well as eight that are unknown for other lineages (Fig. 3); b) the percentage

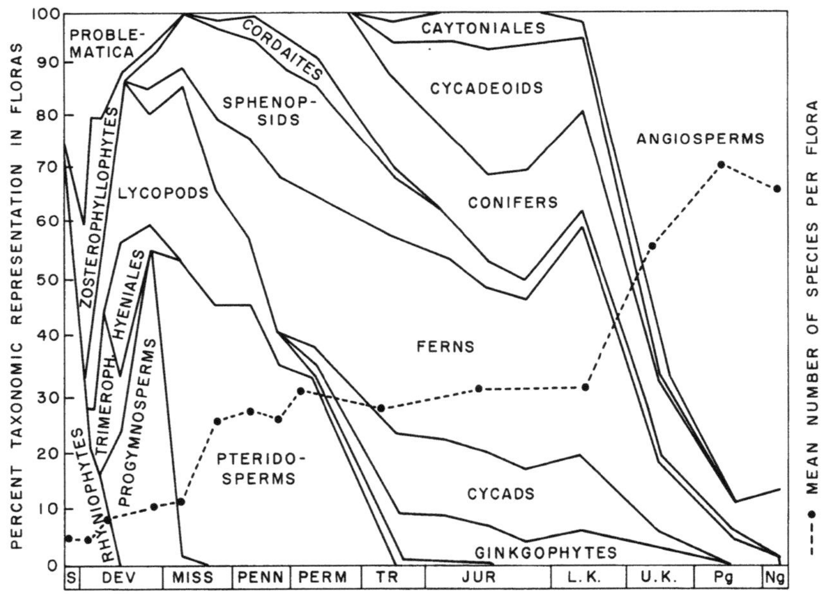

Fig. 2. Changes in the percentage of clades within fossil assemblages ("floras") during the Phanerozoic. Data taken from [15]. Mean number of species within each flora shown as a dashed line

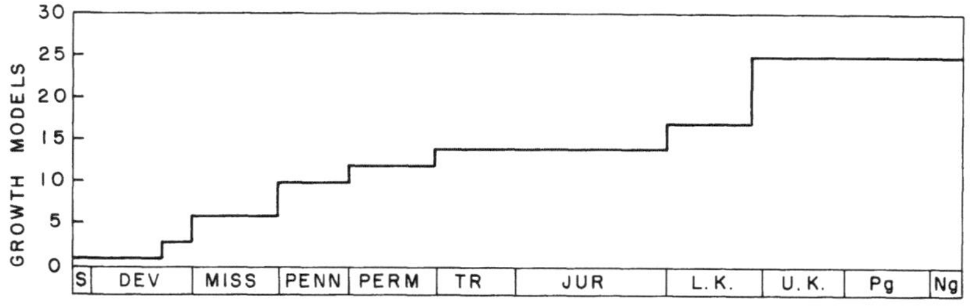

Fig. 3. Changes in the number of growth patterns or Baupläne seen in tracheophytes (a maximum of 25 is reported for living taxa

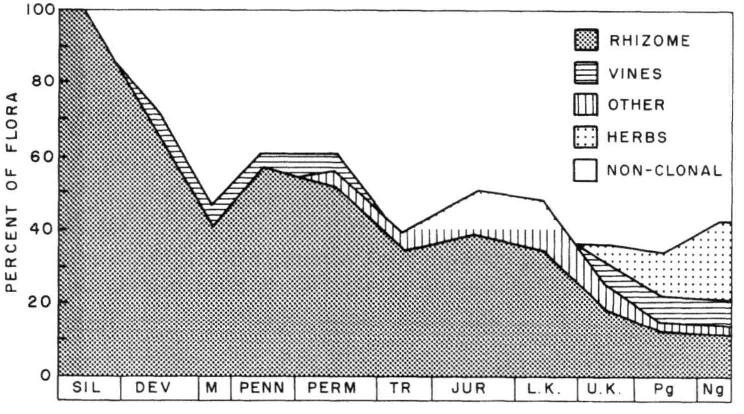

Fig. 4. Changes in the percent of clonal and non-clonal growth habits during the Phanerozoic [32]

of clonal growth forms decreases from the Silurian to the late Lower Cretaceous, followed by a modest increase due to the appearance of herbaceous angiosperms (Fig. 4); c) the number of arborescent forms increases from the early Devonian to the Neogene, with non-arborescent forms lagging behind, particularly after the Lower Cretaceous; d) the increase in arborescent growth forms is accompanied by an increase in seed plants; and e) three major shifts in the relative dominance of sexual reproduction have occurred during the Phanaerozoic (the dominance of free-sporing plants in the Paleozoic was superceded by wind-pollinated seed plants in the Mesozoic, followed by a shift to biotically pollinated flowering plants in the Cenozoic) (Fig. 5). Trend (e) is clearly tied to the taxonomic diversification and flor-

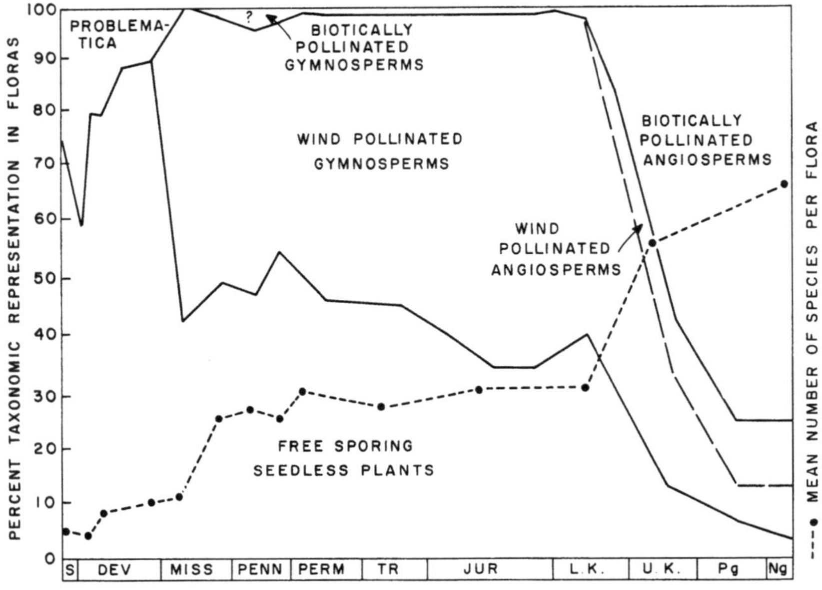

Fig. 5. Changes in the percentage of taxonomic representation in florast that are pteridophytes (free-sporing) or abiotically or biotically pollinated seed-plants (see Fig. 2)

istic global displacement of pteridophytes by gymnosperms, followed by the diversification of the flowering plants (Fig. 1). However, it is clear that "pteridophyte" is a grade level achieved by many separate lineages, as may be the case for gymnosperms and angiosperms [17]. In any case, the shift from an abiotically to a biotically pollinated flora represents a grade- and structure-level change in fossil assemblages.

Compositional and taxonomic changes in floras led to many evolutionary changes in the complexity of plant assemblages, environmental tolerance, and geographic range. The increase in the total (global) species richness, the increase in the number of growth forms in assemblages, and the appearance of arborescence collectively indicate that the number and kinds of species interactions in plant communities had the potential to increase. The physiognomic structures of communities became more complex. The early Paleozoic was characterized by rhizomatous plant communities (=relatively low-growing, vegetatively proliferative plants lacking secondary growth and therefore possessing limited height and stratification) [14, 17, 32]. Vertical stratification and total plant biomass were limited. With the appearance of arborescent growth forms in the late Devonian, both vertical stratification and total biomass increased. The appearance of secondary wood resulted in an increase in the retention of minerals and carbon. Although increases in the complexitiy of assemblages continued at the structural grade, the reproductive grade predominant in any flora generally restricted or limited the species richness and the environmental tolerance of communities. For example, the pteridophyte-dominated floras of the Paleozoic required hydric conditions for sexual reproduction, while efficient sexual reproduction in abiotically pollinated seed plants requires communities with a high density of conspecifics. Consequently, with the exception of some upland "mesic" gymnosperms, most Paleozoic communities were environmentally restricted to humid climates or access to freestanding water. Mesozoic communities, dominated by gymnosperms, could have been more successful in mesic to xeric environments (where competition with pteridophytes was reduced) by virtue of the seed habit. Species richness in the Mesozoic is seen to plateau rapidly. Most likely this is the result of the density-dependence of seed productivity in abiotically pollinated plants [19]. The Mesozoic plateau is followed by a major increase in species richness (both global and within assemblage) which occurs in the Cretaceous with the radiation of angiosperms, the bulk of which were biotically pollinated. Entomophily effectively eliminated the density-dependency of seed-set, thereby allowing a greater variation in species packing and the invasion of angiosperms into gymnosperm-dominated communities [5]. In addition to entomophily, angiosperms exhibit the greatest number of structural grades [8]. This repertoire permits the occupancy of all vegetative niches previously achieved by non-angiosperms, as well as new ones (e.g., marine).

Terrestrial Arthropods

Four principal events highlight the evolution of the most species-diverse terrestrial arthropods (=myriapods and insects). Each was associated with an adaptive radiation (Fig. 6): a) the development of a trachea system that localized respiratory exchanges with the ambient air, which was simultaneous with b) the appearance of a waxy epicuticle capable of conserving water (both of which are found in the most archaic terrestrial arthropods – the myriapods); c) the acquisition of meso- and metathorax extensions that eventually led to the evolution of flying insects (*Pterygota*); and d) the intercalation of a pupal stage (=complete metamorphosis, *Endopterygota*),

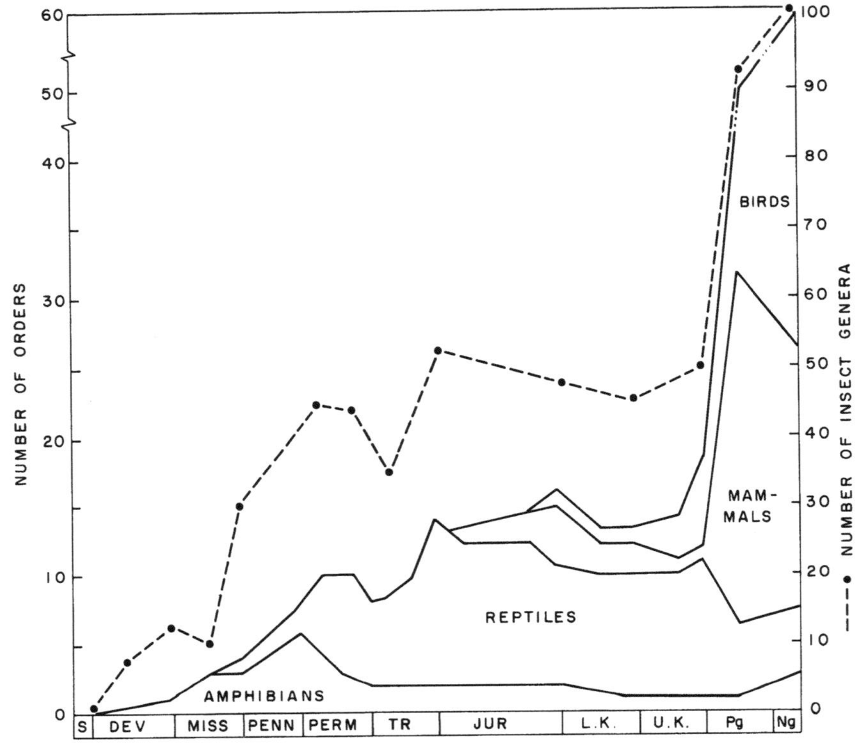

Fig. 6. Ordinal changes in the composition of vertebrates (shown in solid lines) and changes in the numbers of arthropod species (shown in dashed line)

which permitted the dissociation of the structure of the feeding and reproductive/dispersive growth forms. The arthropod solution to the contradictory demands of respiration and water conservation (tracheae and epicuticle), which parallels the evolution of stomata and cuticles in land plants, has placed limits on body size and heat regulation. The endocuticle allows little room for expansion during growth and must be shed periodically (ecdysis). In the post-molt stage before the new cuticle hardens the animal is vulnerable to enemies, heat, and desiccation. In addition, the individual animal is a poiklotherm – forcing air through tracheae or seeking moist, shaded microenvironments are the two methods of lowering body temperature. However, the relatively small body sizes of terrestrial arthropods have had beneficial results. They have partitioned the harvestable productivity of their environments in a remarkably heterogeneous manner. Compete metamorphosis has contributed to this partitioning. From the perspective of species diversity, the *Endopterygota* are the most successful of terrestrial animal groups.

The appearance of a cuticle presumably coincided with the evolution of arthropods from a skin-breathing annelid or onychophoran ancestor (e.g., *Aysheaia pedunculata*, middle Cambrian). The earliest terrestrial arthropods appear to have been myriapods (*Archidesmus loganensis* from the Llandovery/Wenlock and *Necrogammus spp* from the Pridoli); however, some Paleozoic myriapods may have been aquatic or semiaquatic. Most myriapods feed on plant debris (*Diplopoda, Pauropoda*, and *Symphyla*). However, the oldest myriapod-like fossils were carnivores much like the extant centipedes (*Chilopoda*) [28]. The tracheal system and epicuticle probably evolved when carnivorous myriapods sought shelter or prey among the early land plants. Only three Devonian localities have yielded fossils of terrestrial arthropods: a) Alken an der Mosel (Lower Emsian), represented by shallow brackish lagoonal deposits from which trigonotarbid and arthropleurid fragments have been found; b) Rhynie (Siegenian/Emsian), a bog or lake margin depositional environment, yielding trigonotarbids, mites, collembolans, and various chitinous jaws; and c) Gilboa (Givetian), a deltaic environment with no evidence of tidal deposition, which has yielded trigonotarbids, centipedes, mites, spiders, variously unidentified arachnids, and possibly the first true insects. Some complex ecologic relationships between animals and plants had evolved by the Siegenian/Emsian [14]. Most of the fossil arthropods from the Rhynie were very small carnivores. However, plant axes show evidence of puncture wounds that often penetrate into the stele (phloem). These and other necrotic areas on Rhynie plant fossils suggest that some terrestrial arthropods fed on plant sap [14, 28]. Mites and collembola, both present in the fauna, feed on plant debris, while others are fungivoras. Spore represent an important part of their diet. These invertebrates most likely occupied the plant litter that accumulated within mid-Devonian plant communities which had extensive mycorrhizae. Trigonotarpid remains are frequently found within the empty sporangia of *Rhynia major*. These animals may have occupied these sites after feeding on spores [14]. Some Devonian plants, characterized by spines and surficial glands, may have evolved mechanical defenses against terrestrial herbivores. It has also been suggested that by the Upper Devonian some plants evolved specialized spores, dispersed by arthropods, e.g., grapnel-hooks on *Ancyrospora, Hystricosporites*, and *Nikitinsporites* [14]. The presence of mites (*Acari*), springtails (*Collembola*), arthropleurids, centipedes, and arachnids in the Upper Devonian fauna indicate that invertebrates had diversified into carnivorous, fungivorous, herbivorous (xylophagous, saprophagous), and coprophagous taxa well before the adaptive radiation of the true insects [11, 14, 28].

The insects presumably evolved from a myriapod-like ancestor, possible sharing some features with extant *Symphyla* (e.g., *Scutigerella*). The oldest purported insect comes from the Gilboa locality (Givetian), but the designation of these remains (head, eye, and leg fragments) to the *Archaeognatha* is still speculative. These findings would be compatable with the derivation of winged insects (*Pterygota*) from a wingless ancestor (*Apterygota*) [34]. The earliest winged insects (Namurian) could not fold their wings (*Paleoptera*). Extant primitive *Pterygota* (represented by *Ephemeroptera* and *Odonata*) have restricted wing articulation. The remaining *Pterygota*, forming the division *Neoptera*, evolved a special sclerite and associated muscles in the articular wing area that permits wing flexure. Wing folding allows concealment among foliage and under logs or stones, and allows the insertion of head and thorax into smaller spaces for feeding. Insects with complete metamorphosis (*Endopterygota* or "holometabolous" insects) – known from the Permian – are thought to have evolved from generalized *Exopterygota* (or "hemimetabolous") insects. Wing folding and complete metamorphosis were innovations with profound effects on the invertebrate terrestrial fauna. By the Carboniferous, insects and other arthropods may have played a significant role in plant reproduction, and they were certainly the principal herbivores. For example, the pollen of some early seed plants represented in the Carboniferous coal measure swamps may have been too large to be effectively transported by wind (*Monoletes*-type pollen was up to 560 μm). *Monoletes* pollen has been found in arthropod coprolites and on the legs of *Arthropleura*, a myriapod. By contrast, fossil spores show bore-holes in their surfaces. Extensive herbivory, rather than saprophagy, is indicated by numerous plant petrifactions revealing coprolite-filled burrows. Some leaf types of the Lower and Upper Carboniferous have dense mats of glandular hairs. Trichomes such as those found on some Paleozoic plants have the capacity to discourage arthropod herbivores.

The Upper Carboniferous insects include generalized thysanurans of the suborder *Machiloidea* and five palaepterous orders of which three closely related ones were dominant in terms of diversity (*Palaeodictyoptera, Megasecoptera*, and *Diaphanopterodea*). These three orders were characterized by prominent external ovipositors and specialized mouthparts (long beaks with piercing organs, and with an enlarged clypeal region, suggesting a well developed sucking apparatus). All these insects presumably fed on plant sap. The nymphs of the *Palaeodictyoptera* and *Megasecoptera* possessed beaks much like their adult forms and were apparently terrestrial (tracheal gills and other aquatic modifications are lacking); presumably they inhabited moist soil and plant litter. The top carnivores were the *Protodonata*. They were aggressive predators reaching the largest sizes known among insects (wingspans up to 750 mm). Their nymphs were presumably aquatic and carnivorous. The *Protorthoptera* were the largest and most diverse of the *Neoptera Exopterygota*. They had mandibulate mouthparts and probably fed on living and dead plant material. Similarly, the herbivorous *Orthoptera* (*Saltatoria* and *Blattodea*) were well represented by the Upper Carboniferous [2, 3, 11]. In summary, the Carboniferous arthropods were characteristically herbivorous and insectal, just as the mid-Devonian fauna was predominantly carnivorous and myriapod-dominated. This faunistic change was coincidental with a floristic transition and an increase in vertical stratification and plant biomass.

The end of the Paleozoic saw the most diverse insect fauna (in terms of form) in the Phanerozoic (*Palaeodictyoptera, Megasecoptera,* along with new groups such as the *Hemiptera, Mecoptera,* and *Coleoptera*). Nineteen orders of insects are known from the Permian (ten with no fossil history in the Carboniferous). By the Jurassic (seventeen orders, two less than the Permian), the archaic orders of the Paleozoic are absent from the fossil record (the Palaeoptera are extinct) and the insect fauna was essentially modern in appearance. By the mid-Cretaceous, the *Lepidoptera* and *Isoptera* are represented by several primitive families, along with the *Hodotermitidae* are *Formicidae*. Social bees and wasps are known from the Eocene (*Apidae*) and Oligocene (*Vespidae*). The rise of the *Neoptera Endopterygota* in the Permian is coincidental with the ascendence of a seed plant-dominated flora, consisting of many hermaphroditic, potentially biotically pollinated plant groups (cycadeoids and some cycads). Insects that are specialized pollinators make their appearance with the advent of flowering plants in the Cretaceous. There is little doubt that, starting with the first truly terrestrial arthropod fauna in the Devonian, arthropod and plant evolution have been interrelated [2, 3, 5, 11, 28].

The Fossil Record of Vertebrates

Changes in the community composition and species richness of Phanerozoic terrestrial vertebrates resulted from the superimposition of four major radiations and two major reductions in the number of vertebrate orders (Fig. 6) [21]. a) Tetrapods, which differ from their ancestors by possessing pentadactyl limbs, were the first terrestrial vertebrates [1, 4, 22, 25]. They appear in the mid-Devonian in an ichthyostegid fauna and radiate through the Carboniferous and Permian [23]. After the late Jurassic, all but the transitional groups giving rise to contemporary *Lissamphibia* apparently went extinct [33]. b) The amniotes, which possess amniotic eggs, appear in the Pennsylvanian as small, lizard-like animals (proterothyridrids) [10, 27]. Shortly after the appearance of the first amniotes, there was a great split between the *Theropsida* and the *Sauropsida*. The *Theropsida* quickly radiated in the Permian and Triassic through the *Pelycosauria*, and the first mammals appeared by the end of the Triassic [21]. By the Triassic-Jurassic, some reptiles achieved mammal-like dentition (therapsids) and probably endothermy (cynodonts, therocephalians, diademontids, and trithelodontides). The *Sauropsida* split, about the Paleozoic-Mesozoic boundary, into the lineages leading to lizards and snakes (*Lepidosauria*) and to dinosaurs, pterosaurs, and crocodiles. By the Triassic-Jurassic boundary, the essentially modern vertebrate fauna (except for birds and mammals of modern aspect) was in place. c) With the close of the Mesozoic, the mammals radiated from small possibly insectivorous animals in the Triassic into diverse and often large carnivores and herbivores [7, 16, 21]. d) Birds, which are feathered archosaurs, had evolved as part of the great reptilian radiation of the Mesozoic and made their earliest appearance as carnivores in the Jurassic (*Archaeopterygiformes*) [6]. By the Cretaceous, birds were well advanced in appearance, except for the presence of teeth in the jaws of some species (all had fused carpometacarpals, large plate-like sterna, and pygostyles instead of free caudal vertebrate) [6]. Respectively, omnivorous and vegetarian families of birds appear in the Maastrichtian and the Eocene.

The extent to which amphibians, reptiles, or mammals dominated the fauna of a geologic epoch, however, provides limited information about the complexity of species interactions. An examination of the structural features of Paleozoic amphibians and Mesozoic reptiles indicates that these groups had the potential to exploit many of the same niches and habitats currently occupied by mammals and cursorial birds [7, 12, 13, 16, 21]. Unlike sedentary autotrophs, whose niche and habitat are inextricably bound and whose mode of sexual reproduction limited their geographic and environmental tolerances, the reproductive biology of tetrapods appears to have been less critical in defining their ecologic roles. Similarly, since all animals are heterotrophs, the history of terrestrial tetrapods is dictated in large measure by that of terrestrial plants and that of the Paleozoic's major herbivores – the arthropods.

The non-amniote tetrapods were the only terrestrial vertebrates from the mid-Devonian to the mid-Carboniferous, and they were the dominant tetrapods until the early Permian (Fig. 6). Diverse in size and shape, a large number appear to have been aquatic or semiaquatic predators, having basically flattened bodies and large heads with jaws that produced their maximum force when the mouth was open [4, 25]. Another branch of Paleozoic non-amniotes, the seymouriamorphs and cotylosaurs [10, 24, 25], was fully terrestrial and, by the Carboniferous, reptile-like in appearance ("batrachosaurs," a structural grade of tetrapods with terrestrial adults and probably aquatic juveniles). Until the Permian, all cranial and dentitial evidence suggests that the majority of non-amniote tetrapods were exclusively carnivores, feeding on small tetrapods, available fish, and insects. A number of related Carboniferous "amphibians" may have been herbivores – all have close-set, peg-like teeth with chisel ends – they are the adelogyrinids, *Spathicephalus* and Carboniferous relatives of the Permian embolomere *Archeria* [31]. Not until the late Pennsylvanian is there any evidence of herbivorous dentition which suggests specialization toward herbivory (diadectids), e.g., *Scincosaurus*, a coal measure amphibian, had spatulate teeth that may have been capable of masticating plant material. It appears likely, therefore, that for almost 100 million years (the bulk of the Paleozoic record of terrestrial life) that the majority of terrestrial herbivores were arthropods [2, 3, 11–13, 20–28]. Also for most of the Devonian, it appears likely that terrestrial vertebrates were dependent upon the primary productivity of the seas and freshwater lakes.

The evolution of reptiles in the Late Carboniferous coincided with a major radiation of insects. The earliest reptiles were small lizard-like animals that probably fed on insects (Fig. 6), which appear to have had limited species diversity through the Mississippian and were confined to areas near water. However, coincident with the appearance and rise of arborescent plants, insect diversity increased with the expansion of many orders capable of flight [34]. By the Upper Carboniferous, these orders include non wing-folding insects or *Palaeoptera* (*Megasecoptera, Palaeodictyoptera,* and *Diaphanopterodea,* which had sucking mouthparts; *Protodonata* and *Emphemeroptera,* which had mandibles) and wing-folding insects or *Neoptera* (*Blattodea*, omnivores; *Saltatoria*, phytophagous insects; *Protorthoptera, Miomoptera,* and *Caloneurodea*, all active predators). During this time, the feeding mechanisms of tetrapods underwent rapid modification, changing the mechanical function of jaws from quick seizure of prey to the application of compressive force after closure [9]. This was accomplished by the differentiation of the temporalis and pterygoideus muscles from the single jaw muscle mass characteristic of amphibians. The late Pennsylvanian and early Permian mark the first appearances of herbivorous reptiles. It is possible that the large plant biomass of this time interval permitted the diversification of herbivores in late Paleozoic communities (cf. [20, 34]). But perhaps more importantly, global changes in climate occurred, resulting in the loss of extensive hydric environments. The arborescent pteridophyte-dominated structure of the late Pennsylvanian assemblage underwent a major and irreversible decline, presumably due to the extensive elimination of suitable habitats for gametophyte development. Similarly, the *Palaeoptera*-dominated insect fauna declined. These insects had aquatic larvae whose habitat requirements were similar to those of pteridophytic gametophytes.

The edaphosaurids were the first herbivorous pelycosaurs (Upper Carboniferous); these taxa were replaced by caseids by the mid-Permian. Unfortunately, little is known about the numerical abundance of herbivores and carnivores. From the perspective of species diversity, it would appear that carnivores far outnumbered herbivores until the mid-Triassic. The appearance and radiation of the therapsids in the faunas of the mid-Permian to the mid-Triassic may have shifted this relationship. Late Permian deposits consist of a mixture diverse in herbivores and carnivores [12, 13,21]. Triassic cynodonts may have achieved a structural grade level equivalent to that of prototherian mammals and are considered to be the immediate sister group to the mammals. These animals may also have been endothermic. The Permo-Triassic marks a transition in the composition of terestrial faunas. Ancient tetrapods, such as the stereospondylous temnospondyls and procolophonids survived this transition. Herbivorous therapsids, reduced in species diversity (dicynodonts, grazing herbivores), survived into the Triassic, while theriodonts survived and radiated rapidly in the early Triassic. This Permo-Triassic restructuring of faunas coincides with the decline of arborescent pteridophyte communities and the expansion of abiotically pollinated seed plants into mesic and xeric habitats [17–19]. In addition, the taxonomic composition of terestrial insects underwent a major restructuring [5, 11]. In the Permian, the *Paleoptera* (non wing-folding insects) declined, while the *Neoptera* expanded and the *Coleoptera* made their first appearance. Collectively these shifts in the primary producers and herbivores (plants and insects) of terrestrial communities may have had consequent effects on specialized herbivores and carnivores, leading to their extinction.

The Mesozoic reptiles diversified and radiated into most of the ecologic roles occupied by extant vertebrates, even some that no longer exist ("dinosaurs") [21–23, 25]. For example, several groups of marine reptiles dominated the seas, while ornithischians (all herbivores) and saurischians diversified into a tremendous range of terrestrial grades. The pterosaurs occupied many of the roles of contemporary carnivorous birds. Some of the grade levels achieved by archosaurs give insights into the general ecology of specific epochs. For example, some Jurassic and Cretaceous coelurosaurs were small, slightly built carnivorous bipeds. Their apperance suggests that they occupied sites in open plant communities. The stomach contents of herbivorous sauropods have been found to contain twigs and small branches (2.5 cm long; 1 cm wide). It has been suggested that these animals may have been significant in disturbing ecologic succession and restricting the conversion of open countryside to dense forests. The long necks of some sauropods were capable of permitting browsing of branches more than 20 meters above the ground. The late Mesozoic floras were dominated by conifers with understories of cycads and ferns. Many conifers and most cycads are characterized as sclerophyllous and woody. Most sauropods lacked molariform teeth, possessing only anterior, incisor teeth. Gastroliths have been found in Jurassic and Cretaceous sediments.

By the late Cretaceous, the species diversity of mammals (data provided in [21]) equals or exceeds that of archosaurs (Fig. 6). The origin(s) of mammals may be traced back to cynodonts, ictidosaurs, and bauriamorphs, of which the former two may have been endothermic. Physiologically, endothermy is a critical aspect of the mammalian "Gestalt," and it confers a relative independence of climatic conditions. Endothermic therapsids do not appear to have had an advantage over their ectothermic contempories since during the Mesozoic mammal-like therapsids were replaced by increasingly agile archosaurs. This may have been the result of the prevailing equable climate of the Mesozoic and a consequence of the net production of endothermic metabolism. Endotherms use 90% of their energy intake for heat maintenance (1–10% net production of biomass), while the net productivity of amphibians and reptiles is between 30–90%. Under equable conditions, the productivity of ectotherms can usually exceed that of endotherms. However, juvenile ectotherms presumably would have been active only during the day, and the noctural niches would have been sparsely occupied by reptiles or amphibians. With the advent of angiosperms, insect diversity increased with the exploitation of pollen and nectar, as well as the diversity of foliage types [5]. During the night, when many insects are inactive, small, agile noctural endotherms would have had access to a trophic recourse unavailable to adult ectotherms. It is conceivable, therefore, that the early evolutionary patterns of therians, angiosperms, and the insects diversifying on angiosperms were intimately connected.

The Mesozoic was also the time of another major adaptive radiation, the flying archosaurs – birds (Fig. 6). The earliest bird, *Archaeopteryx*, had teeth in both jaws (a basic vertebrate character) and feathers. Small coelurosaurs are the most likely antecedents [6]. Evidence suggests that *Archaeopteryx* represents an almost ideal avian ancestor, unspecialized in many ways and little modified over contemporary small theropod dinosaurs. It is known that by the Cretaceous, a large number of aquatic shorebirds had evolved, e.g., *Gallornis*, a flamingo-like animal. The terrestrial avifauna of the Cretaceous is poorly known. By the end of the Eocene most of the orders of birds had evolved and most of the families represented in these orders consisted of additional water birds and nonpasserine forest dwellers. A second radiation of bird families occurred in the Miocene, consisting mostly of land-dwelling forms, particularly passerines which were adapted to drier, less forested habitats. The Eocene and Miocene radiation of birds are coincident with vegetational changes resulting from global climatic changes. By the end of the Miocene, angiosperms had taken on a relatively modern appearance and the percentage of herbivorous families of birds increased sharply. The basic ecology of Cenozoic birds was probably very similar to that of the contemporary avifauna (carnivores, 46%; omnivores, 30%; herbivores, 24%).

A Trophic Triangle

Large-scale patterns of terrestrial community evolution can be revealed by concentrating on the interactions among three principal biotic components – plants, arthropods, and vertebrates. Although the net productivity of terrestrial plants supports the bulk of the energy requirements of most contemporary tetrapods, early Paleozoic terrestrial vertebrates were most probably largely dependent upon the productivity of aquatic plants, at least until the late Devonian. As the terrestrial flora diversified and radiated into drier upland habitats, the amount of plant biomass capable of supporting a terrestrial tetrapod fauna was achieved. The first and most extensive herbivores, either aquatic or terrestrial, were the arthropods [2–5, 11, 14, 28]. The earliest land plant fossils are associated with a limited assemblage of arthropod fossils. Permineralized plant axes of the early Devonian show evidence of having been pierced and chewed. For over 100 million years plants evolved mechanical and biochemical techniques for reducing arthropod herbivores. Chemical analyses of Paleozoic plant fossils reveal a diversity of compounds possessing anti-herbivorous properties.

With the appearance and expansion of pteridophyte-dominates forests in the Carboniferous, the global ecology changed rather dramatically. In addition to providing a large, species-diverse biomass, tree growth habits produced a vertical stratification both in plant resources and in the distribution of insect populations [20, 34]. In general, the bulk of terrestrial arthropods are found in the most accessible stratum of forest communities directly usable as food – leaves, young twigs and stems, ovules, seeds, and fruits (=the "herbaceous" canopy). Among the late Paleozoic arthropod herbivores, species belonging to the *Palaeodictyoptera* and *Megasecoptera* appear to have occupied the canopies of living plants, while species of the *Orthoptera, Collembola,* and *Thysanura* appear in plant debris interpreted as part of the forest leaf-litter. As the vertical stratification of plant communities increased during the late Devonian and Mississippian, the adult forms of an amphibian-dominated fauna progessively found a more diverse and abundant vegetation as a food source. Similarly, they may have found it more difficult to capitalize on the diverse insect fauna that followed its food-base vertically (*Palaeoptera*). By the early Pennsylvanian, arborescent lycopod-dominated forests reached heights in excess of 40 meters; based on reconstructions of lepidodendrids, the bulk of the leaves in these forests would have been from 15–25 meters above ground-level. Plant communities such as these were dramatically altered with the advent of a general drying trend toward the end of the Paleozoic. The arborescent pteridophyte-dominated flora and the *Palaeoptera*-dominated insect fauna were dramatically reduced in species diversity, both for presumably similar reasons – the loss of suitable reproductive-sites.

The radiation of herbivorous tetrapods was coincidental with the structural and reproductive shifts associated with the amphibian-to-reptile transition – a transition that occurred with the loss of hydric habitats for amphibian reproduction. Of particular interest is the decline of the *Palaeoptera* and the possible influence of this decline on the feeding habits of tetrapods. The *Palaeoptera* have aquatic or semiaquatic larvae – a food resource to the juveniles of amphibians. With a reduction in the available sites for the growth of *Palaeoptera* larvae, it is reasonable to assume that there was a consequent effect on the lower trophic-level biomass available to juvenile amphibians. Similarly, the "dilution" in the primary food source of early Paleozoic tetrapods (insects) was aggravated by the appearance and rapid radiation of flying insects in the mid-Mississippian. In addition to an effective predator-avoidance mechanism, many Paleozoic insect wings appear to have mimicked the morphology (and possibly the color) of plant leaves, e.g., *Phylomylacris villeti* (cockroach) and *Odontopteris callosa* (leaf genus).

By the end of the Carboniferous, plant and insect evolution most probably made vertebrate herbivory energetically possible and even profitable. With the advent of insect flight (Namurian), the efficiency with which insects were used as food by adult tetrapods may have decreased; comparable levels of efficiency were most probably not achieved until the Mesozoic with the appearance of small, agile, Late Triassic insectivorous mammals and, later, with the appearance of birds.

The land plants evolved rapidly and extended their range into a variety of habitats with the advent of the seed plants. The evolution of the seed was in many respects the equivalent to the appearance of the amniotic egg and internal fertilization. Both of these innovations freed their structural grades from a temporary dependency on free-standing water or humid conditions for sexual reproduction. However, unlike animals whose habitat tolerances appear to be largely independent of their mode of reproduction (= niche and habitat are separable ecologic attributes), plants are sedentary and the individual must be capable of vegetative growth and sexual reproduction in the same environment ("ecotrope" concept). An exception to this general rule may be found in many pteridophytes capable of clonal growth. These organisms can maintain themselves in habitats that preclude sexual reproduction. It is likely, therefore, that before the advent of seed plants, some pteridophytes had established an "upland," predominantly clonal, community structure. These communities would have provided food and shelter to various arthropods and smaller tetrapods. With the appearance of seed plants, a major floristic change began, and the geographic and environmental range of plant communities was extended. Many of the structural changes seen in the pteridophyte-to-gymnosperm-assemblage transition can be interpreted both as adaptations to drier, mesic habitats and as adaptations capable of reducing the effects of herbivory (e.g., reduced, sclerotic leaves; thick cuticles; secondary growth).

The close of the Permian saw broad-scale changes in the physiognomic composition and floristics of terrestrial plant assemblages. The Carboniferous pteridophyte-dominated communities underwent a progressive contraction in geographic range as a consequence of the recession of epicontinental oceans and the formation of mid-continent rain shadows resulting from orogenies. Many plant lineages went to extinction and the global species richness of tracheophytes declined. Coincident with plant extinctions, the composition and faunistics of arthropods were altered (Figs. 1 and 6). The *Palaeoptera* declined in species numbers, while the *Neoptera* increased in diversity and abundance. Wind-pollinated plants became the dominant elements of the Mesozoic. Based on the ecology of contemporary gymnosperm communities, the Mesozoic ecosystem must have been characterized by large forests of conspecific gymnosperms with understories of mesic-tolerant pteridophytes. In such communities, the greatest diversity of plant species occurs along lakeshores, riverbanks, and other ecotones. Contrary to conventional wisdom, however, evidence is mounting that suggests that some ancient gymnosperms lineages may have been insect-pollinated, e.g., cycads, Gnetales, cycadeoids. These facultative entomophilous taxa could tolerate relatively low conspecific density levels, thereby providing some measure of community species heterogeneity. A few bisporangiate plants show evidence of protecting ovules while at the same time producing large quantities of pollen grains (possibly as a reward to pollinating insects such as the *Coleoptera*). Nevertheless, within fossil assemblages, species diversity leveled off from the early Permian to the late Jurassic, perhaps due to the predominant abiotic mode of pollination. Until the Jurassic the species diversity of vertebrate carnivores exceeded that of herbivores. The morphologic diversification of carnivores may have been related to a relatively high incidence of higher-level trophic interaction (e.g., carnivore-carnivore predation), while the relatively low diversity of herbivore morphology may reflect the homogeneity of plant community compositions. By the end of the Jurassic, the herbivore species diversity equalled or exceeded that of carnivores. The late Mesozoic marks the appearance of angiosperms, which could tolerate low numbers of conspecifics and exploit a variety of niches.

The early angiosperms quickly radiated, until by the Upper Cretaceous they comprised over 70% of the within-assemblage species diversity. The success of the angiosperms has been traditionally ascribed to their reproductive mode. Effectively, angiosperms are the "mammals" of the plant world. And with biotic pollination, flowering plants have achieved copulatory powers of species recognition. Based on neontologic studies, the primitive angiopsperm flower was most probably radially symmetrical, many-parted and hermaphroditic, with ovules protected in carpellary tissues and pollen produced in a fairly well displayed (to pollinators) androecium. In many respects, this primitive flower has a morphology analogous to the bisporangiate structures of Mesozoic gymnosperms thought to be insect-pollinated – cycadeoids. The structural grade changes seen in angiosperms are not to be minimized, however. The appearance of vessels and sieve-tube members confer many physiologic advantages to angiosperms, while the developmental plasticity of angiosperms has resulted in the greatest compendium of Baupläne seen in one reproductive grade [8].

Coincident with the ecologic rise of angiosperms, insect diversity increased during the late Mesozoic and early Cenozoic. The angiosperms by virtue of their biotic pollination had evolved a highly canalized partitioning of various toxins to inhibit herbivores and attractants to stimulate pollination. Insect diversity rose with the appearance of pollen- and nectar-gathering insects and the radiation of non-pollinating insects that utilized the flowering plants as food and shelter. These specialized insects, most of which were capable of flight, were accessible as food to nocturnal animals, which fed upon them when they were torpid or simply resting. Endothermy, a physiologic criterion for the mammalian, avian, and may be other archosaur grades, permits an independency between activity and climate. Based on morphologic data, endothery had evolved in a number of synapsid lineages during the Mesozoic. Yet this innovation did not appear to confer much success in terms of dominance. The mammals did not evolve into large-bodied, diurnal forms until long after the extinction of most large dinosaurs and only at the same time as the ecologic rise and success of the flowering plants. Since the earliest mammals may have relied heavily on insects for food, it is not inconceivable that the evolutionary patterns seen in early angiopserms, the various specialized insect groups dependent upon angiosperms, and the late Mesozoic-early Cenozoic mammals were interdependent.

Summary

The species numbers of both tracheophytes and terrestrial vertebrates have increased throughout the Phanerozoic with but a few, nonetheless major, episodes of decline (Figs. 1 and 6). This increase appears to result from the superimposition of major adaptive radiations characterized by the appearance of evolutionarily novel, structural and/or reproductive grades. The appearance of a new grade usually coincides with or is shortly followed by a decline in the species numbers of the preceding, presumably ancestral clade. In some instances, there is evidence that this may result from a large scale competitive displacement, e.g., the angiosperms appear to have competed with the gymnosperms for similar habitats. In other instances, abiotic factors, such as climate change and continental plate movements, appear to be responsible, e.g., the elimination of arborescent pteridophyte-dominated communities, and the expansion of gymnosperms. Patterns in plant evolution influence those of vertebrates, since animals are dependent upon autotrophs for food and shelter. Thus, a greater variety of food and shelter was available to animals as plant communities evolved. The tracheophytes show a pattern of habitat expansion in the Paleozoic analogous to that seen in hydrarch community succession, and another, analogous to xerarch succession, in the Mesozoic. Coincident with this "geographic" expansion, plant communities show evidence of progressive stratification through much of the Phanerozoic. By the end of the Paleozoic, plant biomass was large and fully supported a completeley terrestrial fauna, which by the Mesozoic achieved a diversity of ecologic roles comparable for the most part to the present day. During the Paleozoic, vertebrate evolution was intimately connected to plant evolution via the dependence of the former on insects for a large portion of their food resource. The earliest reptiles and mammals may have been heavily dependent upon insects for food. The history of global community structure has some long periods of compositional stasis during which taxonomic diversity changed at varying rates. For example, the architecture or "geometry" of early and later Mesozoic plant communities were relatively similar, but the taxonomic structure or "topology" of these communities was significantly different. No period is characterized by taxonomic stasis. Changes in the topology of communities appear consistently to enrich community structures, i.e., even with the virtual elimination of a lineage, its survivors (and evolutionarily new groups) occupy and augment the number of ecologic roles. During many periods of relatively low taxonomic turnover, there is a perceptible trend toward increased specialization ("niche partitioning"). With the appearance of novel grades there is a decrease in the species numbers of the antecendent specialists, while new lineages expand as generalists. Among plants, the survivors of archaic lineages appear to be generalists.

References

1. Andrews SM (1973) Interrelationships of crossopterygians. In: Interrelationships of fishes, eds Greenwood PH, Miles RS, Patterson C. Londond: Linnaean Society, pp 138–177
2. Burnham L (1981) Fossil insects from Montceau-les-Mines (France): a preliminary report. Bull Soc Hist Nat Autun 100:5–12
3. Burnham L (1983) Studies on Upper Carboniferous insects: 1. the Geraridae (Order *Protorthoptera*). Psyche 90:1–57
4. Carroll RI (1977) Patterns of amphibian evolution: an extended example of the incompleteness of the fossil record. In: Patterns of evolution as illustrated by the fossil record, ed Hallam A. Amsterdam: Elsevier Publishing Company, pp 405–438
5. Crepet WL (1979) Insect pollination: a paleontological perspective. BioScience 29:102–108
6. Feduccia A (1980) The age of birds. Cambridge, MA: Harvard University Press

7. Gingerich PD (1977) Patterns of evolution in the mammalian fossil record. In: Patterns of evolution as illustrated by the fossil record, ed Hallam A. Amsterdam: Elsevier Publishing Company, pp 469–500
8. Hallé F, Oldeman RAA, Tomlinson PB (1978) Tropical trees and forests. Berlin: Springer-Verlag
9. Heaton MJ (1979) Cranial anatomy of primitive captorhinid reptiles from the late Pennsylvanian and early Permian Oklahoma and Texas. Bull Okla Geol Surv 127:1–84
10. Heaton MJ (1980) The Cotylosauria: a reconsideration of a group of archaic tetrapods. In: The terrestrial environment and the origin of land vertebrates, ed Panchen AL. London: Academic Press, pp 497–551
11. Hughes NF, Smart J (1967) Plant-insect relationships in Paleozoic and later time. In: The fossil record, ed Harland WB et al. London: Geological Society London, pp 107–117
12. Kemp TS (1982) Mammal-like reptiles and the origin of mammals. London: Academic Press
13. Kemp TS (1983) The relationship of mammals. Zool J Linn Soc 77:353–384
14. Kevan PG, Chaloner WG, Savile DBO (1975) Interrelationships of early terrestrial arthropods and plants. Palaeontology 18:391–417
15. Knoll AH (1985) Patterns of changes in plant communities. In: Community ecology, eds Diamond J, Case T. New York: Harper and Row, in press
16. Lillegraven JA (1972) Ordinal and familial diversity of Cenozoic mammals. Taxon 21:261–274
17. Niklas KJ, Tiffney BH, Knoll AH (1980) Apparent changes in the diversity of fossil plants. Evol Biol 12:1–89
18. Niklas KJ, Tiffney BH, Knoll AH (1983) Patterns in vascular land plant diversification. Nature 303:614–616
19. Niklas KJ, Tiffney BH, Knoll AH (1985) Patterns in vascular land plant diversification: an analysis at the species level. In: Phanerozoic diversity patterns: profiles in macroevolution, ed Valentine JW. Princeton, NJ: Princeton University Press, pp 97–128
20. North FJ (1931) Insect-life in the coal forests, with special reference to South Wales. Trans Cardiff Nat Soc 62:16–44
21. Padian K, Clemens WA (1985) Terrestrial vertebrate diversity: episodes and insights. In: Phanerozoic diversity patterns: profiles in macroevolution, ed Valentine JW. Princeton, NJ: Princeton University Press, pp 41–96
22. Panchen AL (1972) The interrelationships of the earliest tetrapods. In: Studies in vertebrate evolution, eds Joysey KR, Kemp TS. Edinburgh: Oliver and Boyd, pp 65–87
23. Panchen AL (1975) A new genus and species of anthracosaur amphibian from the Lower Carboniferous Scotland. Phil Trans Roy Soc Lond B 269:581–640
24. Panchen AL (1980) The origin and relationships of the anthracosaur amphibia from the late Palaeozoic. In: The terrestrial environment and the origin of land vertebrates, ed Panchen AL. London: Academic Press, pp 319–350
25. Panchen AL (ed) (1980) The terrestrial environment and the origin of land vertebrates. Syst Assoc Spec, vol 15. London: Academic Press
26. Panchen AL (1985) On the amphibian *Crassigyrinus scoticus* Watson from the Carboniferous of Scotland. Phil Trans Roy Soc Lond B 309:505–568

27. Reisz RR (1980) A protothyridid captorhinomorph reptile from the Lower Permian of Oklahoma. Life Sci Contr Roy Ontario Mus 121:1–16
28. Rolfe WDI (1980) Early invertebrate terrestrial faunas. In: The terrestrial environment and origin of land vertebrates, ed Panchen AL. London: Academic Press, pp 117–157
29. Sadler PM (1981) Sediment accumulation rates and the completeness of stratigraphic sections. J Geol 89:569–584
30. Smithson TR (1980) A new labyrinthodont amphibian from the Carboniferous of Scotland. Palaeontology 23:915–923
31. Smithson TR (1980) An early tetrapod fauna from the Namurian of Scotland. In: The terrestrial environment and the origin of land vertebrates, ed Panchen AL. London: Academic Press, pp 407–438
32. Tiffney BH, Niklas KJ (1986) Clonal growth in land plants – a paleobotanical perspective. In: Population biology and evolution of clonal organisms, eds Jackson JBV, Buss CW, Cook RE. New Haven: Yale University Press, pp 35–66
33. Warren AA, Hutchinson MN (1983) The last labyrinthodont? A new brachyopoid (*Amphibia, Temnospondyli*) from the early Jurassic Evergreen formation of Queensland, Australia. Phil Trans Roy Soc Lond B 303:1–62
34. Wootton RJ (1976) The fossil record and insect flight. In: Insect flight, ed Rainey RC. Symp Roy Ent Soc 7:235–254

Patterns and Processes in the History of Life,
eds. D. M. Raup and D. Jablonski, pp. 407–428. Dahlem Konferenzen 1986
Springer-Verlag Berlin, Heidelberg

Phanerozoic Marine Communities

R. K. Bambach

Dept. of Geological Sciences
Virginia Polytechnic Institute and State University
Blacksburg, VA 24061, USA

Abstract. Marine paleocommunities are time-averaged assemblages of primarily skeletal remains. Nonetheless, they do represent the communities from which they were derived. Patterns of change in marine communities reflect the major changes in faunal composition and ecospace utilization for the whole marine biota. Although species turnover is frequent, marine shelf community types persist for long intervals of time. Reef communities show great fluctuation in importance superimposed on two phases of long-term development of complexity. Community composition has been strongly influenced by an onshore to offshore displacement of faunal types through time. The onshore origins of new faunal types contrasts with the higher rates of evolutionary turnover in offshore settings. Attributes of community structure such as succession and trophic grouping can be monitored at points in the fossil record but no continuity of trends is yet well established. The elaboration of predatory life habits and the development of both active infaunal and multilayered epifaunal modes of life (tiering) characterize the changes in community structure during the Phanerozoic. Emergent properties of communities and coordinated patterns of evolutionary change (coevolution?) suggest that communities are more than epiphenomena even though their basic distribution is always dictated by adaptive responses along environmental gradients.

Introduction

The Nature of Marine Paleocommunities

Definitions of paleocommunities vary [10], but all agree that paleocommunities are represented by fossil assemblages derived from the organisms that lived in only one habitat. The assemblages must be untransported (even though they may be disturbed). Fossil communities differ from living communities because preservation is biased strongly in favor of well skeletonized organisms and "time-averaging" is always involved in accumulating a fossil assemblage. It should also be emphasized that complete reconstruction of marine environments, including biochemical conditions and other ambient features, is impossible.

Organisms lacking hard parts are seldom preserved, thus making "holistic" paleocommunity analysis speculative [46]. However, between 25% and 35% of the whole fauna in almost any environment [10, 69] is readily preserved. Therefore, paleocommunities do give a consistent representation of the available fauna and its diversity. This is also documented by "Lagerstätten," such as the Cambrian Burgess Shale [19] and Carboniferous Francis Creek Shale [70]. About two thirds of the preserved fauna in such deposits is softbodied and includes unusual organisms, but the remainder of the fauna was skeletonized and could have been preserved in less unusual circumstances.

No fossil assemblage can precisely record standing crop abundances. Most assemblages contain materials that accumulated over years or even decades. If a deposit is formed by storm reworking, long buried shells may be redeposited with recently killed organisms [22]. Even in a mass kill event, such as an ash fall, any dead shells would be preserved along with recently killed material. Although much information on seasonal fluctuations in living communities is lost [62], these time-averaged assemblages accurately represent the preservable fauna that lived in the community [98]. Such assemblages may even preserve the details of the influence of environmental gradients on the fauna (A. Miller, in preparation).

Despite the limitations of preservation, unmixed and untransported assemblages are the only record of past communities that we have. Therefore, we must study patterns of change in these assemblages to learn about community evolution. Assemblages of fossils form in many ways. There is an extensive literature on the terminology applied to assemblages with different histories of accumulation [4, 10]. For the purposes of this paper I have tried to utilize literature on only those assemblages that can be regarded as thanatocoenoses (a grouping of fossils from a single biotope – the fossilizable part of a paleobiocoenose).

The Framework of Biotic Change Through the Phanerozoic

In the last five years (1981–1985), the data accumulated over the past 150 years on marine fossils have been compiled in a series of studies that provide an overview of the history of the marine biosphere. This history is the basic framework within which community evolution has taken place. Diversity at all lower taxonomic levels (species, genus, family) has increased in three steps (Early and Middle Cambrian, Middle and Late Ordovician, Triassic to Paleogene), each followed by an interval of relatively little diversity change [75]. Different groups of class-level taxa have contributed to each stage of diversity increase (trilobites in the Cambrian; articulate brachiopods, crinoids, ostracodes, cephalopods, anthozoa, and stenolaemate bryozoa in the Middle and Late Paleozoic; and gastropods, bivalves, malacostracans, echinoids, osteichthyes, and gymnolaemate bryozoa in the Mesozoic and Cenozoic) [74]. Within class-level taxa, diversity increase is achieved by particular suborder-to-order-level groups which develop new morphologic features that permit them to invade previously unexploited ecospace [7]. Increasing diversity in the marine biosphere has been reflected in increasing species richness within communities [5]. The increase in diversity within communities has not been achieved by packing more species into already established modes of life (guilds). Instead, the number of modes of life (guilds) has increased as diversity has increased [6].

There are some extinct higher taxa of early Paleozoic age whose functional morphology and ecospace requirements are not well understood (such as hyolithids and conodonts), but both their total diversity at particular times and their diversity within communities is known. None of them is so diverse or dominant that the general conclusions noted here are invalidated.

The implications of these changes for the evolution of marine communities are numerous. The step and plateau pattern of overall diversity increase implies that there may be variation in the rates and style of community change through time [11]. The different groups of dominant classes at different times mean that the taxonomic dominants in communities have changed over time. The addition of new adaptive strategies and new guilds as diversity increased means that community structure has changed over time. The following sections of this paper will review the patterns of change in the distribution and structure of marine communities through the Phanerozoic.

The Composition and Distribution of Marine Communities Through the Phanerozoic

The patterns of increase in diversity and change in taxonomic dominance in marine communities over time are clearly visualized form the illustrations in the book *Ecology of Fossils* [57]. Although these reconstructions have imperfections (inadequate representation of lithologic fabrics, unrealistic population densities), they give a good general impression of relative change. The changes in the expression of epifaunal and infaunal modes of life are also noticeable. Few infaunal organisms are shown in the lower Paleozoic, a variety of sedentary epifauna dominate Middle and Late Paleozoic communities, there is a mixture of infaunal and epifaunal organisms through much of the Mesozoic and a predominance of infauna in the Cenozoic. The increasing importance of nektonic and nektobenthic organisms, especially from the Devonian on, is also represented. The details involved in these changes of composition and distribution are the first-order apsects of community evolution. Questions have been raised about the possibility of a shift toward more skeletonized organisms through time which would imply that reconstructions based primarily on the skeletonized fauna might be inadequate for visualizing trends, but analyses of Lagerstätten and modern faunas (see section on The Nature of Marine Paleocommunities, above) and of the history of bioturbation and infaunal tiering (see section on Modes of Life, below) indicate no obvious change in relative proportions of skeletal and non-skeletal organisms during the Phanerozoic.

Communities on Soft Substrates

Communities are not discrete entities ("superorganisms") but are made by the response of the biota to environmental gradients [99]. Multivariate statistical techniques can reveal these gradient arrays in fossil communities [16]. Although assemblages of dead shells are time-averaged, subtle environmental change, such as the continuous decrease in vegetation cover in a transition from ellgrass to bare sediment, may be clearly recorded in the accumulating fossil assemblage (A. Miller, in preparation).

In general onshore-offshore distribution of marine shelf communities has been apparent in the fossil record since the first summaries of fossil communities were compiled in the late 1960s [12]. Gradient arrays of communities are observed both in whole faunas and in limited groups of organisms such as trace fossils [59]. Well documented community gradients are known from every geologic period. In the Late Cambrian the variety of trilobites distributed along lithofacies gradients was associated in as many ecologic subdivisions as later, presumably more complex faunas [52]. The classic fivefold pattern of Lower Silurian communities first recognized by Ziegler [104] has been shown to be an intergrading gradient pattern [18]. Devonian community patterns in New York vary with gradients in rates of sedimentation and depositional topography [86]. Low diversity molluscan shelf faunas of the Late Triassic display gradient relationships [49]. Salinity gradients are reflected by some Jurassic communities [27]. Trophic relationships have been reconstructed in considerable detail for Miocene communities distributed along environmental gradients [35].

Comparative studies of Ordovician to Devonian communities reveal that there are remarkable stabilities of some patterns along environmental gradients, although there is considerable taxonomic turnover between communities widely separated in time and even within communities in local settings. The major pattern of community groups in more stable environments is maintained throughout the Ordovician [51]. Similar community structures and even similar brachiopod morphologies and taxa characterize Caradoc and Ludlow shelf communities [38]. Intertidal and shallow shelf communities of the Middle Ordovician and Lower Devonian are remarkedly congruent in structure [96]. In the Late Devonian the community array is very similar (with only differences from different topographic gradients) in two areas of the Appalachians [55, 56]. Similar community structure is maintained despite the evolutionary change that occurs in these Late Devonian communities [54]. Wallace [97] has even documented remarkable homeomorphy between Late Devonian communities from different biogeographic provinces. It is as if the response of the Middle Paleozoic fauna to the same environmental conditions at different times was channelled by either selecton pressures or phylogenetic constraints on the higher taxa available to produce the same morphologic suites. Such convergence or homeomorphy is not expected in nonselectionist models of community development. In fact, it is not expected in any current community evolution models and suggests that there are stronger pressures to develop particular community structures than have previously been appreciated.

Few comparable studies are available from Cenozoic rocks, but one recent study suggests that community congruence over time is still achieved. Hickman [34] points out that six community types of deep-water mollusks develop in the Late Eocene and can be traced through the Oligocene of the northeastern Pacific margin. She claims that all six community types have species turnover but persist in time and that modern counterparts exist for all six as well.

Many environmental factors are arrayed along gradients (salinity, turbidity, productivity, disturbance frequency, and substratum type). Many gradients in benthic paleocommunities relate to substratum lithology, especially the granulometry of the sediment. Older studies that claimed little relation between lithology and community were done before the impact of storm reworking on shell bed formation became known. Most modern studies that include evaluation of sedimentology show community relationships with the substratum. The contention that community distribution is often depth-related has been disputed. Depth is not usually the direct control, but many factors do correlate with depth. Communities on soft substrates generally display onshore-offshore gradients at all points through the Phanerozoic.

Communities on Hard Substrates

Communities on hard substrates are those on rocky shores and hardgrounds. Few rocky shorelines are preserved in the fossil record because they occur in areas of erosion, not deposition. Submarine hardgrounds are frequently preserved, however. Palmer [61] has surveyed the changes in hardground communities from the Cambrian to the Cretaceous. He noted that diversity increased from the Cambrian to the Middle Ordovician but, while fluctuating, did not increase significantly thereafter. Various feeding levels (tiering) also developed in the early Paleozoic and have simply been maintained ever since. Diversity and feeding levels did not change for almost 400 million years, whereas there was considerable taxonomic turnover noted, even from period to period. Change in other aspects did happen during this interval, however, as the proportion of boring organisms, encrusting organisms with true exoskeletons, and organisms inhabiting cavities and overhangs in hardgrounds all increased in the late Paleozoic and Mesozoic. Thus the Jurassic hardground communities investigated by Fürsich [26] are much more complex than the Cambrian communities studied by Brett et al. [15].

Reefs

Carbonate buildups are abundant throughout the geologic record [33, 101]. They are usually said to represent warm, shallow marine environments. Recently, however, Ziegler et al. [106] have shown that abundant carbonates are related to low latitudes and do not expand their range to higher latitudes during times of warmer conditions. Because carbonate production is based on algal productivity and photosynthesis, it appears that light penetration into the water (a function of latitude) is more critical than higher temperatures for carbonate production.

I am restricting the idea of reefs to those carbonate buildups with rigid skeletal frameworks which built up some significant topographic relief off the seafloor. Major reviews of the history of reefs include those by Newell [60], Heckel [33], and Wilson [101]. Complex, high diversity reef structures have had a remarkably episodic history of fluctuation [78]. Despite this episodic history, the structural complexity of reefs has consistently increased during each of the two major phases of reef development.

Fluctuations in reef communities have been brought about by extinctions (the disappearance of the *archaeocyathids* in the Middle Cambrian, the successive declines in *Tabulata, Rugosa,* and *Stomatoporoidea* during the Upper Devonian, and the loss of the rudistid bivalves at the end of the Createceous) and by interruptions in the preserved record such as the sea level lows in the Late Ordovician and Late Permian to Early Triassic. These interruptions also probably accelerated evolutionary change in reef ecosystems because of the stress of the accompanying environmental perturbations.

The most ancient reef-like buildups were stromatolite structures near shelf margins in the Coronation Mobile Belt about 2 billion years ago. The Early Cambrian had some rather large reef structures made by *archaeocyathids*. The two major Phanerozoic phases of reef development were from the Middle Ordovician to the Late Devonian and from the Late Triassic to the Recent. Large-scale reef structures equivalent to modern atolls and barrier reefs are preserved only in the Devonian and the Neogene. This might be a result of bias in the preservational record because large reef systems are primarily located in oceanic settings rather than on continental platforms. However, the pattern of change in community makeup during each large phase of reef development suggests that large reefs were possible only at the times they actually appear in the record.

The Ordovician-Devonian phase of reef development starts with bryozoan-based systems in which early tabulates are also found. By the Silurian, tabulate corals are the dominant framebuilders with *Rugosa* subsidiary and reef caps encrusted by stromatoporoids. In the Devonian, large reef structures such as those in central Canada or in the Canning Basin of Australia are comprised of colonial *Rugosa* as the dominant forms with *Tabulata* still important and stromatoporoids still forming the rough-water caps. This complex reef system died out in the Late Devonian. It did not decline in a single event, however. The *Tabulata* declined first (in the Eifelian), followed by the *Rugosa* in the Givetian and then the stromatoporoids in the Frasnian.

Although small algal-sponge-brachiopod mounds are found in Carboniferous, Permian, and Middle Triassic rocks, they form rather modest structures. The famous Permian Reef complex in West Texas is primarily a diagenetically altered bank edge with small bioherms, not a massive organic reef. Middle Triassic reef-like mounds have constituents and structure similar to Permian structures, despite the intervening Permian extinction.

Scleractinian-based reefs begin their development in the Late Triassic [28, 82]. These were patch reef structures with only one or two frame-building species at first, even though they were accompanied by a varied molluscan fauna. Scleractinian reefs were still patch reefs in the Late Jurassic [31]. During the Cretaceous, coral patch reefs served also as the site of development of rudistid bivalves [73]. By the Late Creataceous the rudistids had become major reef-forming organisms and corals were subsidiary [17, 45]. After the extinction of the rudistids at the end of the Cretaceous, there were few reefs until the Late Eocene, when hermatypic corals expanded greatly in diversity [24]. By the Oligocene the reef systems in the Caribbean were at their acme with about seventy species of hermatypic corals in forty genera. Oligocene reef ecolgoic zonation was comparable to modern reefs. Diversity in the Caribbean dropped at the end of the Oligocene, but the species turnover was not sudden. Twenty of the living species of Caribbean corals are known to have existed by the Late Miocene. Modern reef structures are of quite recent origin. Fast growing hermatypic corals such as *Acropora* are much less diverse in both the Caribbean and Indo-Pacific in the Miocene and Pliocene than in the Pleistocene and Holocene [24]. No trace of sclerosponges are found in the deep portions of Miocene and Pliocene reefs, although they are important in the deep parts of Pleistocene and Holocene reefs.

Large-Scale Pattern of Community Displacement

The class level taxa that dominate the three successive stages of diversification in the marine realm also appear associated within different community settings at different times throughout the Paleozoic. Cambrian faunal elements (primarily trilobites) that had dominanted all environments from the shoreline across the shelf to the slope were displaced by the faunal dominants of the Middle and Upper Paleozoic beginning in nearshore communities in the Late Cambrian-Early Ordovician and spreading to outer shelf settings by the Late Ordovician [77]. In the Late Ordovician the classes that participated in the Mesozoic-Cenozoic diversity increase displaced the typically Paleozoic classes from nearshore communities, and trilobite-dominated communities persisted primarily only in slope and basin settings. Sepkoski and Miller [76] have extended observations on this onshore-offshore spectrum of class level domination of communities throughout the Paleozoic. They observe the waning of the trilobite-dominated offshore communities and, beginning in the Devonian, the extension (with great fluctuation and frequent shift back to only nearshore settings) of the Mesozoic-Cenozoic dominant classes into shelf communities.

During the Cretaceous, ecologic types common in the Paleozoic, such as epifaunal suspension feeders, which were still widespread in the Jurassic, were displaced by ecologic types that typify modern faunas, such as infaunal suspension feeders [40]. This displacement also began in nearshore settings and shifted across the spectrum of shelf environments, so that epifaunal suspension feeders dominated soft sediment communities only in offshore, deep-water settings by the end of the Cretaceous. The ecologic types in the "archaic" community types of the Cretaceous include members of the order-level taxa that had diversified in the Paleozoic within the classes that continue to diversify in the Mesozoic-Cenozoic interval, whereas the new ecologic types that first appear in the nearshore assemblages are representatives of the order-level taxa within these same classes that contribute virtually all of the expansion of diversity in the Mesozoic and Cenozoic [7]. This displacement is not a displacement of one coherent faunal grouping by another but instead is caused by the origination of future faunal dominant groups in near-shore settings and their expansion through time into offshore environments. Some former dominants do remain in near-shore associations (such as sponges) or in refugia such as cryptic habitats in reefs.

The origination of new ecologic types (major adaptive strategies) may be a special feature of evolution in nearshore environments [42]. Nearshore, shallow-water settings seem to have produced even the initial radidation of metazoans as documented by Crimes [21] for trace fossils and for the earliest skeletal faunas (Mount and Signor, in preparation). Mount and Signor note, however, that the earliest skeletal organisms are primarily in subtidal, low-stress environments and not in shoreline settings. This phenomenon of evolutionary innovation in nearshore habitats is of interest because rates of speciation and taxonomic turnover in communities have often been observed to be higher in offshore settings [14, 23, 39]. It should be remembered that this phenomenon is one of origination of new higher taxa and not just speciation. The evolution of new species has been common throughout the ocean depths. Causes for this paradoxical relationship are yet to be determined. Jablonski et al. [42] propose several possibilities. Many features, such as larval ecology [41], may play roles in speciation rates. Speciation rates may not always equate with the likelihood of producing evolutionary novelty. Novelty may depend more on conditions that might favor major adaptive shifts.

The Structure of Marine Communities Through Time

Change in the composition and distribution of communities raises questions about change in the interactions of organisms in communities. These interactions result in aspects of community structure such as succession and niche partitioning.

Succession

The influence of succession in fossil communities was emphasized by Johnson [44] and Walker and Alberstadt [95]. Distinct patterns of autogenic succession are associated with recolonization of the seafloor after disturbance [53, 66] such as the dumping of dredge spoil [53, 66] or severe pollution from oil spills [29]. Such sequences must also have accompanied recolonization after storm disturbances, which have been common at all times in shallow marine habitats [22]. These successional sequences may have changed as faunas have changed. Unfortunately, this type of succession is not preserved in the formation of most fossil assemblages, and its history cannot be traced [25, 53]. This is once consequence of time-averaging in paleocommunities.

Successional sequences on hard substrates and in reefs may be preserved [30, 32, 95]. Hardground succession still awaits systematic study through time, however. Four phases of reef succession (stabilization, colonization, diversification, and dominantion) are documented from many parts of the geologic record [95]. These phases seem to be common to the development of reefs at all times. No detailed studies follow shifts in the structure of these successional phases as reef systems change, either through the Paleozoic or in the Mesozoic-Cenozoic.

The successional sequences noted above are produced by the interactions of living organisms (autogenic succession). Walker and Alberstadt [95] noted a type of community development in Ordovician shelf settings in which the accumulation of dead shells that had colonized muddy substrates provided a new substratum that permitted other organisms to become established. This is similar to the stabilization and colonization phases of reef development. Kidwell and Jablonski [47] have pointed out that this is not a pure autogenic succession and have named the influence of accumulated dead skeletal materials on living organisms (either facilitating their appearance in a community or inhibiting it) as taphonomic feedback. This is an important aspect of the formation of fossil assemblages of all ages and has been documented in the Ordovician [95], Carboniferous [102], and Miocene [47]. Again, detailed understanding of the changes in taphonomic feedback as faunas change awaits future study. This will be one of the most important methods for working out the meaning of time-averaging in paleocommunities.

The examples of succession discussed above relate to biological interactions as the control on community change. The use of the term succession for community change caused by change in the physical environment (allogenic succession) is synonomous with the phenomenon of community response to environmental gradients. Studies such as those by Bretsky and Bretsky [13] and Copper and Grawbarger [20] are not analyses of biologically mediated succession.

The Utilization of Ecospace

The niche structure of a community reveals how resources are utilized by the biota. As marine faunas have changed, the ways in which resources have been partitioned in communities has also changed. The major trend in the change of community structure through time has been toward the utilization of more total ecospace. Each taxon that increases in diversity also invades previously unexploited ecospace [7]. As diversity within communities has increased, the number of guilds in communities has increased [6]. Changes in trophic structure, especially well documented by the rise of predation, and changes in modes of life, both through the increase of infaunal habits and the elaboration of epifaunal stratification, reflect the secular increase in ecospace utilization that marks community change through time. Some aspects of biotic interaction such as commensalism and symbiosis are important and clearly influence the structure of communities such as coral reefs. Unfortunately, evidence of most of these interactions are only rarely preserved, so it is not possible to discuss changes in mutualistic interactions through time or their impact on community change through time.

Trophic Structure

Although trophic analysis has been proposed as an important tool for paleocommunity study [71, 72, 94], questions have been raised about its reliability because of biases of preservation and the effects of time-averaging [69, 83]. In some cases, however, convincing trophic reconstruction of part or most of a community has been achieved. Although no continuity of studies through time is yet available, it is clear that the overall trophic structure of communities has increased in complexity from simple Paleozoic patterns [6, 94] to complex trophic webs in the Cenozoic [37, 84]. This contrasts with the stability in stratification in the deposit feeding segment of communities between the Silurian and the Recent [50].

The most completely documented aspect of change in trophic patterns is the rise of predation. This is because some taxa can be recognized as predators, either from their morphology or by analogy with living groups, and because many styles of predation leave traces in shell boring or breakage [81, 91]. Predation is one of the factors that may control community structure. Changes in predation have been suggested as instrumental in restructuring marine communities in the mid-paleozoic [80] and in the Mesozoic [90].

The trend toward increase in and severity of predation is unambiguous. Few predators are known from the Cambrian. Starfish and cephalopods diversify in the Ordovician, but no boring or shell-crushing predation is documented from either the Ordovician or Silurian. For example, I found no borings or crushed valves in over 14,000 Silurian bivalve specimens collected from the Llandovery to Gedinnian section at Arisaig, Nova Scotia. Shell breakage associated with predation has been documented from the later Paleozoic [1, 68], however, accompanying the evolution of chondrichthyes [80]. The first documented shell boring predation is in the Early Devonian [79]. Smith et al. (in preparation) have recently discovered the first borings of Paleozoic age in which incomplete holes have a raised central boss similar to naticid drill holes in the Mesozoic and Cenozoic. Although evidence of shell-boring and -crushing predation is found in Devonian and younger Paleozoic fossils, it is not nearly as common as it is in Cretaceous and younger assemblages. Changes in morphology related to resistance to shell breakage are noted in post-Silurian communities [80], but is is only during the Mesozoic that less predation-resistant morphologies are virtually eliminated. It is also only in the Late Mesozoic that evidence of predation in preserved faunas rises to the frequencies observed commonly through the Cenozoic [92, 93].

Modes of Life

Benthic marine community structure has also changed as adaptive strategies for various modes of life have changed with time. Tiering (utilization of different levels above and below the sediment-water interface) has changed [3], as has and bioturbation, both in depth (tiering) and rate [87]. Although many of the adaptive reasons for these changes relate to elaboration of feeding activities (suspension feeding high in the water mass, deep deposit feeding in the sediment), the resulting physical interactions of the organisms with the sediment and with each other have had major consequences beyond trophic structure alone [2, 65, 87].

Epifaunal tiering has developed and changed as various groups evolved that could extend up from the bottom into the water mass [3]. On soft sediments these heights were limited to about 10 cm in the Cambrian (eocrinoids and sponges) but ranged up to 1 m (crinoids) by the Silurian and remained so through the Paleozoic. Bryozoa, blastoids, and others filled the 10–20 cm range tier after the Cambrian during the rest of the Paleozoic. In the early Mesozoic, crinoids remained at the highest tier but were replaced by alcyonarians and sponges in the later Mesozoic and Cenozoic. A lower tier ($\sim$20 cm) has also persisted in the Mesozoic and Cenozoic, inhabited by a variety of taxa. The growth of reefs (discussed above) has also had a history of topographic change with small, low relief mounds at the beginning of each phase of reef development and large masses capable of extending tens to hundreds of meters off the seafloor and developing a complex ecological zonation culminating each of the two phases of reef development.

Infaunal tiering has also increased through the Phanerozoic, as deep deposit feeding, infaunal suspension feeding using siphons, and infaunal predation have developed [3]. Thayer's major compilation of the sequence of infaunal innovations discusses their likely effects on benthic communities [87]. Larsen and Rhoads [48] have demonstrated the change in degree of bioturbation in a particular habitat from the Ordovician to the Devonian and summarize other observations on the impact of increased bioturbation through time on sedimentary fabrics. In Cambrian shelf settings bioturbation seldom extended more than a few centimeters into the sediments. During the Ordovician and most of the Silurian, shelf sediments have bioturbated fabrics in the upper parts of thin (3–10 cm thick) storm-reworked beds. Thin storm-rewoked beds are often thoroughly bioturbated by the Devonian, although homogenization of beds on a meter scale is only common in low sedimentation rate settings and in very nearshore environments (where it had occurred even in Late Cambrian time). In Cretaceous and Cenozoic shelf strata, bioturbated fabrics extending over meter-thick beds are typical, not unusual. The rates as well as the depths of bioturbation have increased during the Phanerozoic [87]. All the organisms living in communities with this greater burrowing activity have had to adapt to those conditions.

Coevolutionary change is involved in the change in community structure through time. Because of the preservational biases of the fossil record and because of the difficulty of demonstrating explicit cause-effect relationships in ancient settings (through lack of experimental verification), it is hard to specify how the interactive events have occurred. Nonetheless, the changes in community structure that are known, both in trophic organization and space utilization, demonstrate that coevolutionary response has been a significant aspect of community evolution.

Possible External Influences on Marine Communities

Secular change in geography, in oxygen levels, and in nutrient supplies during the Phanerozoic have been documented or suggested. They may have had an impact on the changes in the biosphere and in the changes seen in communities. The Termiers [85] have argued that secular change in paleogeography coupled with changes in the ecosystem and in biosphere-mediated features such as the atmosphere have influenced the course of evolution.

Valentine [88] suggested that plate tectonic events have had an impact on world marine diversity as change in geography caused fluctuation in the potential for endemism. Valentine et al. [89] also argue for changes in provinciality as an influence to add to the within-community changes in diversity observed by Bambach [5]. Ziegler et al. [105] document changes in biogeography and provinciality related to plate tectonic-driven changes in geography, paleooceanography, and climatology during the Paleozoic. The discipline of vicariance biogeography is based on similar changes that have taken place in more recent geologic time.

Oxygen concentration influences marine benthic faunas [67]. Rhoads and Morse [67] have speculated that change in the marine biosphere has been regulated by increase in oxygen availability through the Phanerozoic. Progressive ventilation of the oceans and fluctuations in rates of ventilation driven by changes in thermo-haline deep circulation have been suggested as influencing marine environments and the communities inhabiting them [9, 100].

Nutrients supplied from the land to ocean waters influence productivity in nearshore ecosystems. Over time they replace nutrients lost from the marine ecosystem by sedimentation. Change in both organic productivity and erosion on land through time would have affected the rate of supply of nutrient materials from the land to the sea. For example, human influence has increased natural nutrient levels greatly in fresh waters. Total dissolved phosphorous and nitrogen are elevated two to fifty times above natural levels in the runoff of rivers from inhabited regions [58]. Changes would have occurred at times in the distant past, too, as the terrestrial ecosystem became inhabited and productivity increased on land. Berner and Raiswell [8] report changes in the burial of organic carbon and pyritic sulfur in marine sediments over the Phanerozoic that may reflect changes in the supply of organic detritus from the land. Such changes could influence activities such as deposit feeding in marine communities, as well as community diversity.

Establishing specific connections between evolutionary events and large-scale changes in geography, oxygen availability, and nutrients in marine ecosystems is difficult. The magnitude of such changes is difficult to establish, too. Whether there has been a direct interaction between such changes and changes in marine communities is not yet clear. However, the evolution of marine communities has occurred in the context of changes in these basic factors through the Phanerozoic.

The Status of Community Evolution

Hoffman [36] contended that "ecological communities are merely an epiphenomenon of the overlap in distributional patterns of various organisms." I fell he made two errors in reaching this harsh judgement. One is the equating of the ecological hierarchy with the taxonomic hierarchy and the other was the confusion of community membership with the community itself.

Hoffman assumes that communities are composed of species and that they are a part of an ecological hierarchy equivalent to the taxonomic hierarchy. In the taxonomic hierarchy all smaller groups are included within each larger group, and all are genealogically related. The ecological hierarchy (individual, population, community, province, realm) is not a clear-cut nested sequence. Criteria differ at each level. For example, communities are assemblages of populations within particular habitats, whereas provinces are geographic regions typified by endemic species (not assemblages of communities).

Hoffman also claimed that communities were reducible to the properties of their components unless group selection could be demonstrated as a community property. Yet he ignored or passed over many irreducible emergent properties of the assemblage of organisms in their environment. It may be true that the constituents of a community can co-occur because they are each adapted to particular environmental conditions and therefore end up in assemblages along environmental gradients. But no amount of autecologic information alone can predict abundances and the concomitant suite of chracteristics of community structure. For example, competitive interactions are being identified in living marine benthic communities that may determine abundances, diversity, and growth rates [43, 63, 64, 103]. The properties of community structure are emergent properties, not reducible to the individual components alone. Community composition may be epiphenomenal, but community structure is not.

The results of the emergent properties of communities are seen in the fossil record as repetitive dominance-diversity relations, the phenomenon of community displacement associated with faunal changes and coevolutionary relationships associated with the expansion of the utilization of ecospace over time.

If communities were not integrated responses of the fauna to the environment, then why should repetitive dominance-diversity associations be so common? Communities or community types persist over immense lengths of time (tens of millions of years). It may even be that random short-term fluctuations must be evened out by time-averaging before the emergent long-term influence of relatively weak interactions can be detected consistently. In persistent community types, the same general suite of morphologies is often maintained despite species and even generic turnover. It is as if the interaction of the environment and evolution are producing selection for particular morphotypes.

Newly developed major faunal groups do not just enter the community gradient spectrum in a random fashion. The displacement of old community types by newly developed associations as major faunal change occurs suggests that community types have some coevolutionary integrity.

The changes in ecospace utilization within communities through time [6, 7] required many coevolutionary responses. These must have included responses to "by-product effects" (such as interference from increased rates of bioturbation) as well as responses to direct interactions such as predation.

Conclusions

Communities do not exist independently of their constituent organisms. Communities have no genetic continuity, no geneology. It is debatable whether communities evolve or whether they change as evolution operates. Under any circumstances communities are the context in which all evolution occurs.

Communities reflect the integration of the available adaptive strategies at particular points within the environmental spectrum. Homeomorphs, products of convergent evolution, are frequently found in succeeding communities in the same environmental setting. Community types and community structures are maintained over long intervals of time. It is almost as if communities are responding to a process equivalent to stabilizing selection as long as the fauna retains the same basic suite of adaptive strategies in its constituent taxa.

When new adaptive strategies appear, community structure as well as community composition is altered. The major times of increase in diversity have also been times of increase in ecospace utilization. This has resulted in the restructuring and displacement of communities over relatively short intervals of time. The increase in utilization of ecospace, represented by increasing numbers of guilds, increases in the complexity of food webs, and increase in tiering and its effects, has typified the history of marine communities throughout the Phanerozoic.

Acknowledgements. I wish to thank the organizing committee of this Dahlem Workshop for the invitation to participate and thereby providing me the stimulating opportunity to review and synthesize the various facets of the history of marine communities. C. Babin read the manuscript and made constructive suggestions to clarify several points. The Dahlem editorial staff also improved some of the more clumsy statements. Numerous colleagues provided material still in preparation and that generous sharing of information was of great use.

References

1. Alexander RR (1981) Predation scars preserved in Chesterian brachiopods: probable culprits and evolutionary consequences for the articulates. J Paleontol 55:192–203
2. Aller RC (1982) The effects of macrobenthos on chemical properties of marine sediment and overlying water. In: Animal-sediment relations, eds McCall PM, Tevesz MJS. New York: Plenum, pp 52–102

3. Ausich WI, Bottjer DJ (1982) Tiering in suspension-feeding communities on soft substrata throughout the Phanerozoic. Science 216:173–174
4. Babin C (1980) Elements of palaeontology. Chichester: John Wiley and Sons
5. Bambach RK (1977) Species richness in marine benthic habitats through the Phanerozoic. Paleobiology 3:152–167
6. Bambach RK (1983) Ecospace utilization and guilds in marine communities through the Phanerozoic. In: Biotic interactions in recent and fossil benthic communities, eds Tevesz MJS, McCall PL. New York: Plenum, pp 719–746
7. Bambach RK (1985) Classes and adaptive variety: the ecology of diversification in marine faunas through the Phanerozoic. In: Phanerozoic diversity patterns: profiles in macroevolution, ed Valentine JW, Ch 6. Princeton: Princeton, pp 191–253
8. Berner RA, Raiswell R (1983) Burial of organic carbon and pyrite sulfur in sediments over Phanerozoic time: a new theory. Geochem Cos 47:855–862
9. Berry WBN, Wilde P (1978) Progressive ventilation of the oceans – an explanation for the distribution of the Lower Paleozoic black shales. Am J Sci 278:257–275
10. Boucot AJ (1981) Principles of benthic marine paleoecology. New York: Academic Press
11. Boucot AJ (1983) Does evolution take place in an ecological vacuum? II. J Paleontol 57:1–30
12. Bretsky PW (1969) Evolution of Paleozoic benthic marine communities. Palaeogeog Palaeoclimatol Palaeoecol 6:45–59
13. Bretsky PW, Bretsky SS (1975) Succession and repetition of Late Ordovician fossil assemblages from the Nicolet River valley, Quebec. Paleobiology 1:225–237
14. Bretsky PW, Lorenz DM (1970) Adaptive response to environmental stability: a unifying concept in paleoecology. Proceedings of the Norm American Paleontology Convention, Part E. Lawrence, KS: Allen Press, pp 522–550
15. Brett CE, Liddell WD, Derstler KL (1983) Late Cambrian hard substrate communities from Montan/Wyoming: the oldest known hardground encrusters. Lethaia 16:281–289
16. Cisne JL, Rabe BD (1978) Coenocorrelation: gradient analysis of fossil communities and its applications in stratigraphy. Lethaia 11:341–364
17. Coates AG (1977) Jamaican coral-rudist frameworks and their geologic setting. In: Reefs and related carbonates – ecology and sedimentology, eds Frost SH, Weiss MP, Saunders JB. Am Assoc Petrol Geol Stud 4:83–91
18. Cocks LRM, McKerrow WS (1984) Review of the distribution of the commoner animals in Lower Silurian marine benthic communities. Palaeontology 27:663–670
19. Conway Morris S (1979) The Burgess Shale (Middle Cambrian) fauna. Ann Rev Ecol Syst 10:327–349
20. Copper P, Grawbarger DJ (1978) Paleoecological succession leading to a Late Ordovician biostorme on Manitoulin Island, Ontario. Can Earth Sci 15:1987–2005
21. Crimes TP (1974) Colonisation of the early ocean floor. Nature 248:328–330
22. Einsele G, Seilacher A (eds) (1982) Cyclic and event stratification. New York: Springer-Verlag
23. Fortey RA (1980) Generic longevity in Lower Ordovician trilobites: relation to environment. Paleobiology 6:24–31
24. Frost SH (1977) Cenozoic reef systems of Caribbean – prospects for paleoecologic synthesis. In: Reefs and related carbonates – ecology and sedimentology, eds Frost SH, Weiss MP, Saunders JB. Am Assoc Petrol Geol Stud Geol 4:93–110

25. Fürsich FT (1978) The influence of faunal condensation and mixing on the preservation of fossil benthic communities. Lethaia 11:243–250
26. Fürsich FT (1979) Genesis, environments and ecology of Jurassic hardgrounds. N Geol Paläontol Abh 158:1–63
27. Fürsich FT (1981) Salinity-controlled benthic associations from the Upper Jurassic of Portugal. Lethaia 14:203–223
28. Fürsich FT, Wendt J (1977) Biostratinomy and paleoecology of the Cassian Formation (Triassic) of the southern alps. Palaeogeog Palaeoclimatol Palaeoecol 22:257–323
29. Glemarec M, Hussenot E, Le Moal Y (1982) Utilization of biological indications in hypertrophic sedimentary areas to describe dynamic process after the AMOCO Cadiz oil-spill. In: International symposium on utilization of coastal ecosystems. Planning, pollution, and productivity. Rio Grande, Brazil: Index to Scientific and Technical Proceedings
30. Goldring R, Kazmierczak J (1974) Ecological succession in intraformational hardground formation. Palaeontology 17:949–962
31. Hallam A (1975) Coral patch reefs in the Bajocian (Middle Jurassic) of Loraine. Geol Mag 112:383–392
32. Halleck MS (1973) Crinoids, hardgrounds, and community succession: The Silurian Laurel-Waldron contact in southern Indiana. Lethaia 6:239–252
33. Heckel PH (1974) Carbonate buildups in the geologic record: a review. In: Reefs in time and space, ed Laporte LF. EPM Spec Publ 18:90–154
34. Hickman CS (1984) Composition, structure, ecology, and evolution of six Cenozoic dep-water mollusk communities. J Paleontol 58:1215–1234
35. Hoffman A (1977) Synecology of macrobenthic assemblages of the Korytnica clays (Middle Miocene; Holy Cross Mountains, Poland). Acta Geol Polon 27:227–280
36. Hoffman A (1979) Community paleooecology as an epiphenomenal science. Paleobiology 5:357–379
37. Hoffman A, Pisera A, Studenci W (1978) Reconstruction of a Miocene kelp-associated macrobenthic ecosystem. Acta Geol Polon 28:377–387
38. Hurst JM, Watkins R (1981) Lower Paleozoic clastic, level-bottom community organization and evolution based on Caradoc and Ludlow comparisons. In: Communities of the past, eds Gray J, Boucot AJ, Berry WBN. Stroudsburg: Hutchinson Ross, pp 69–100
39. Jablonski D (1980) Apparent versus real biotic effects of transgressions and regressions. Paleobiology 6:397–407
40. Jablonski D, Bottjer DJ (1983) Soft-bottom epifaunal suspension-feeding assemblages in the Late Cretaceous. In: Biotic interaction in recent and fossil benthic communities, eds Tevesz MJS, McCall PM. New York: Plenum, pp 747–812
41. Jablonski D, Lutz RA (1983) Larval ecology of marine benthic invertebrates: paleobiological implications. Biol Rev 58:21–89
42. Jablonski D, Sepkoski JJ Jr, Bottjer DJ, Sheehan PM (1983) Onshore-offshore patterns in the evolution of Phanerozoic shelf communities. Science 222:1123–1125
43. Jackson JBC (1983) Biological determinants of present and past sessile animal distributions. In: Biotic interactions in recent and fossil benthic communities, eds Tevesz MJS, McCall PL. New York: Plenum, pp 39–120
44. Johnson RG (1972) Conceptual models of benthic marine communities. In: Models in Paeobiology, ed Schopf TJM. San Francisco: Freeman, Cooper and Company, pp 148–159

45. Kauffman FG (1974) Structure, succession and evolution of Antillean Cretaceous "Reefs": rudistid frameworks. In: Principles of benthic community analysis, Sedimenta IV, eds Ziegler AM et al. University of Miami: Comparative Sedimentology Laboratory, pp 12.14–12.25
46. Kauffman EG, West RW (1976) Basic concepts of community ecology and paleoecology. In: Structure and classification of paleocommunities, eds Scott RW, West RR. Stroudsburg: Dowden, Hutchinson and Ross, pp 1–28
47. Kidwell SM, Jablonski D (1983) Taphonomic feedback: ecological consequences of shell accumulation. In: Biotic interactions in recent and fossil benthic communities, eds Tevesz MJS, McCall PM. New York: Plenum, pp 195–248
48. Larson DW, Rhoads DC (1983) The evolution of infaunal communities and sedimentary fabrics. In: Biotic interactions in recent and fossil benthic communities, eds Tevesz MJS, McCall PM. New York: Plenum, pp 627–248
49. Laws RA (1982) Late Triassic depositional environments and molluscan associations from west-central Nevada. Paleogeog Palaeoclimatol Palaeoecol 37:131–148
50. Levinton JS, Bambach RK (1975) A comparative study of Silurian and Recent deposit-feeding bivalve communities. Paleobiology 1:97–124
51. Lockley MG (1983) A review of brachiopod dominated palaeocommunities from the type Ordovician. Palaeontology 26:111–145
52. Ludvigsen R, Westrop SR (1983) Trilobite biofacies of the Cambrian-Ordovician boundary interval in northern North America. Alcheringa 7:301–319
53. McCall PM, Tevesz MJS (1983) Soft bottom succession and the fossil record. In: Biotic interactions in recent and fossil benthic communities, eds Tevesz MJS, McCall PM. New York: Plenum, pp 157–194
54. McGhee GR Jr (1981) Evolutionary replacement of ecological equivalents in Late Devonian benthic marine communities. Palaeogeog Palaeoclimatol Palaeoecol 34:267–283
55. McGhee GR Jr, Sutton RG (1981) Late Devonian marine ecology and zoogeography of the central Appalachians and New York. Lethaia 14:27–43
56. McGhee GR Jr, Sutton RG (1983) Evolution of Late Frasnian (Late Devonian) marine environments in New York and the central Appalachians. Alcheringa 7:9–21
57. McKerrow WS (ed) (1978) The ecology of fossils. Cambridge: M.I.T. Press
58. Meybeck M (1982) Carbon, nitrogen and phosphorus transport by world rivers. Am Sci 282:401–450
59. Miller MF, Ekdale AA, Picard MD (eds) (1984) Trace fossils and paleoenvironments: marine carbonate, marginal marine terrigenous and continental terrigenous settings. J Paleontol 58:283–597
60. Newell ND (1971) An outline history of tropical organic reefs. Am Museum Novit 2465:1–37
61. Palmer T (1982) Cambrian to Cretaceous changes in hardground communities. Lethaia 15:309–323
62. Peterson CH (1977) The paleoecological significance of undetected short-term temporal variability. J Paleontol 51:976–981
63. Peterson CH (1979) Predation, competitive exclusion, and diversity in the soft-sediment benthic communities of estuaries and lagoons. In: Ecological processes in coastal and marine systems, ed Livingston RJ. New York: Plenum, pp 233–264
64. Peterson CH, Andre SV (1980) An experimental analysis of interspecific competition among marine filter feeders in a soft-sediment environment. Ecology 61:129–139

65. Rhoads DC, Boyer LF (1982) The effects of marine benthos on physical properties of sediments. In: Animal-sediment relations, eds McCall PM, Tevesz MJS. New York: Plenum, pp 3–52
66. Rhoads DC, McCall PL, Yingst JY (1978) Disturbance and production on the estuarine seafloor. Am Sci 66:577–586
67. Rhoads DC, Morse JW (1971) Evolutionary and ecologic significance of oxygen-deficient marine basins. Lethaia 4:413–428
68. Schindel DE, Vermeij GJ, Zipser E (1982) Frequencies of repaired shell fractures among the Pennsylvanian gastropods of north-central Texas. J Paleontol 56:729–740
69. Schopf TJM (1978) Fossilization potential of an intertidal fauna: Friday Harbor, Washington. Paleobiology 4:261–270
70. Schram FR (1979) The Mazon Creek biotas in the context of a Carboniferous faunal continuum. In: Mazon Creek fossils, ed Nitecki MH. New York: Academic Press, pp 159–190
71. Scott RW (1976) Trophic classification of benthic communities. In: Structure and classification of paleocommunities, eds Scott RW, West RR. Stroudsburg: Dowden, Hutchinson and Ross, pp 29–66
72. Scott RW (1978) Approaches to trophic analysis of paleocommunities. Lethaia 11:1–14
73. Scott RW (1981) Biotic relations in Early Cretaceous coral-algal-rudist reefs, Arizona. J Paleontol 55:463–478
74. Sepkoski JJ Jr (1981) A factor analytic description of the Phanerozoic marine fossil record. Paleobiology 7:36–53
75. Sepkoski JJ Jr, Bambach RK, Raup DM, Valentine JW (1981) Phanerozoic marine diversity and the fossil record. Nature 293:435–437
76. Sepkoski JJ Jr, Miller AJ (1985) Evolutionary faunas and the distribution of Paleozoic benthic communities in space and time. In: Phanerozoic diversity patterns: profiles in macroevolution, ed Valentine JW. Princeton: Princeton pp 153–190
77. Sepkoski JJ Jr, Sheehan PM (1983) Diversification, faunal change, and community replacement during the Ordovician radiations. In: Biotic interactions in recent and fossil benthic communities, eds Tevesz MJS, McCall PM. New York: Plenum, pp 673–717
78. Sheehan PM (1985) Reefs are not so different – they follow the evolutionary pattern of level-bottom communities. Geology 13:46–49
79. Sheehan PM, Lesperance PJ (1978) Effect of predation on the population dynamics of a Devonian brachiopod. J Paleontol 52:812–817
80. Signor PW III, Brett CE (1984) The mid-Paleozoic precursor to the Mesozioc marine revolution. Paleobiology 10:229–245
81. Sohl NF (1969) The fossil record of shell boring by snails. Am Zool 9:725–734
82. Stanley GD Jr (1979) Paleoecology, structure and distribution of Triassic coral buildups in western North America. Univ Kansas Paleontol Contr Art 65:1–58
83. Stanton RJ Jr, Dodd JR (1976) The application of trophic structure of fossil communities in paleoenvironmental reconstruction. Lethaia 9:327–342
84. Stanton RJ Jr, Nelson PC (1980) Reconstruction of the trophic web in paleontology: community structure in the Stone City Formation (Middle Eocene, Texas). J Paleontol 54:118–135
85. Termier H, Termier G (1984) Recherches sur le role des ecosystemes dans la differenciation de guelgues grands phylums. Geobios Mem Spec No 8:167–174 (Lyon)

86. Thayer CW (1974) Marine paleoecology in the Upper Devonian of New York. Lethaia 7:121–155
87. Thayer CW (1983) Sediment-mediated biological disturbance and the evolution of marine benthos. In: Biotic interactions in recent and fossil benthic communities, eds Tevesz MJS, McCall PM. New York: Plenum, pp 479–625
88. Valentine JW (1971) Plate tectonics aand shallow marine diversity and endemism, an actualistic model. Syst Zool 20:253–264
89. Valentine JW, Foyne TC, Pert D (1978) A provincial model of Phanerozoic marine diversity. Paleobiology 4:55–66
90. Vermeij GJ (1977) The Mesozoic marine revolution: evidence from snails, predators and grazers. Paleobiology 3:245–258
91. Vermeij GJ (1983) Shell-breaking predation through time. In: Biotic interactions in recent and fossil benthic communities, eds Tevesz MJS, McCall PM. New York: Plenum, pp 649–669
92. Vermeij GJ, Zipser E, Dudley EC (1980) Predation in time and space: peeling and drilling in terebrid gastropods. Paleobiology 6:352–364
93. Vermeij GJ, Zipser E, Zardini R (1982) Breakage-induced shell repair in some gastropods from the Upper Triassic of Italy. J Paleontol 56:233–235
94. Walker KR (1972) Trophic analysis: a method for studying the function of ancient communities. J Paleontol 46:82–93
95. Walker KR, Alberstadt LP (1975) Ecological succession as an aspect of structure in fossil communities. Paleobiology 1:238–257
96. Walker KR, Laporte LF (1970) Congruent fossil communities from Ordovician and Devonian carbonates of New York. J Paleontol 44:928–944
97. Wallace P (1978) Homeomorphy between Devonian brachiopod communities in France and Iowa. Lethaia 11:259–272
98. Warme JE, Ekdale AA, Ekdale SF, Peterson CH (1976) Raw material of the fossil record. In: Structure and classification of paleocommunities, eds Scott RW, West RR. Stroudsburg: Dowden, Hutchinson and Ross, pp 143–170
99. Whittaker RH (ed) (1973) Ordination and classification of communities. The Hague: W Junk
100. Wilde P, Berry WBN (1984) Destabilization of the oceanic density structure and its significance to marine "extinction" events. Palaeogeog Palaeoclimatol Palaeoecol 48:143–162
101. Wilson JL (1975) Carbonate facies in geologic history. New York: Springer-Verlag
102. Wilson MA (1982) Origin of brachipod-bryozoan assemblages in an Upper Carboniferous limestone: importance of physical and ecological controls. Lethaia 15:263–273
103. Woodin SA (1983) Biotic interactions in recent marine sedimentary environments. In: Biotic interactions in recent and fossil benthic communities, eds Tevesz MJS, McCall PL. New York: Plenum, pp 3–30
104. Ziegler AM (1965) Silurian marine communities and their environmental significance. Nature 207:270–272
105. Ziegler AM, Bambach RK, Parrish JT, Barrett SF, Gierlowski EH, Parker WZ, Raymond A, Sepkoski JJ Jr (1981) Paleozoic biogeography and climatology. In: Paleobotany, paleoecology and evolution, ed Niklas KJ, vol 2. New York: Praeger, pp 231–266
106. Ziegler AM, Hulver ML, Lottes AL, Schmachtenberg WF (1984) Uniformitarianism and paleoclimates: inferences from the distribution of carbonate rocks. In: Fossils and climate, ed Brenchley PJ. New York: John Wiley and Sons, pp 3–25

Appendix: Geological Time Scales

Three recent geological time scales for the Phanerozoic Eon. The time for beginning of each interval of geological time is given in millions of years before present. Time scale divisions after Harland et al. [1], except Carboniferous, which follows Snelling [3].

Interval	Time of Beginning		
	(1)	(2)	(3)
Cenozoic Era			
Quaternary Sub-era			
Holocene Epoch	0.01	0.01	0.01
Pleistocene Epoch	2.0	2.0	1.6
Tertiary Sub-era			
Neogene Period			
Pliocene Epoch			
Piacenzian Stage	–	–	3.4
Zanclian Stage	5.1	5.5	5.3
Miocene Epoch			
Messinian Stage	–	–	6.5
Tortonian Stage	11.3	–	10.5
Serravallian Stage	–	–	15.1
Langhian Stage	–	–	16.3
Burdigalian Stage	–	–	21.8
Aquitanian Stage	24.6	23.0	23.7
Paleogene Period			
Oligocene Epoch			
Chattian Stage	32.8	27.0	30.0
Rupelian Stage	38.0	34.0	36.6
Eocene Epoch			
Priabonian Stage	42.0	37.0	40.0
Bartonian Stage	–	39.0	43.6
Lutetian Stage	50.5	45.0	52.0
Ypresian Stage	54.9	53.0	57.8
Paleocene Epoch			
Thanetian Stage	60.2	59.0	62.3
Danian Stage	65.0	65.0	66.4
Mesozoic Era			
Cretaceous Period			
Maastrichtian Stage	73	72	72
Campanian Stage	83	83	83
Santonian Stage	87.5	86	86
Coniacian Stage	88.5	88	88

Interval	Time of Beginning		
	(1)	(2)	(3)
Turonian Stage	91	91	91
Cenomanian Stage	97.5	95	95
Albian Stage	113	107	107
Aptian Stage	119	112	114
Barremian Stage	125	114	116
Hauterivian Stage	131	119	120
Valanginian Stage	138	126	128
Berriasian Stage	144	130	135
Jurassic Period			
Tithonian Stage	150	135	139
Kimmeridgian Stage	156	140	144
Oxfordian Stage	163	150	152
Callovian Stage	169	158	159
Bathonian Stage	175	170	170
Bajocian Stage	181	178	176
Aalenian Stage	188	181	180
Toarcian Stage	194	189	188
Pliensbachian Stage	200	195	195
Sinemurian Stage	201	201	201
Hettangian Stage	213	204	205
Triassic Period			
Rhaetian Stage	219	210	210
Norian Stage	225	220	220
Carnian Stage	231	229	230
Ladinian Stage	238	233	235
Anisian Stage	243	239	240
Scythian Epoch (includes Spathian, Smithian, Dienerian, and Griesbachian Stages)	248	245	250
Paleozoic Era			
Permian Period			
Tatarian (=Dzhulfian) Stage	253	250	255
Kazanian Stage	–	–	260
Ufimian Stage	258	258	–
Kungurian Stage	263	265	270
Artinskian Stage	268	273	–
Sakmarian Stage	–	280	280
Asselian Stage	286	290	290
Carboniferous Period			
Pennsylvanian Sub-period			
Stephanian Epoch	296	–	300

Interval	Time of Beginning		
	(1)	(2)	(3)
Westphalian Epoch	315	–	310
Namurian Epoch (B+C)	320	–	–
Mississippian Sub-period			
Namurian Epoch (A) (=Serpukhovian Epoch)	333	320	325
Visean Epoch	352	355	–
Tournaisian Epoch	360	360	355
Devonian Period			
Famennian Stage	367	–	–
Frasnian Stage	374	375	375
Givetian Stage	380	–	–
Eifelian Stage	387	385	390
Emsian Stage	394	–	–
Siegenian Stage	401	–	–
Gedinnian Stage	408	400	405
Silurian Period			
Pridoli Epoch	414	–	–
Ludlow Epoch	421	–	420
Wenlock Epoch	428	–	425
Llandovery Epoch	438	418	435
Ordovician Period			
Ashgill Epoch	448	425	440
Caradoc Epoch	458	438	455
Llandeilo Epoch	468	455	460
Llanvirn Epoch	478	470	–
Arenig Epoch	488	475	490
Tremadoc Epoch	505	495	510
Cambrian Period			
Late Cambrian	525	–	–
Middle Cambrian	540	–	–
Lenian Stage	–	–	–
Atdabanian Stage	–	–	550
Tommotian Stage	590	530	570

References

(1) Harland, W. B.; Cox, A. V.; Llewellyn, P. G.; Pickton, C. A. G.; Smith, A. G.; and Walters, R. 1982. Geologic Time Scale. Cambridge: Cambridge University Press.
(2) Odin, G. S., ed. 1982. Numerical Dating in Stratigraphy (2 Vols.). Chichester: Wiley.
(3) Snelling, N. J., ed. 1985. The Chronology of the Geological Record. Geol. Soc. Lond. Mem. *10*. Oxford: Blackwell.

List of Participants with Fields of Research

BABIN, C.
Laboratoire de Paleontologie
et Stratigraphie du Paleozoique
Université de Bretagne Occidentale
U.E.R. Sciences
Avenue le Gorgeu
29283 Brest Cedex
France

Paleozoic Mollusca Bivalvia: evolution, biostratigraphy

BAMBACH, R. K.
Dept. of Geological Sciences
Virginia Polytechnic Institute
and State University
Blacksburg, VA 24061
USA

Community paleoecology, ecospace utilization and patterns of diversity change through time

BANDEL, K.
Institut für Paläontologie
Universität Erlangen-Nürnberg
Loewenichstrasse 28
8520 Erlangen
Federal Republic of Germany

Paleontology, paleobiology, evolution of molluscs

CHARLESWORTH, B.
Dept. of Biology
University of Chicago
1103 E. 57th St.
Chicago, IL 60637
USA

Population genetics and evolutionary theory

CONNOR, E. F.
Dept. of Environmental Sciences
Clark Hall
University of Virginia
Charlottesville, VA 22903
USA

Population and community ecology, biogeography, and insect-plant interactions

DZIK, J.
Zaklad Paleobiologii PAN
Aleja Zwirki i Wigury 93
02-089 Warsaw
Poland

Evolutionary paleontology

ERBEN, H. K.
Institut für Paläontologie
Universität Bonn
Nussallee 8
5300 Bonn 1
Federal Republic of Germany

Phyletic extinctions (esp. „mass extinctions"), general patterns of organismic evolution in Phanerozoic time

FISHER, D. C.
Museum of Paleontology
University of Michigan
Ann Arbor, MI 48109
USA

Paleontology – functional morphology, phylogenetic inference, taphonomy

FLESSA, K. W.
Dept. of Geosciences
University of Arizona
Tucson, AZ 85721
USA

Paleobiology, biogeography

FLÜGEL, E.
Institut für Paläontologie
Universität Erlangen – Nürnberg
Loewenichstrasse 28
8520 Erlangen
Federal Republic of Germany

Evolution of fossil reefs

FÜRSICH, F. T.
Institut für Paläontologie
und Historische Geologie
Universität München
Richard-Wagner-Strasse 10
8000 München 2
Federal Republic of Germany

Paleoecology of mesozoic benthic marine systems

FUTUYMA, D. J.
Dept. of Ecology and Evolution
State University of New York
Stony Brook, NY 11794
USA

Evolution of associations among plants and herbivorous insects, from genetic, ecological, and phylogenetic perspectives; coevolution

GOULD, S. J.
Museum of Comparative Zoology
Harvard University
Cambridge, MA 02138
USA

Evolutionary biology

HALLAM, A.
Dept. of Geological Sciences
University of Birmingham
P.O. Box 363
Birmingham B15 2TT
England

Evolutionary paleobiology, paleobiogeography

HSÜ, K. J.
Geologisches Institut
ETH-Zentrum
8092 Zürich
Switzerland
Sedimentology, geochemistry, paleoenvironmental analyses

HÜSSNER, H. M.
Institut für Paläontologie
Universität Erlangen – Nürnberg
Loewenichstrasse 28
8520 Erlangen
Federal Republic of Germany
Carbonate sedimentology, evolution of reefs, mass extinctions

JABLONSKI, D.
Dept. of Geophysical Sciences
University of Chicago
5734 S. Ellis Avenue
Chicago, IL 60637
USA
Paleobiology and macroevolution

JÄRVINEN, O.
Dept. of Zoology
University of Helsinki
P. Rautatiekatu 13
00100 Helsinki 10
Finland
Community ecology of birds; ecological zoogeography; population biology of endangered populations

LaBARBERA, M.
Dept. of Anatomy
University of Chicago
1025 East 57th Street
Chicago, IL 60637
USA
Functional morphology and biomechanics of living and fossil marine invertebrates

LEVINTON, J. S.
Dept. of Ecology and Evolution
State University of New York
Stony Brook, NY 11794
USA
Marine ecology, evolutionary biology

MEETER, D. A.
Dept. of Statistics
Florida State University
Tallahassee, FL 32306
USA
Ecological statistics: time series analysis, design of experiments and tests of hypotheses

MOSBRUGGER, V.
Institut für Paläontologie
Universität Bonn
Nussallee 8
5300 Bonn 1
Federal Republic of Germany
Paleobotany, evolutionary biology

MÜLLER, G.
Institut für Anatomie
Währingerstrasse 13
1090 Vienna
Austria

Experimental embryology, vertebrate limb morphogenesis, evolutionary biology

NAGL, W.
FB Biologie
Universität Kaiserslautern
Postfach 3049
6750 Kaiserslautern
Federal Republic of Germany

Genome organization and evolution of species, DNA evolution in cell cultures, cell differentiation and malignant transformation

NIKLAS, K. J.
Section of Plant Biology
Cornell University
Ithaca, NY 14850
USA

Plant development and evolution

PANCHEN, A. L.
Dept. of Zoology
The University
Newcastle upon Tyne NE1 7RU
England

Vertebrate paleontology: origin of land vertebrates and early tetrapods; taxonomic theory

RAUP, D. M.
Dept. of Geophysical Sciences
University of Chicago
5734 S. Ellis Avenue
Chicago, IL 60637
USA

Theoretical paleobiology and evolutionary biology

REIF, W.-E.
Institut und Museum für Geologie und Paläontologie
Universität Tübingen
Sigwartstrasse 10
7400 Tübingen
Federal Republic of Germany

Evolutionary biology, functional morphology

de RICQLES, A. J.
Laboratoire d'Anatomie Comparée
Université Paris VII
2, Place Jussieu
75005 Paris
France

Evolutionary biology, paleohistology

RIEGER, R. M.
Institut für Zoologie
Universität Innsbruck
Universitätsstrasse 4
6020 Innsbruck
Austria

The origin and radiation of the bilateria as deduced from neonological information (with special reference to the new ultrastructural information)

RUNNEGAR, B.
Dept. of Geology and Geophysics
University of New England
Armidale, NSW 2351
Australia
Evolution of the mollusca, crystallography of carbonate biominerals, protein evolution

SEILACHER, A.
Institut und Museum für
Geologie und Paläontologie
Universität Tübingen
Sigwartstrasse 10
7400 Tübingen 1
Federal Republic of Germany
Paleobiology

SELANDER, R. K.
Dept. of Biology
University of Rochester
River Campus
Rochester, NY 14627
USA
Molecular population genetics

SEPKOSKI, Jr., J. J.
Dept. of Geophysical Sciences
University of Chicago
5734 S. Ellis Avenue
Chicago, IL 60637
USA
Paleontology, paleobiology

SIMBERLOFF, D.
Dept. of Biological Science
Florida State University
Tallahassee, FL 32306
USA
Ecology, evolution, biogeography

SOULÉ, M. E.
118 Little Oaks Rd.
Encinitas, CA 92024
USA
Evolutionary genetics and phenetics; conservation biology; environmental ethics

SOUSA, W. P.
Dept. of Zoology
University of California
Berkeley, CA 94720
USA
Disturbance and successional dynamics in marine intertidal communities; ecology of host-parasite interactions

STEARNS, S. C.
Zoologisches Institut
Rheinsprung 9
4051 Basel
Switzerland
Evolutionary theory, evolutionary ecology, life-history evolution

STINNESBECK, W.
Institut für Paläontologie
Universität Bonn
Nussallee 8
5300 Bonn 1
Federal Republic of Germany
Cretaceous-tertiary boundary events, upper cretaceous molluscs, paleoenvironmental analyses

TURNER, J. R. G.
Dept. of Genetics
University of Leeds
Leeds LS2 9JT
England

Evolutionary genetics

UNDERWOOD, A. J.
Dept. of Zoology
School of Biological Sciences
University of Sydney
Sydney NSW 2006
Australia

Experimental intertidal ecology; epistemology of ecological experiments

URBANEK, A. J.
Zaklad Paleobiologii PAN
Zwirki i Wigury 93
02-089 Warsaw
Poland

Paleontology and evolutionary biology, morphogenesis and evolution of colonial organisms (graptolites, pterobranchs)

VALENTINE, J. W.
Dept. of Geological Sciences
University of California
Santa Barbara, CA 93106
USA

Origin of phyla

VERMEIJ, G. J.
Dept. of Zoology
University of Maryland
College Park, MD 20742
USA

Paleobiology, malacology

WAGNER, G. P.
Institut für Zoologie
der Universität Wien
Althanstrasse 14
1090 Vienna
Austria

Quantitative genetic theory of phenotypic evolution, evolution of development

WAKE, D. B.
Museum of Vertebrate Zoology
2593 Life Sciences Building
University of California
Berkeley, CA 94720
USA

Evolutionary biology, with emphasis on patterns and processes of evolution in urodeles; functional and evolutionary morphology, genetics in relation to speciation and geographical ecology

WEIDICH, K. F.
Institut für Paläontologie
und Historische Geologie
Universität München
Richard-Wagner-Str. 10/II
8000 München 2
Federal Republic of Germany

Mesozoic microfaunas; cretaceous foraminifera

Author Index

Subject Index

Dahlem Workshop Reports

Life Sciences Research Reports (LS)

LS 1 The Molecular Basis of Circadian Rhythms
Editors: J. W. Hastings, H.-G. Schweiger

LS 2 Appetite and Food Intake
Editor: T. Silverstone

LS 3 Hormone and Antihormone Action at the Target Cell
Editors: J. H. Clark et al

LS 4 Organization and Expression of Chromosomes
Editors: V. G. Allfrey et al

LS 5 Recognition of Complex Acoustic Signals
Editor: T. H. Bullock

LS 6 Function and Formation of Neural Systems
Editor: G. S. Stent

LS 7 Neoplastic Transformation: Mechanisms and Consequences
Editor: H. Koprowski

LS 8 The Bases of Addiction
Editor: J. Fishman

LS 9 Morality as a Biological Phenomenon
Editor: G. S. Stent

LS 10 Abnormal Fetal Growth: Biological Bases and Consequences
Editor: F. Naftolin

LS 11 Transport of Macromolecules in Cellular Systems
Editor: S. C. Silverstein

LS 12 Light-Induced Charge Separation in Biology and Chemistry
Editors: H. Gerischer, J. J. Katz

LS 13 Strategies of Microbial Life in Extreme Environments
Editor: M. Shilo

LS 14 The Role of Intercellular Signals: Navigation, Encounter, Outcome
Editor: J. G. Nicholls

LS 15 Biomedical Pattern Recognition and Image Processing
Editors: K. S. Fu, T. Pavlidis

LS 16 The Molecular Basis of Microbial Pathogenicity
Editors: H. Smith et al

LS 17 Pain and Society
Editors: H. W. Kosterlitz, L. Y. Terenius

LS 18 Evolution of Social Behavior: Hypotheses and Empirical Tests
Editor: H. Markl

LS 19 Signed and Spoken Language: Biological Constraints on Linguistic Form
Editors: U. Bellugi, M. Studdert-Kennedy

Physical and Chemical Sciences Research Reports (PC)

PC 1 The Nature of Seawater (out of print)

PC 2 Global Chemical Cycles and Their Alteration by Man
Editor: W. Stumm

Distributor for LS 1–19 and PC 1 + 2:
Verlag Chemie, Pappelallee 3, 6940 Weinheim,
Federal Republic of Germany

Springer

MIX
Papier aus verantwortungsvollen Quellen
Paper from responsible sources
FSC® C105338

If you have any concerns about our products,
you can contact us on
ProductSafety@springernature.com

In case Publisher is established outside the EU,
the EU authorized representative is:
Springer Nature Customer Service Center GmbH
Europaplatz 3, 69115 Heidelberg, Germany

Printed by Libri Plureos GmbH
in Hamburg, Germany